Neural and Neuroendocrine Mechanisms in Host Defense and Autoimmunity

Neural and Neuroendocrine Mechanisms in Host Defense and Autoimmunity

Edited by

C. Jane Welsh

Texas A&M University
College Station, TX USA

Mary W. Meagher

Texas A&M University
College Station, TX USA

Esther M. Sternberg

National Institutes of Mental Health
Bethesda, MD, USA

 Springer

C. Jane Welsh
Departments of Veterinary Integrative Biosciences
and Veterinary Pathobiology
College of Veterinary Medicine and Biomedical Sciences
Texas A&M University
College Station, TX 77843-4458
jwelsh@cvm.tamu.edu

Mary W. Meagher
Department of Psychology
College of Liberal Arts
Texas A&M University
College Station, TX 77843-4225
m-meagher@tamu.edu

Esther M. Sternberg
Section on Neuroendocrine Immunology and Behavior
National Institute of Mental Health
National Institutes of Health
Rockville, MD 20892-9401
ems@codon.nih.gov

Library of Congress Control Number: 2005939172

ISBN-10: 0-387-31411-3
ISBN-13: 978-0387-31411-2

Printed on acid-free paper.

Printed in the United States of America.   (BS/IBT)

9 8 7 6 5 4 3 2 1

springer.com

To my parents Frank and Agnes Welsh for enlightenment and my family Dr. Colin R. Young, James and Robert Young for all their support and understanding, as well as to Dr. Sharon Boston and the Brazos Valley Multiple Sclerosis Support Group who instigated our research interest in stress and multiple sclerosis.

Preface

Susceptibility to infections and autoimmunity is profoundly affected by neural and neuroendocrine factors that in turn mediate the psychological status of the organism. Physiological stressors have provided one important experimental approach to investigate the interactive bidirectional communication between the immune and nervous systems. The central nervous system alters the immune response through activation of (1) the hypothalamic-pituitary-adrenal (HPA) axis resulting in the production of glucocorticoids, (2) the sympathetic nervous system resulting in the production of catecholamines, and (3) the parasympathetic nervous system and release of acetylcholine from the vagus nerve. In this context, the immune system functions as a "sensory system" alerting the central nervous system to the presence of pathogenic intruders via cytokine secretion. The immune system's response to neuroendocrine factors released by activation of the CNS may result in increased immunity or immunosuppression depending on the type and timing of the stressor in relation to exposure to the immune agent. Stress-induced immunosuppression causes increased susceptibility to infection resulting in more severe diseases or may also allow for the establishment of persistent infections leading to autoimmunity. Conversely, inhibition or failure of HPA axis function predisposes to more severe inflammation in certain conditions. This book focuses on the role of stress and neuroendocrine factors, in particular the HPA axis and adrenergic responses, in the pathogenesis of infectious diseases, autoimmunity, and inflammation. Although much recent work has highlighted the importance of cholinergic mechanisms in regulating inflammation, these will not be addressed here. An in-depth understanding of the cytokine and neuroendocrine networks involved in these interactions may lead to the development of targeted therapeutics for a variety of diseases.

This book includes selected authorities in research covering the role of stress in cell trafficking (Dhabhar), and stress effects on sympathetic nervous system influences on the immune system (Fleshner and Sanders). The impact of anthrax lethal toxin on the glucocorticoid system (Webster) and gender dimorphism and the effects of sex hormones on various conditions (Chaudry) are also covered. The neuroendocrine influences on influenza (Bailey), herpes (Bonneau), and human immunodeficiency (Sloan) viral infections are a major focus of this book. Multiple sclerosis has been selected as an example of autoimmune disease. The clinical aspects of stress and multiple sclerosis (Mohr) and the effects of stress on experimental models of multiple sclerosis (Meagher, Stephan, and Welsh) are discussed.

This book developed as a result of a symposium held by the International Society for NeuroImmunoModulation (ISNIM) entitled "Neural and Neuroendocrine Factors in Susceptibility and Resistance to Infectious Diseases" at the meeting of the American Association of Immunologists, Denver, Colorado, in May 2003.

Acknowledgments

We are indebted to Andrea Macaluso for initiating this book and for all her work making it a reality. We would also like to acknowledge Virginia T. Ladd, President and Executive Director of the American Autoimmune Related Diseases Association Inc., for financial support with the original meeting and Craig Smith, M.Sc., for assistance in organizing the meeting.

Contributors

Michael T. Bailey, Ph.D.
College of Dentistry Section of Oral Biology
The Ohio State University
Columbus, OH 43218-2357

Robert H. Bonneau, Ph.D.
Department of Microbiology and Immunology
The Pennsylvania State University College of Medicine
Milton S. Hershey Medical Center
Hershey, Pennsylvania 17033

Irshad H. Chaudry, Ph.D.
Center for Surgical Research
University of Alabama at Birmingham
Birmingham, AL 35294-0019

Mashkoor A. Choudhry, Ph.D.
Center for Surgical Research
University of Alabama at Birmingham
Birmingham, AL 35294-0019

Steve Cole, Ph.D.
Department of Medicine, Division of Hematology-Oncology
David Geffen School of Medicine at UCLA
Los Angeles CA 90095-1678

Alicia Collado-Hidalgo, Ph.D.,
Department of Psychiatry
UCLA Semel Institute for Neuroscience and Human Behavior
Los Angeles CA 90095-7076

Firdaus S. Dhabhar, Ph.D.,
Associate Professor
Department of Psychiatry & Behavioral Sciences Stanford University
School of Medicine 401 Quarry Road, C231
Stanford, CA 94305-5718

Monika Fleshner, Ph.D.
Department of Integrative Physiology
University of Colorado
Boulder, CO 80309-0354

Elisabeth Good, B.S.
Department of Psychology
College of Liberal Arts
Texas A&M University
College Station, TX 77843-4225

Ashley Hammons, B.S.
Department of Veterinary Integrative Biosciences
College of Veterinary Medicine and Biomedical Sciences
Texas A&M University
College Station, TX 77843-4458

Stephan von Hörsten, M.D.
Franz-Penzoldt-Center
University Erlangen
Section Experimental Therapy
91054 Erlangen, Germany

John Hunzeker, Ph.D.
Department of Microbiology and Immunology
The Pennsylvania State University College of Medicine
Milton S. Hershey Medical Center
Hershey, Pennsylvania 17033

Robin R. Johnson, MS.,
Department of Psychology
College of Liberal Arts
Texas A&M University
College Station, TX 77843-4225

Mary W. Meagher, Ph.D.
Department of Psychology
College of Liberal Arts
Texas A&M University
College Station, TX 77843-4225

Wentao Mi, Ph.D.,
Department of Veterinary Integrative Biosciences
College of Veterinary Medicine and Biomedical Sciences
Texas A&M University
College Station, TX 77843-4458

Mahtab Moayeri, PhD
National Institute of Allergy and Infectious Diseases
National Institutes of Health
Bethesda, MD 20892-4349

David C. Mohr, Ph.D.
UCSF Departments of Psychiatry and Neurology
Veterans Affairs Medical Center
Mental Health Service
San Francisco, CA 94121

David A. Padgett, Ph.D.
College of Dentistry Section of Oral Biology
The Ohio State University
Columbus, OH 43218-2357

Thomas Prentice, M.S.
Department of Psychology
College of Liberal Arts
Texas A&M University
College Station, TX 77843-4225

Virginia Sanders, Ph.D.
Department of Molecular Virology
Immunology & Medical Genetics
The Ohio State University
Columbus, OH 43210

John F. Sheridan, Ph.D.
College of Dentistry Section of Oral Biology
The Ohio State University
Columbus, OH 43218-2357

Amy Sieve, Ph.D.
Department of Immunology & Microbiology
The University of North Texas Health Science Center
Fort Worth, Texas 76107

Erica K. Sloan, Ph.D.
Department of Medicine, Division of Hematology-Oncology
David Geffen School of Medicine at UCLA
Los Angeles CA 90095-1678

Andrew Steelman, MBiot.
Department of Veterinary Integrative Biosciences
College of Veterinary Medicine and Biomedical Sciences
Texas A&M University
College Station, TX 77843-4458

Thomas Skripuletz, MD
Department of Functional and Applied Anatomy
Medical School of Hannover
30625 Hannover,
Germany

Michael Stephan, MD
Department of Functional and Applied Anatomy
Medical School of Hannover
30625 Hannover,
Germany

Esther M. Sternberg, M.D.
Section on Neuroendocrine Immunology and Behavior
National Institute of Mental Health
National Institutes of Health
Rockville, MD 20892-9401 U.S.A.

Ralph Storts, DVM, Ph.D.
Department of Veterinary Pathobiology
College of Veterinary Medicine and Biomedical Sciences
Texas A&M University
College Station, TX 77843-4467

Jeanette Webster, Ph.D.,
Section on Neuroendocrine Immunology and Behavior
National Institute of Mental Health
National Institutes of Health
Rockville, MD 20892-9401

C. Jane Welsh, Ph.D.
Departments of Veterinary Integrative Biosciences
and Veterinary Pathobiology
College of Veterinary Medicine and Biomedical Sciences
Texas A&M University
College Station, TX 77843-4458

Thomas Welsh, Ph.D.,
Department of Animal Science,
College of Agriculture
Texas A&M University
College Station, TX 77843-2471

Colin R. Young, Ph.D.,
Departments of Veterinary Integrative Biosciences and Psychology
College of Veterinary Medicine and Biomedical Sciences
Texas A&M University
College Station, TX 77843-4458

Contents

**11. Social Stress Alters the Severity of an Animal Model of
Multiple Sclerosis**   216
Mary W. Meagher, Robin R. Johnson, Elisabeth Good, and
C. Jane Welsh

**12. Early Postnatal Nongenetic Factors Modulate Disease
Susceptibility in Adulthood: Examples from Disease Models
of Multiple Sclerosis, Periodontitis, and Asthma**   241
Michael Stephan, Thomas Skripuletz, and Stephan von Hörsten

1
Introduction

ESTHER M. STERNBERG

The study of host susceptibility to infectious disease has shifted focus over the decades from one primarily centered on the organism and its characteristics to one centered on the host's immune-defense mechanisms. Recent research has made clear, however, that there is a two-way relationship between host and invading microorganism in which factors released by the host in response to the microorganism alter infectivity and course of infection and in turn by which factors released by the microorganism alter host defense mechanisms to infection. The purpose of this volume is to highlight these sometimes symbiotic and sometimes detrimental bidirectional influences of host-on-pathogen and pathogen-on-host responses. In particular, this volume focuses on extraimmune host responses of the central nervous and neuroendocrine systems that are activated during stress and during infection.

This interaction is particularly relevant to infectious disease, because an infected host is "stressed" for multiple reasons. Infection itself activates the stress response, through bacterial or viral products and through host cytokines released in response to invading microorganisms. In addition, infection is also associated with psychological and physiological stressors that further activate host stress-response mechanisms. Cytokines activate central neuroendocrine stress responses directly, by crossing the blood-brain barrier at leaky sites and via active transport, or indirectly by activating second messenger enzyme systems in cerebral endothelial cells with resultant release of second messengers such as nitric oxide and prostaglandins (reviewed in Marques-Deak et al., 2005). In addition, cytokines activate central vagal autonomic pathways by binding to receptors on vagal paraganglia cells (Watkins and Maier, 1999). This volume will not focus on these afferent pathways, however, but will focus on the effects on infection of the hormonal and neurochemical products released when these stress-response systems are activated.

When considering the effects of neuroendocrine and neural systems on host defense to infection, one can analyze the relationship at a systems and anatomical level as well as at a cellular and molecular level. Much research has been done in this area at a systems level, examining the effects of the various effector arms of the central nervous system on host susceptibility and resistance to infection. These include the hypothalamic-pituitary-adrenal (HPA) axis and the sympathetic, parasympathetic, and peripheral nervous systems. It is becoming increasingly clear that there is a specificity of pattern of activation of stress-response pathways, depending upon the particular type of stressor to which the host is exposed. Thus, physiological stress such as hypovolemia and psychological stress such as pain have been shown to activate different brain regions. Furthermore, patterns of activation of central stress-response pathways differ according to route, dose, and particular pathogen products to which the host is exposed. Finally, the location at which neurotransmitters and neurohormones are released, whether locally at sites of inflammation or infection, regionally in lymph nodes or immune organs, or systemically in the bloodstream, also determine the ultimate effects of these molecules on host responses to infectious agents.

This volume will address such systemic physiological responses, including the role of stress activation of the neuroendocrine stress response (HPA axis) on an important aspect of innate first-line host defense—immune cell trafficking (Dhabhar)—and on viral infection (Theiler's virus: Welsh *et al.*). It will address the effects of activation of the sympathetic nervous system on cellular and molecular aspects of innate and cell-mediated immunity (Fleshner; Sanders) and the effects of the autonomic nervous system on HIV infection (Sloan *et al.*). In addition, the effects on viral infection of psychological activation of these stress-response systems by stressful and early life events and naturalistic stressors will be addressed (HSV: Bonneau and Hunzeker; influenza: Bailey, Padgett, and Sheridan; Theiler's virus: Meagher). New research also reviewed in this volume is increasingly showing that different types of stress, such as restraint versus social stress, have different effects not only on stress-response activation patterns but also on susceptibility to and course of infection. In addition, such stressors differentially affect the course and severity of infection with specific microorganisms, whether specific viruses (HSV, influenza) or bacteria.

It is also important to consider these relationships at a cellular and molecular level not only in order to fully understand the pathogenesis of the effects of physiological stresses on the infected host but also as potential therapeutic targets. Thus the effects of bacterial toxins on repressing trans-activation of nuclear hormone receptors such as the glucocorticoid receptor is another level of interaction between microorganism and host that may play an important role in altering host defenses. In this context, the microorganism could potentially induce a state of glucocorticoid resistance that would interfere with the host's ability to suppress inflammation through release of glucocorticoids from the adrenals—an event that ordinarily occurs during infection as a result of HPA axis stimulation by cytokines or bacterial products and should counter excessive host inflammatory responses. Defining the precise molecular mechanism of such effects could lead to development of new therapeutic approaches that bypass the block and reinstate appropriate glucocorticoid modulation of inflammation in the context of infection.

Such interactions also underscore the concept that pathology in the context of infection may result from either inadequate host immune/ inflammatory defenses that are unable to contain the invading microorganism or from excessive immune/inflammatory host responses that result in tissue damage from inflammation. An appropriate inflammatory/immune response is necessary to control and clear infectious agents from the host. However, during stress, excessive release of glucocorticoids reduces immune-defense mechanisms, resulting in worse infection. In contrast, glucocorticoid resistance induced by bacterial products may prevent control of inflammatory responses to the pathogen, thus worsening tissue damage from excessive host responses. In addition, activation of different arms of the central nervous system stress response may also alter the balance of appropriate inflammatory responses to pathogens. In some contexts, glucocorticoids or adrenergic factors may enhance and in others they may suppress inflammatory/immune responses. Also addressed in this text is the important role of sex hormones, in particular estrogens, on this balance of inflammation and infection.

An additional level at which host defense responses may alter infection is through the direct effects of neurohormones and neurotransmitters on microorganism replication. In some cases, mediators released during stress, such as NO or Hsp72, may facilitate recovery from infection (*E. coli*: Fleshner), while in others neurotransmitters such as catecholamine accelerate viral replication (HIV-1: Sloan *et al.*). Not addressed in this volume, but also a direct effect of host mediators on pathogens, is the antibacterial effect of chromogranins released from adrenal medullary cells during adrenergic activation (Briolat *et al.*, 2005).

Given the complexity of the many hormones and neurotransmitters released during infection and the changing inflammatory/immune responses that occur over time as the infection evolves, it is not surprising that host responses to infectious agents vary greatly depending on host and environmental and pathogen factors. In some cases, environmental factors such as psychological stress may contribute to this variability, and in others internal physiological variables such as physiological stress may contribute. Interactions of pathogen products with these host response pathways at a system's cellular and molecular level also alter host defenses. Similarly, mediators released by the host in the course of infection alter the infectivity of pathogens. Careful dissection of these pathogen-host interactions at all levels will provide insights into pathogenesis of infection and the many environmental and host factors that alter the course of infection and could ultimately yield important new therapeutic approaches and targets for treatment of infection.

4   E.M. Sternberg

References

Briolat, J., Wu, S.D., Mahata, S.K., Gonthier, B., Bagnard, D., Chasserot-Golaz, S., Helle, K.B., Aunis, D., and Metz-Boutigue, M.H. (2005). New antimicrobial activity for the catecholamine release-inhibitory peptide from chromogranin A. Cell. Mol. Life Sci. 62:377–385.

Marques-Deak, A., Cizza, G., and Sternberg, E. (2005). Brain-immune interactions and disease susceptibility. *Mol. Psychiatry* 10:239–250.

Watkins, L.R., and Maier, S.F. (1999). Implications of immune-to-brain communication for sickness and pain. *Proc. Natl. Acad. Sci. U. S. A.* 96:7710–7713.

Part I
Neural and Neuroendocrine Mechanisms in Host Defense: Molecular and Cellular Mechanisms

2
Stress-induced Changes in Immune Cell Distribution and Trafficking: Implications for Immunoprotection versus Immunopathology

FIRDAUS S. DHABHAR

1. Introduction

Effective immunoprotection requires rapid recruitment of leukocytes into sites of surgery, wounding, infection, or vaccination. Immune cells circulate continuously on surveillance pathways that take them from the blood, through various organs, and back into the blood. This circulation is essential for the maintenance of an effective immune defense network (Sprent and Tough, 1994). The numbers and proportions of leukocytes in the blood provide an important representation of the state of distribution of leukocytes in the body and of the state of activation of the immune system. A stress-induced change in leukocyte distribution within different body compartments is perhaps one of the most underappreciated effects of stress and stress hormones on the immune system.

Because the blood is the most accessible and commonly used compartment for human studies, it is important to carefully evaluate how changes in blood immune parameters might reflect *in vivo* immune function in the context of the specific experiments or study at hand. Moreover, because most blood collection procedures involve a certain amount of stress, because all patients or subjects will have experienced acute and chronic stress, and because many studies of psychophysiological effects on immune function focus on stress, it is important to keep in mind the effects of stress on blood leukocyte distribution.

Numerous studies have shown that stress and stress hormones induce significant changes in absolute numbers and relative proportions of leukocytes in the blood. In fact, changes in blood leukocyte numbers were used as an indirect measure for changes in plasma cortisol before methods were available to directly assay the hormone (Hoagland *et al.*, 1946), and numerous studies have shown that glucocorticoid hormones induce significant changes in blood leukocyte distribution (Fauci and Dale, 1974, 1975; Dhabhar *et al.*, 1996). Studies have also delineated the rapid and significant effects of catecholamine hormones in mediating stress-induced changes in blood leukocyte distribution (Benschop *et al.*, 1993, 1996; Carlson *et al.*, 1997; Mills *et al.*, 1998, 2001; Redwine *et al.*, 2003).

Dhabhar *et al.* were the first to propose that stress-induced changes in blood leukocyte distribution may represent an adaptive response (Dhabhar *et al.*, 1994; Dhabhar and McEwen, 1999a). They suggested that acute stress–induced changes in blood leukocyte numbers represent a redistribution of leukocytes from the blood to other organs such as the skin and lining of the gastrointestinal and urinary-genital tracts and draining sentinel lymph nodes (Dhabhar and McEwen, 1996a, 2001). They hypothesized that such a leukocyte redistribution may enhance immune function in those compartments to which immune cells traffic during stress. In agreement with this hypothesis, it was demonstrated that a stress-induced redistribution of leukocytes from the blood to the skin is accompanied by a significant enhancement of skin immunity (Dhabhar and McEwen, 1996, 1999a; Dhabhar *et al.*, 2000). Studies also showed that acute stress initially increases trafficking of all leukocyte subpopulations to a site of surgery or immune activation (Viswanathan and Dhabhar, 2005). Although all leukocyte subpopulations traffic to a site of immune activation in greater numbers during stress, tissue damage, antigen-, or pathogen-driven chemoattractants synergize with acute stress to determine which specific subpopulations are recruited more vigorously (Viswanathan and Dhabhar, 2005). Thus, depending on the primary chemoattractants driving an immune response, acute stress may selectively mobilize specific leukocyte subpopulations into sites of surgery, wounding, or immune activation. Such a stress-induced increase in leukocyte trafficking may be an important mechanism by which acute stressors alter the course of different (innate versus adaptive, early versus late, or acute versus chronic) immune responses.

It is important to keep in mind the Yin-Yang nature of a stress-induced increase in leukocyte trafficking to sites of immune activation. Such an increase in leukocyte trafficking may be beneficial for promoting immunoprotection during surgery, wound healing, vaccination, infection, or localized cancer. However, a stress-induced increase in leukocyte trafficking may have harmful consequences during stress-induced exacerbations of inflammatory (e.g., cardiovascular disease, gingivitis) and autoimmune (e.g., psoriasis, arthritis, multiple sclerosis) diseases (Amkraut *et al.*, 1971; Al'Abadie *et al.*, 1994; Garg *et al.*, 2001; Ackerman *et al.*, 2002) or graft-rejection (Kok-van Alphen and Volker-Dieben, 1983). Whereas decades of research have examined the pathological effects of stress on immune function and on health, the study of salubrious or health-promoting effects of stress is relatively new (Dhabhar *et al.*, 1995a; Dhabhar and McEwen, 2001). Much work remains to be done to elucidate the mechanisms mediating these bidirectional effects of stress on health and to translate basic findings regarding the adaptive effects of stress from bench to bedside. Here we discuss stress-induced and stress hormone–induced changes in blood leukocyte distribution and examine their functional consequences.

2. Stress

Although the word *stress* generally has negative connotations, stress is a familiar aspect of life, being a stimulant for some but a burden for others. Numerous definitions have been proposed for the word *stress*. Each definition focuses on aspects of an internal or external challenge, disturbance, or stimulus; on perception of a stimulus by an organism; or on a physiological response of the organism to the stimulus (Goldstein and McEwen, 2002; McEwen, 2002; Sapolsky, 2004). Physical stressors have been defined as external challenges to homeostasis and psychological stressors as the "anticipation justified or not, that a challenge to homeostasis looms" (Sapolsky, 2005). An integrated definition states that stress is a constellation of events, consisting of a stimulus (stressor), which precipitates a reaction in the brain (stress perception), which activates physiologic fight-or-flight systems in the body (stress response) (Dhabhar and McEwen, 1997). The physiologic stress response results in the release of neurotransmitters and hormones that serve as the brain's alarm signals to the rest of the body. It is often overlooked that a stress response has salubrious adaptive effects in the short run (Dhabhar *et al.*, 1995a; Dhabhar and McEwen, 1996, 2001) although stress can be harmful when it is longlasting (Irwin *et al.*, 1990; McEwen, 1998; Sapolsky, 2004). An important distinguishing characteristic of stress is its duration and intensity. Thus, *acute stress* has been defined as stress that lasts for a period of a few minutes to a few hours and *chronic stress* as stress that persists for several hours per day for weeks or months (Dhabhar and McEwen, 1997). The intensity of stress may be gauged by the peak levels of stress hormones, neurotransmitters, and other physiological changes such as increases in heart rate and blood pressure, and by the amount of time for which these changes persist during and after the cessation of stress.

It is important to bear in mind that there exist significant individual differences in the manner and extent to which stress is perceived, processed, and coped with. These differences become particularly relevant in case of human subjects because stress perception, processing, and coping mechanisms can have significant effects on the kinetics and peak levels of circulating stress hormones and on the duration for which these hormone levels are elevated. The magnitude and duration of catecholamine and glucocorticoid hormone exposure in turn can have significant effects on leukocyte distribution and function (Dhabhar and McEwen, 2001; Pruett, 2001; Schwab *et al.*, 2005).

3. Stress-induced Changes in Blood Leukocyte Numbers

The phenomenon of acute stress–induced changes in blood leukocyte numbers is well-known. Changes in blood leukocyte numbers were used as an indirect measure for changes in plasma cortisol levels long before methods were available to directly assay the hormone (Hoagland *et al.*, 1946). Stress-induced changes in blood leukocyte numbers have been reported in fish (Pickford *et al.*, 1971), hamsters (Bilbo *et al.*, 2002), mice (Jensen, 1969), rats (Dhabhar *et al.*, 1994, 1995a, 1996; Rinder *et al.*, 1997), rabbits (Toft *et al.*, 1993), horses (Snow *et al.*, 1983), non-human primates (Morrow-Tesch *et al.*, 1993), and humans (Herbert and Cohen, 1993; Schedlowski *et al.*, 1993a; Mills *et al.*, 1998; Bosch *et al.*, 2003; Redwine *et al.*, 2004). This suggests that the phenomenon of stress-induced leukocyte redistribution has a long evolutionary lineage and that perhaps it has important functional significance.

Studies in rodents have shown that stress-induced changes in blood leukocyte numbers are characterized by a significant decrease in numbers and percentages of lymphocytes and monocytes and by an increase in numbers and percentages of neutrophils (Dhabhar *et al.*, 1994, 1995a). Flow cytometric analyses have revealed that absolute numbers of peripheral blood T cells, B cells, NK cells, and monocytes all show a rapid and significant decrease (40% to 70% lower than baseline) during stress (Dhabhar *et al.*, 1995a). Moreover, it has been shown that stress-induced changes in leukocyte numbers are rapidly reversed upon the cessation of stress (Dhabhar *et al.*, 1995a). In apparent contrast with animal studies, human studies have shown that stress can increase rather than decrease blood leukocyte numbers (Naliboff *et al.*, 1991; Schedlowski *et al.*, 1993a; Brossschot *et al.*, 1994; Mills *et al.*, 1995; Bosch *et al.*, 2003). Factors that help resolve this apparent contradiction are discussed later in this chapter.

4. Hormones Mediating Stress-induced Changes in Blood Leukocyte Numbers

The catecholamines epinephrine and norepinephrine and the glucocorticoid hormones have been identified as the major endocrine mediators of stress-induced changes in leukocyte distribution (Fauci and Dale, 1974; Schedlowski *et al.*, 1993a, 1993b; Dhabhar *et al.*, 1995a; Benschop *et al.*, 1996; Dhabhar *et al.*, 1996). Studies have revealed that stress-induced changes in leukocyte distribution are mediated by hormones released by the adrenal gland (Dhabhar *et al.*, 1996; Dhabhar and McEwen, 1999b). Adrenalectomy (which eliminates the corticosterone and epinephrine stress response) has been shown to reduce the magnitude of the stress-induced changes in blood leukocyte numbers. (Dhabhar *et al.*, 1996; Dhabhar and McEwen, 1999b). Cyanoketone treatment, which virtually eliminates the corticosterone stress response, also virtually eliminates the stress-induced decrease in blood lymphocyte numbers, and significantly enhances the stress-induced increase in blood neutrophil numbers (Dhabhar *et al.*, 1996). Several other studies have shown that glucocorticoid treatment induces changes in leukocyte distribution in mice (Dougherty and White, 1945; Spain and Thalhimer, 1951;

Cohen, 1972; Zatz, 1975), guinea pigs (Fauci, 1975), rats (Ulich *et al.*, 1988; Miller *et al.*, 1994; Dhabhar *et al.*, 1996), rabbits (Miller *et al.*, 1994), and humans (Fauci and Dale, 1974; Fauci, 1976; Onsrud and Thorsby, 1981).

Because adrenal steroids act at two distinct receptor subtypes that show a heterogeneity of expression in immune cells and tissues (Spencer *et al.*, 1990, 1991; Miller *et al.*, 1990, 1991, 1992, 1993; Dhabhar *et al.*, 1993, 1995b, 1996), Dhabhar *et al.* investigated the role played by each receptor subtype in mediating changes in leukocyte distribution (Dhabhar *et al.*, 1996). Acute administration of aldosterone (a specific type I adrenal steroid receptor agonist) to adrenalectomized animals did not have a significant effect on blood leukocyte numbers. In contrast, acute administration of corticosterone (the endogenous type I and type II receptor agonist) or RU28362 (a specific type II receptor agonist) to adrenalectomized animals induced changes in leukocyte distribution that were similar to those observed in intact animals during stress. These results suggest that corticosterone, acting at the type II adrenal steroid receptor, is a major mediator of the stress-induced decreases in blood lymphocyte and monocyte distribution. Taken together, these studies show that stress and glucocorticoid hormones induce a significant decrease in blood lymphocyte numbers when administered under acute or chronic conditions.

Apart from glucocorticoids, studies have also demonstrated the importance of catecholamine hormones in mediating stress-induced changes in blood leukocyte distribution in rodents and humans (Benshop *et al.*, 1993, 1996; Schedlowski *et al.*, 1993b; Carlson *et al.*, 1997; Mills *et al.*, 1998; Mills *et al.*, 2001; Redwine *et al.*, 2003; Engler *et al.*, 2004). In apparent contrast with glucocorticoid hormones, catecholamine hormones have been shown to increase blood leukocyte numbers in rats (Harris *et al.*, 1995) and humans (Landmann *et al.*, 1984). On closer examination, it is observed that after adrenaline or noradrenaline administration, neutrophil and NK cell numbers increase rapidly and dramatically whereas T- and B-cell numbers decrease (Tonnesen *et al.*, 1987; Landmann, 1992; Schedlowski *et al.*, 1993b; Benschop *et al.*, 1996). Carslon *et al.* have shown that catecholamine pretreatment results in increased accumulation of lymphocytes in the spleen and lymph nodes (Carlson *et al.*, 1997), which would be in agreement with a catecholamine-induced decrease in lymphocytes in the blood. By acutely administering epinephrine, norepinephrine, selective α and β adrenergic receptor agonists, or corticosterone to adrenalectomized animals, researchers have shown that increases in blood granulocyte numbers may be mediated by the α_1 and β adrenergic receptors and are counteracted by corticosterone acting at the type II adrenal steroid receptor (Dhabhar and McEwen, 1999b). Increases in lymphocytes may be mediated by the α_2 receptor, whereas decreases in lymphocytes may be mediated by β adrenergic and type II adrenal steroid receptors (Dhabhar and McEwen, 1999b).

Therefore, the absolute number of specific blood leukocyte subpopulations may be significantly affected by the ambient concentrations of epinephrine, norepinephrine, and corticosterone. Differences in concentrations and combinations of these hormones may explain reported differences in blood leukocyte numbers during different stress conditions (e.g., short- versus long-duration acute stress, acute versus chronic stress) and during exercise.

5. A Stress-induced Decrease in Blood Leukocyte Numbers Represents a Redistribution Rather Than a Destruction or Net Loss of Blood Leukocytes

From the above discussion it is clear that stress and glucocorticoid hormones induce a rapid and significant decrease in blood lymphocyte, monocyte, and NK cell numbers. This decrease in blood leukocyte numbers may be interpreted in two possible ways. It could reflect a large-scale destruction of circulating leukocytes. Alternatively, it could reflect a redistribution of leukocytes from the blood to other organs in the body. In favor of the latter explanation, experiments were conducted to test the hypothesis that acute stress induces a redistribution of leukocytes from the blood to other compartments in the body (Dhabhar et al., 1995a).

The first series of experiments examined the kinetics of recovery of the stress-induced reduction in blood leukocyte numbers. It was hypothesized that if the observed effects of stress represented a redistribution rather than a destruction of leukocytes, one would see a relatively rapid return of leukocyte numbers back to baseline upon the cessation of stress. Results showed that all leukocyte subpopulations that showed a decrease in absolute numbers during stress showed a complete recovery with numbers reaching pre-stress baseline levels within 3 h after the cessation of stress (Dhabhar et al., 1995a). Plasma levels of lactate dehydrogenase (LDH), which is a marker for cell damage, were also monitored in the same experiment. If the stress-induced decrease in leukocyte numbers were the result of a destruction of leukocytes, one would expect to observe an increase in plasma levels of LDH during or after stress. However, no significant changes in plasma LDH were observed, further suggesting that a redistribution rather than a destruction of leukocytes was primarily responsible for the stress-induced decrease in blood leukocyte numbers (Dhabhar et al., 1995a). Further studies have shown that lymph nodes, bone marrow, and skin are target organs of a stress-induced redistribution of leukocytes within the body (Dhabhar, 1998; Stefanski, 2003; Viswanathan and Dhabhar, 2005).

It is important to bear in mind that although glucocorticoids are known to induce leukocyte apoptosis under certain conditions (Cohen, 1992), glucocorticoid hormones have also been shown to induce changes in various immune parameters (Munck et al., 1984), including immune cell distribution (Dhabhar et al., 1995a; Dhabhar and McEwen, 2001), in the absence of cell death. It has been suggested that some species may be "steroid-resistant" and others may be "steroid-sensitive," and that glucocorticoid-induced changes in blood leukocyte numbers represent changes in leukocyte redistribution in steroid-resistant species (humans and guinea pig), and leukocyte lysis in steroid-sensitive species (mouse and rat) (Claman, 1972). However, it is now accepted that even in species previously thought to be steroid-sensitive, changes in adrenal steroids similar to those described here produce changes in leukocyte distribution rather than an increase in leukocyte destruction (Cohen, 1972).

6. Target Organs of a Stress-induced Redistribution of Blood Leukocytes

Based on the above discussion, the obvious question one might ask is, "Where do blood leukocytes go during stress?" Numerous studies using stress or stress hormone treatments have investigated this issue. Using gamma imaging to follow the distribution of adoptively transferred radiolabeled leukocytes in whole animals, Toft et al. have shown that stress induces a redistribution of leukocytes from the blood to the mesenteric lymph nodes (Toft et al., 1993). It has been reported that anesthesia stress, as well as the infusion of ACTH and prednisolone in rats results in decreased numbers of labeled lymphocytes in the thoracic duct, and the cessation of drug infusion results in normal circulation of labeled lymphocytes (Spry, 1972). This suggests that ACTH and prednisolone (which would produce hormonal changes similar to those observed during stress) may cause the retention of circulating lymphocytes in different body compartments thus resulting in a decrease in lymphocyte numbers in the thoracic duct and a concomitant decrease in numbers in the peripheral blood (Spry, 1972). It has also been reported that a single injection of hydrocortisone, prednisolone, or ACTH results in increased numbers of lymphocytes in the bone marrow of mice (Cohen, 1972), guinea pigs (Fauci, 1975), and rats (Cox and Ford, 1982). Fauci et al. have suggested that glucocorticoid-induced decreases in blood leukocyte numbers in humans may also reflect a redistribution of immune cells to other organs in the body (Fauci and Dale, 1974). Finally, corticosteroids have been shown to induce the accumulation of lymphocytes in mucosal sites (Walzer et al., 1984), and the skin has been identified as a target organ to which leukocytes traffic during stress (Dhabhar and McEwen, 1996; Viswanathan and Dhabhar, 2005). In vitro catecholamine treatment has also been shown to direct leukocyte traffic to spleen and lymph nodes (Carlson et al., 1997). Studies have identified lymph nodes, bone marrow, and skin as the target organs of a stress-induced redistribution of leukocytes within the body (Dhabhar, 1998; Stefanski, 2003; Viswanathan and Dhabhar, 2005).

It is important to note that in these studies, a return to basal glucocorticoid levels was followed by a return to basal blood lymphocyte numbers, further supporting the hypothesis that decrease in blood leukocyte numbers is the result of a glucocorticoid-induced redistribution rather than a glucocorticoid-induced destruction of blood leukocytes. The above discussion shows that a stress-induced decrease in blood leukocyte numbers reflects a redistribution or redeployment of leukocytes from the blood to other organs (lymph nodes, bone marrow, and skin) in the body.

7. Acute Stress-induced Changes in Blood Leukocyte Numbers: Contradicting Results or a Biphasic Response?

As stated before, stress has been shown to induce a significant decrease in blood leukocyte numbers in a range of different species. However, studies have also shown that stress can increase rather than decrease blood leukocyte numbers in humans (Naliboff et al., 1991; Schedlowski et al., 1993a; Brosschot et al., 1994; Mills et al., 1995; Bosch et al., 2003). This apparent contradiction is resolved when three important factors are taken into account: first, stress-induced increases in blood leukocyte numbers are observed after stress conditions that primarily result in the activation of the sympathetic nervous system. These stressors are often of a short duration (few minutes) or relatively mild (e.g., public speaking) (Naliboff et al., 1991; Schedlowski et al., 1993a; Brosschot et al., 1994; Mills et al., 1995). Second, the increase in leukocyte numbers may be accounted for by stress- or catecholamine-induced increases in granulocytes and NK cells (Naliboff et al., 1991; Schedlowski et al., 1993a; Brosschot et al., 1994; Mills et al., 1995; Benschop et al., 1996). Because granulocytes form a large proportion of circulating leukocytes in humans (60–80% granulocytes), an increase in granulocyte numbers is reflected as an increase in total leukocyte numbers in contrast with rats and mice (10–20% granulocytes). Third, stress or pharmacologically induced increases in glucocorticoid hormones induce a significant decrease in blood lymphocyte and monocyte numbers (Hoagland et al., 1946; Stein et al., 1951; Schedlowski et al., 1993a; Dhabhar et al., 1996). Thus, stress conditions that result in a significant and sustained activation of the hypothalamic-pituitary-adrenal (HPA) axis will result in a decrease in blood leukocyte numbers.

In view of the above discussion, it may be proposed that acute stress induces an initial increase followed by a decrease in blood leukocyte numbers. Stress conditions that result in activation of the sympathetic nervous system, especially conditions that induce high levels of norepinephrine, may induce an increase in circulating leukocyte numbers. These conditions may occur during the very beginning of a stress response, very short duration stress (order of minutes), mild psychological stress, or during exercise. In contrast, stress conditions that result in the activation of the HPA axis induce a decrease in circulating leukocyte numbers. These conditions often occur during the later stages of a stress response, long-duration acute stressors (order of hours), or during severe psychological, physical, or physiological stress. An elegant and interesting example in support of this hypothesis comes from Schedlowski *et al.* who measured changes in blood T cell and NK cell numbers as well as plasma cate-cholamine and cortisol levels in parachutists (Schedlowski *et al.*, 1993a). Measurements were made 2 h before, immediately after, and 1 h after the jump. Results showed a significant increase in T cell and NK cell numbers immediately (minutes) after the jump that was followed by a significant decrease 1 h after the jump. An early increase in plasma catecholamines pre-ceded early increases in lymphocyte numbers, whereas the more delayed rise in plasma cortisol preceded the late decrease in lymphocyte numbers (Schedlowski *et al.*, 1993a). Importantly, changes in NK cell activity and antibody-dependent cell-mediated cytotoxicity closely paralleled changes in blood NK cell numbers, thus suggesting that changes in leukocyte numbers may be an important mediator of apparent changes in leukocyte "activity." Similarly, Rinner *et al.* have shown that a short stressor (1-min handling) induced an increase in mitogen-induced proliferation of T and B cells obtained from peripheral blood, whereas a longer stressor (2-h immo-bilization) induced a decrease in the same proliferative responses (Rinner *et al.*, 1992). In another example, Manuck *et al.* showed that acute psycho-logical stress induced a significant increase in blood Cytolytic T Lympho-cyte (CTL) numbers only in those subjects who showed heightened catecholamine and cardiovascular reactions to stress (Manuck *et al.*, 1991).

Thus, an acute stress response may induce biphasic changes in blood leukocyte numbers. Soon after the beginning of stress (order of minutes) or during mild acute stress or exercise, catecholamine hormones and neu-rotransmitters induce the body's "soldiers" (leukocytes) to exit their "bar-racks" (spleen, lung, marginated pool, and other organs) and enter the "boulevards" (blood vessels and lymphatics). This results in an increase in blood leukocyte numbers, the effect being most prominent for NK cells and granulocytes. As the stress response continues, activation of the HPA axis results in the release of glucocorticoid hormones that induce leukocytes to exit the blood and take position at potential "battle stations" (skin, mucosal lining of gastrointestinal and urinary-genital tracts, lung, liver, and lymph nodes) in preparation for immune challenges that may be imposed by the actions of the stressor (Dhabhar *et al.*, 1995a; Dhabhar and McEwen, 1996, 2001). Such a redistribution of leukocytes results in a decrease in blood leukocyte numbers, the effect being most prominent for T and B lympho-cytes, NK cells, and monocytes. Thus, acute stress may result in a redistrib-ution of leukocytes from the barracks, through the boulevards, and to potential battle stations within the body.

8. Stress-induced Redistribution of Blood Leukocytes: Molecular Mechanisms

It is likely that the observed changes in leukocyte distribution are mediated by changes in either the expression, or affinity, of adhesion molecules on leukocytes and/or endothelial cells. It has been suggested that after stress or after glucocorticoid treatment, specific leukocyte subpopulations (being transported by blood and lymph through different body compartments) may be selectively retained in those compartments in which they encounter a stress- or glucocorticoid-induced "adhesion match" (Dhabhar and McEwen, 1999a). As a result of this selective retention, the proportion of specific leukocyte subpopulations would decrease in the blood, whereas it increases in the organ in which they are retained (e.g., the skin) (Viswanathan and Dhabhar, 2005).

Support for this hypothesis comes from studies that show acute psychological stressors such as public speaking can induce significant changes in leukocyte adhesion molecules (Mills and Dimsdale, 1996; Goebel and Mills, 2000; Bauer et al., 2001; Redwine et al., 2003, 2004; Shephard, 2003). Prednisolone has also been shown to induce the retention of circulating lymphocytes within the bone marrow, spleen, and some lymph nodes thus resulting in a decrease in lymphocyte numbers in the thoracic duct and a concomitant decrease in numbers in the peripheral blood (Spry, 1972; Cox and Ford, 1982). Moreover, glucocorticoid hormones also influence the production of cytokines (Danes and Araneo, 1989) and lipocortins (Hirata, 1989), which in turn can affect the surface adhesion properties of leukocytes and endothelial cells. Further investigation of the effects of endogenous glucocorticoids (administered in physiologic doses and examined under physiologic kinetic conditions) on changes in expression/activity of cell surface adhesion molecules and on leukocyte–endothelial cell adhesion is necessary.

9. Stress-induced Redistribution of Blood Leukocytes: Functional Consequences

It has been proposed that a stress-induced decrease in blood leukocyte numbers may represent an adaptive response (Dhabhar et al., 1994; Dhabhar and McEwen, 1999a, 2001) reflecting a redistribution of leukocytes from the blood to other organs such as the skin, lining of gastrointestinal and urinary-genital tracts, lung, liver, and lymph nodes, which may serve as potential "battle stations" should the body's defenses be breached. Furthermore, such a leukocyte redistribution may enhance immune function in those compartments to which leukocytes traffic during stress (Dhabhar et al., 1994; Dhabhar and McEwen, 1996, 1999a).

Thus, an acute stress response may direct the body's "soldiers" (leukocytes), to exit their "barracks" (spleen and bone marrow), travel the "boulevards" (blood vessels), and take position at potential "battle stations" (skin, lining of gastrointestinal and urinary-genital tracts, lung, liver, and lymph nodes) in preparation for immune challenge (Dhabhar and McEwen, 1996, 1997, 1999a). In addition to sending leukocytes to potential "battle stations," stress hormones may also better equip them for "battle" by enhancing processes like antigen presentation, phagocytosis, and antibody production. Thus, a hormonal alarm signal released by the brain upon detecting a stressor may "prepare" the immune system for potential challenges (wounding or infection) that may arise due to the actions of the stress-inducing agent (e.g., a predator or attacker).

The above hypothesis has profound functional and therapeutic implications. Acute psycho-physiological stress may act as an endogenous adjuvant to bolster the effects of natural or therapeutic immunization. Indeed, studies have shown that acute stress administered at the time of primary (Dhabhar and Viswanathan, 2005; Viswanathan et al., 2005) or secondary (Dhabhar and McEwen, 1996) immunization induces a significant increase in indices of innate (Baumann et al., 1983; Pos et al., 1988; Lyte et al., 1990), cell-mediated (Dhabhar and McEwen, 1996, 1999a; Dhabhar et al., 2000; Bilbo et al., 2002; Saint-Mezard et al., 2003), and humoral immunity (Zalcman et al., 1991, 1993). Stress-induced trafficking of leukocytes to sites of immune activation or surgery may also enhance the rate of wound healing and recovery (Viswanathan and Dhabhar, 2005). It is important to keep in mind that a stress-induced increase in leukocyte trafficking to sites of immune activation is like a double-edged sword: it may be beneficial for promoting immunoprotection during surgery, wound healing, vaccination, infection, or localized cancer. However, it may also mediate stress-induced exacerbations of inflammatory (e.g., cardiovascular disease, gingivitis) and autoimmune (e.g., psoriasis, arthritis, multiple sclerosis) diseases (Amkraut et al., 1971; Al Abadie et al., 1994; Garg et al., 2001; Ackerman et al., 2002).

When interpreting data showing stress-induced changes in functional assays such as lymphocyte proliferation or NK activity, it may be important to bear in mind the effects of stress on the leukocyte composition of the compartment in which an immune parameter is being measured. For example, it has been shown that acute stress induces a redistribution of leukocytes from the blood to the skin and that this redistribution is accompanied by a significant enhancement of a skin cell mediated immune (CMI) response (Dhabhar and McEwen, 1996). In what might at first glance appear to be contradicting results, acute stress has been shown to suppress splenic and peripheral blood responses to T-cell mitogens (Cunnick et al., 1990) and splenic IgM production (Zalcman and Anisman, 1993). However, it is important to note that in contrast with the skin that is enriched in leukocytes during acute stress, peripheral blood and spleen are relatively depleted of leukocytes during acute stress. This stress-induced decrease in blood and spleen leukocyte numbers may contribute to the acute stress–induced suppression of immune function in these compartments.

Moreover, in contrast with acute stress, chronic stress has been shown to suppress skin CMI, and a chronic stress–induced suppression of blood leukocyte redistribution is thought to be one of the factors mediating the immunosuppressive effect of chronic stress (Dhabhar and McEwen, 1997). Again, in what might appear to be contradicting results, chronic stress has been shown to enhance mitogen-induced proliferation of splenocytes (Monjan and Collector, 1997) and splenic IgM production (Zalcman and Anisman, 1993). However, the spleen is relatively enriched in T cells during chronic glucocorticoid administration, suggesting that it may also be relatively enriched in T cells during chronic stress (Miller *et al.*, 1994), and this increase in spleen leukocyte numbers may contribute to the chronic stress–induced enhancement of immune parameters measured in the spleen.

It is also important to bear in mind that the heterogeneity of the stress-induced changes in leukocyte distribution (Dhabhar *et al.*, 1995a) suggests that using equal numbers of leukocytes in a functional assay may not account for stress-induced changes in relative percentages of different leukocyte subpopulations in the cell suspension being assayed. For example, samples that have been equalized for absolute numbers of total blood leukocytes from control versus stressed animals may still contain different numbers of specific leukocyte subpopulations (e.g., T cells, B cells, or NK cells). Such changes in leukocyte composition may mediate the effects of stress even in functional assays using equalized numbers of leukocytes from different treatment groups. This possibility needs to be taken into account before concluding that a given treatment changes an immune parameter on a "per cell" rather than a "per population" basis.

10. Conclusion

It is important to recognize that the relationship between the psychological and physiological manifestations of stress and immune function is complex. Whereas decades of research have examined the pathological effects of stress on immune function and on health, the study of salubrious or health-promoting effects of stress is relatively new (Dhabhar *et al.*, 1995a). Much work remains to be done to elucidate the mechanisms mediating these bidirectional effects of stress on health and to translate basic findings regarding the adaptive effects of stress from bench to bedside.

An important function of endocrine mediators released under conditions of acute stress may be to ensure that appropriate leukocytes are present in the right place and at the right time to respond to an immune challenge that might be initiated by the stress-inducing agent (e.g., attack by a predator, invasion by a pathogen, etc.). The modulation of immune cell distribution by acute stress may be an adaptive response designed to enhance immune surveillance and increase the capacity of the immune system to respond to challenge in immune compartments (such as the skin and the epithelia of lung, gastrointestinal and urinary-genital tracts) that serve as major defense barriers for the body. Thus, neurotransmitters and hormones released during stress may increase immune surveillance and help enhance immune preparedness for potential (or ongoing) immune challenge. Such stress-induced increases in leukocyte trafficking may enhance immunoprotection during surgery, vaccination, or infection but may also exacerbate immunopathology during inflammatory (cardiovascular disease, gingivitis) or autoimmune (psoriasis, arthritis, multiple sclerosis) diseases.

Acknowledgments. I wish to thank Kanika Ghai for reviewing this manuscript and for her helpful comments and suggestions. This work was supported in part by grants from the NIH (AI48995 and CA107498) and The Dana Foundation.

References

Ackerman, K.D., Heyman, R., Rabin, B.S., Anderson, B.P., Houck, P.R., Frank, E., and Baum, A. (2002). Stressful life events precede exacerbations of multiple sclerosis. *Psychosom. Med.* 64:916–920.

Al'Abadie, M.S., Kent, G.G., and Gawkrodger, D.J. (1994). The relationship between stress and the onset and exacerbation of psoriasis and other skin conditions. *Br. J. Dermatol.* 130:199–203.

Amkraut, A.A., Solomon, C.F., and Kraemer, H.C. (1971). Stress, early experience and adjuvant-induced arthritis in the rat. *Psychosom. Med.* 33:203–214.

Bauer, M.E., Perks, P., Lightman, S.L., and Shanks, N. (2001). Are adhesion molecules involved in stress-induced changes in lymphocyte distribution? *Life Sci.* 69: 1167–1179.

Baumann, H., Firestone, G.L., Burgess, T.L., Gross, K.W., Yamamoto, K.R., and Held, W.A. (1983). Dexamethasone regulation of a1-acid glycoprotein and other acute phase reactants in rat liver and hepatoma cells. *J. Biol. Chem.* 258:563–570.

Benschop, R.J., Oostveen, F.G., Heijnen, C.J., and Ballieux, R.E. (1993). Beta 2-adrenergic stimulation causes detachment of natural killer cells from cultured endothelium. *Eur. J. Immunol.* 23:3242–3247.

Benschop, R.J., Rodriguez-Feuerhahn, M., and Schedlowski, M. (1996). Catecholamine-induced leukocytosis: Early observations, current research, and future directions. *Brain Behav. Immun.* 10:77–91.

Bilbo, S.D., Dhabhar, F.S., Viswanathan, K., Saul, A., Yellon, S.M., and Nelson, R.J. (2002). Short day lengths augment stress-induced leukocyte trafficking and stress-induced enhancement of skin immune function. *Proc. Natl. Acad. Sci. U.S.A.* 99: 4067–4072.

Bosch, J.A., Berntson, G.G., Cacioppo, J.T., Dhabhar, F.S., and Marucha, P.T. (2003). Acute stress evokes selective mobilization of T cells that differ in chemokine receptor expression: A potential pathway linking immunologic reactivity to cardiovascular disease. *Brain Behav. Immun.* 17:251–259.

Brosschot, J.F., Benschop, R.J., Godaert, G.L., Olff, M., De Smet, M., Heijnen, C.J., and Ballieux, R.E. (1994). Influence of life stress on immunological reactivity to mild psychological stress. *Psychosom. Med.* 56:216–224.

Carlson, S.L., Fox, S., and Abell, K.M. (1997). Catecholamine modulation of lymphocyte homing to lymphoid tissues. *Brain Behav. Immun.* 11:307–320.

Claman, H.N. (1972). Corticosteroids and lymphoid cells. *N. Engl. J. Med.* 287: 388–397.

Cohen, J.J. (1972). Thymus-derived lymphocytes sequestered in the bone marrow of hydrocortisone-treated mice. *J. Immunol.* 108:841–844.

Cohen, J.J. (1992). Glucocorticoid-induced apoptosis in the thymus. *Semin. Immunol.* 4:363–369.

Cox, J.H., and Ford, W.L. (1982). The migration of lymphocytes across specialized vascular endothelium. IV. Prednisolone acts at several points on the recirculation pathway of lymphocytes. *Cell Immunol.* 66:407–422.

Cunnick, J.E., Lysle, D.T., Kucinski, B.J., and Rabin, B.S. (1990). Evidence that shock-induced immune suppression is mediated by adrenal hormones and peripheral beta-adrenergic receptors. *Pharmacol. Biochem. Behav.* 36:645–651.

Danes, R.A., and Araneo, B.A. (1989). Contrasting effects of glucocorticoids on the capacity of T cells to produce the growth factors interleukin 2 and interleukin 4. *Eur. J. Immunol.* 19:2319–2325.

Dhabhar, F.S. (1998). Stress-induced enhancement of cell-mediated immunity. In S.M. McCann, J.M. Lipton, E.M. Sternberg, G.P. Chrousos, P.W. Gold, and C.C. Smith (eds.), *Neuroimmunomodulation: Molecular, Integrative Systems, and Clinical Advances.* New York: New York Academy of Sciences, pp. 359–372.

Dhabhar, F.S., and McEwen, B.S. (1996). Stress-induced enhancement of antigen-specific cell-mediated immunity. *J. Immunol.* 156:2608–2615.

Dhabhar, F.S., and McEwen, B.S. (1997). Acute stress enhances while chronic stress suppresses immune function in vivo: A potential role for leukocyte trafficking. *Brain Behav. Immun.* 11:286–306.

Dhabhar, F.S., and McEwen, B.S. (1999a). Enhancing versus suppressive effects of stress hormones on skin immune function. *Proc. Natl. Acad. Sci. U.S.A.* 96: 1059–1064.

Dhabhar, F.S., and McEwen, B.S. (1999b). Changes in blood leukocyte distribution: interactions between catecholamine & glucocorticoid hormones. *Neuroimmunomodulation.* 6:213.

Dhabhar, F.S., and McEwen, B.S. (2001). Bidirectional effects of stress & glucocorticoid hormones on immune function: Possible explanations for paradoxical observations. In R. Ader, D.L. Felten, and N. Cohen (eds.), *psychoneuroimmunology,* Third ed. San Diego: Academic Press, pp. 301–338.

Dhabhar, F.S., and Viswanathan, K. (2005). Short-term stress experienced at the time of immunization induces a long-lasting increase in immunological memory. *Am. J. Physiol. Regul. Integr. Comp. Physiol.* 289:R738–744.

Dhabhar, F.S., McEwen, B.S., and Spencer, R.L. (1993). Stress response, adrenal steroid receptor levels, and corticosteroid-binding globulin levels—a comparison between Sprague Dawley, Fischer 344, and Lewis rats. *Brain Res.* 616:89–98.

Dhabhar, F.S., Miller, A.H., Stein, M., McEwen, B.S., and Spencer, R.L. (1994). Diurnal and stress-induced changes in distribution of peripheral blood leukocyte subpopulations. *Brain Behav. Immun.* 8:66–79.

Dhabhar, F.S., Miller, A.H., McEwen, B.S., and Spencer, R.L. (1995a). Effects of stress on immune cell distribution—dynamics and hormonal mechanisms. *J. Immunol.* 154:5511–5527.

Dhabhar, F.S., Miller, A.H., McEwen, B.S., and Spencer, R.L. (1995b). Differential activation of adrenal steroid receptors in neural and immune tissues of Sprague Dawley, Fischer 344, and Lewis rats. *J. Neuroimmunol.* 56:77–90.

Dhabhar, F.S., Miller, A.H., McEwen, B.S., and Spencer, R.L. (1996). Stress-induced changes in blood leukocyte distribution—role of adrenal steroid hormones. *J. Immunol.* 157:1638–1644.

Dhabhar, F.S., Satoskar, A.R., Bluethmann, H., David, J.R., and McEwen, B.S. (2000). Stress-induced enhancement of skin immune function: A role for IFNγ. *Proc. Natl. Acad. Sci. U.S.A.* 97:2846–2851.

Dougherty, R.F., and White, A. (1945). Functional alterations in lymphoid tissue induced by adrenal cortical secretion. *Am. J. Anatomy* 77:81–116.

Engler, H., Dawils, L., Hoves, S., Kurth, S., Stevenson, J.R., Schauenstein, K., and Stefanski, V. (2004). Effects of social stress on blood leukocyte distribution: The role of alpha- and beta-adrenergic mechanisms. *J. Neuroimmunol.* 156:153–162.

Fauci, A.S. (1975). Mechanisms of corticosteroid action on lymphocyte subpopulations. I. Redistribution of circulating T and B lymphocytes to the bone marrow. *Immunology* 28:669–680.

Fauci, A.S. (1976). Mechanisms of corticosteroid action on lymphocyte subpopulations. II. Differential effects of in vivo hydrocortisone, prednisone, and dexamethasone on in vitro expression of lymphocyte function. *Clin. Exp. Immunol.* 24:54–62.

Fauci, A.S., and Dale, D.C. (1974). The effect of in vivo hydrocortisone on subpopulations of human lymphocytes. *J. Clin. Invest.* 53:240–246.

Fauci, A.S., and Dale, D.C. (1975). The effect of hydrocortisone on the kinetics of normal human lymphocytes. *Blood* 46:235–243.

Garg, A., Chren, M.M., Sands, L.P., Matsui, M.S., Marenus, K.D., Feingold, K.R., and Elias, P.M. (2001). Psychological stress perturbs epidermal permeability barrier homeostasis: Implications for the pathogenesis of stress-associated skin disorders. *Arch. Dermatol.* 137:53–59.

Goebel, M.U., and Mills, P.J. (2000). Acute psychological stress and exercise and changes in peripheral leukocyte adhesion molecule expression and density. *Psychosom. Med.* 62:664–670.

Goldstein, D.S., and McEwen, B. (2002). Allostasis, homeostats, and the nature of stress. *Stress* 5:55–58.

Harris, T.J., Waltman, T.J., Carter, S.M., and Maisel, A.S. (1995). Effect of prolonged catecholamine infusion on immunoregulatory function: Implications in congestive heart failure. *J. Am. Coll. Cardiol.* 26:102–109.

Herbert, T.B., and Cohen, S. (1993). Stress and immunity in humans: A meta-analytic review. *Psychosom. Med.* 55:364–379.

Hirata, F. (1989). The role of lipocortins in cellular function as a second messenger of glucocorticoids. In R.P. Schleimer, H.N. Claman, and A. Oronsky (eds.),

Anti-inflammatory Steroid action—Basic and Clinical Aspects. San Diego: Academic Press, pp. 67–95.

Hoagland, H., Elmadjian, F., and Pincus, G. (1946). Stressful psychomotor performance and adrenal cortical function as indicated by the lymphocyte reponse. *J. Clin. Endocrinol.* 6:301–311.

Irwin, M., Patterson, T., Smith, T.L., Caldwell, C., Brown, S.A., Gillin, C.J., and Grant, I. (1990). Reduction of immune function in life stress and depression. *Biol. Psychiatry* 27:22–30.

Jensen, M.M. (1969). Changes in leukocyte counts associated with various stressors. *J. Reticuloendothelial Soc.* 8:457–465.

Kok-van Alphen, C.C., and Volker-Dieben, H.J. (1983). Emotional stress and rejection, cause and effect. *Doc. Ophthalmol.* 56:171–175.

Landmann, R. (1992). Beta-adrenergic receptors in human leukocyte subpopulations. *Eur. J. Clin. Invest.* 22(Suppl. 1):30–36.

Landmann, R., Muller, F.B., Perini, C.H., Wesp, M., Erne, P., and Buhler, F.R. (1984). Changes of immunoregulatory cells induced by psychological and physical stress: Relationship to plasma catecholamines. *Clin. Exp. Immunol.* 58:127–135.

Lyte, M., Nelson, S.G., and Thompson, M.L. (1990). Innate and adaptive immune responses in a social conflict paradigm. *Clin. Immunol. Immunopathol.* 57:137–147.

Manuck, S.B., Cohen, S., Rabin, B.S., Muldoon, M.F., and Bachen, E.A. (1991). Individual differences in cellular immune response to stress. *Psychol. Sci.* 2:111–115.

McEwen, B.S. (1998). Protective and damaging effects of stress mediators: Allostasis and allostatic load. *N. Engl. J. Med.* 338:171–179.

McEwen, B.S. (2002). The end of stress as we know it. Washington, DC: Dana Press, p. 239.

Miller, A.H., Spencer, R.L., Stein, M., and McEwen, B.S. (1990). Adrenal steroid receptor binding in spleen and thymus after stress or dexamethasone. *Am. J. Physiol.* 259:E405–E412.

Miller, A.H., Spencer, R.L., Stein, M., and McEwen, B.S. (1991). Adrenal steroid receptor activation in vivo and immune function. *Am. J. Physiol.* 261:E126–E131.

Miller, A.H., Spencer, R.L., Hasset, J., Kim, C.H., Husain, A., McEwen, B.S., and Stein, M. (1992). Type I and type II adrenal steroid receptor agonists have selective effects on peripheral blood immune cells in the rat. Annual Meeting of the Society for Neuroscience Publishers: Society for Neuroscience, Washington, DC Abst. 424.9, Annaheim, CA, p. 1010 (Abstract).

Miller, A.H., Spencer, R.L., Husain, A., Rhee, R., McEwen, B.S., and Stein, M. (1993). Differential expression of type I adrenal steroid receptors in immune tissues is associated with tissue-specific regulation of type II receptors by aldosterone. *Endocrinology* 133:2133–2139.

Miller, A.H., Spencer, R.L., Hasset, J., Kim, C., Rhee, R., Cira, D., Dhabhar, F.S., McEwen, B.S., and Stein, M. (1994). Effects of selective type I and type II adrenal steroid receptor agonists on immune cell distribution. *Endocrinology* 135: 1934–1944.

Mills, P.J., and Dimsdale, J.E. (1996). The effects of acute psychologic stress on cellular adhesion molecules. *J. Psychosom. Res.* 41:49–53.

Mills, P.J., Berry, C.C., Dimsdale, J.E., Ziegler, M.G., Nelesen, R.A., and Kennedy, B.P. (1995). Lymphocyte subset redistribution in response to acute experimental stress: Effects of gender, ethnicity, hypertension, and the sympathetic nervous system. *Brain Behav. Immun.* 9:61–69.

Mills, P.J., Ziegler, M.G., Rehman, J., and Maisel, A.S. (1998). Catecholamines, catecholamine receptors, cell adhesion molecules, and acute stressor-related changes in cellular immunity. *Adv. Pharmacol.* 42:587–590.

Mills, P.J., Meck, J.V., Waters, W.W., D'Aunno, D., and Ziegler, M.G. (2001). Peripheral leukocyte subpopulations and catecholamine levels in astronauts as a function of mission duration. *Psychosom. Med.* 63:886–890.

Monjan, A.A., and Collector, M.I. (1977). Stress-induced modulation of the immune response. *Science* 196:307–308.

Morrow-Tesch, J.L., McGlone, J.J., and Norman, R.L. (1993). Consequences of restraint stress on natural killer cell activity, behavior, and hormone levels in Rhesus Macaques (*Macaca mulatta*). *Psychoneuroendocrinology* 18:383–395.

Munck, A., Guyre, P.M., and Holbrook, N.J. (1984). Physiological functions of glucocorticoids in stress and their relation to pharmacological actions. *Endocr. Rev.* 5:25–44.

Naliboff, B.D., Benton, D., Solomon, G.F., Morley, J.E., Fahey, J.L., Bloom, E.T., Makinodan, T., and Gilmore, S.L. (1991). Immunological changes in young and old adults during brief laboratory stress. *Psychosom. Med.* 53:121–132.

Onsrud, M., and Thorsby, E. (1981). Influence of in vivo hydrocortisone on some human blood lymphocyte subpopulations. *Scand. J. Immunol.* 13:573–579.

Pickford, G.E., Srivastava, A.K., Slicher, A.M., and Pang P.K.T. (1971). The stress response in the abundance of circulating leukocytes in the Killifish, Fundulus heteroclitus. I The cold-shock sequence and the effects of hypophysectomy. *J. Exp. Zool.* 177:89–96.

Pos, O., Van Dijk, W., Ladiges, N., Linthorst, C., Sala, M., Van Tiel, D., and Boers, W. (1988). Glycosylation of four acute-phase glycoproteins secreted by rat liver cells in vivo and in vitro. Effects of inflammation and dexamethasone. *Eur. J. Cell Biol.* 46:121–128.

Pruett, S.B. (2001). Quantitative aspects of stress-induced immunomodulation. *Int. Immunopharmacol.* 1:507–520.

Redwine, L., Snow, S., Mills, P., and Irwin, M. (2003). Acute psychological stress: Effects on chemotaxis and cellular adhesion molecule expression. *Psychosom. Med.* 65:598–603.

Redwine, L., Mills, P.J., Sada, M., Dimsdale, J., Patterson, T., and Grant, I. (2004). Differential immune cell chemotaxis responses to acute psychological stress in Alzheimer caregivers compared to non-caregiver controls. *Psychosom. Med.* 66: 770–775.

Rinder, C.S., Mathew, J.P., Rinder, H.M., Tracey, J.B., Davis, E., and Smith, B.R. (1997). Lymphocyte and monocyte subset changes during cardiopulmonary bypass: Effects of aging and gender [see comments]. *J. Lab. Clin. Med.* 129:592–602.

Rinner, I., Schauenstein, K., Mangge, H., Porta, S., and Kvetnansky, R. (1992). Opposite effects of mild and severe stress on in vitro activation of rat peripheral blood lymphocytes. *Brain Behav. Immun.* 6:130–140.

Saint-Mezard, P., Chavagnac, C., Bosset, S., Ionescu, M., Peyron, E., Kaiserlian, D., Nicolas, J.F., and Berard, F. (2003). Psychological stress exerts an adjuvant effect on skin dendritic cell functions in vivo. *J. Immunol.* 171(8):4073–4080.

Sapolsky, R.M. (2004). Why zebras don't get ulcers. New York: W.H. Freeman and Company, p. 560.

Sapolsky, R.M. (2005). The influence of social hierarchy on primate health. *Science* 308:648–652.

Schedlowski, M., Jacobs, R., Stratman, G., Richter, S., Hädike, A., Tewes, U., Wagner, T.O.F., and Schmidt, R.E. (1993a). Changes of natural killer cells during acute psychological stress. *J. Clin. Immunol.* 13:119–126.

Schedlowski, M., Falk, A., Rohne, A., Wagner, T.O.F., Jacobs, R., Tewes, U., and Schmidt, R.E. (1993b). Catecholamines induce alterations of distribution and activity of human natural killer (NK) cells. *J. Clin. Immunol.* 13:344–351.

Schwab, C.L., Fan, R., Zheng, Q., Myers, L.P., Hebert, P., and Pruett, S.B. (2005). Modeling and predicting stress-induced immunosuppression in mice using blood parameters. *Toxicol. Sci.* 83:101–113.

Shephard, R.J. (2003). Adhesion molecules, catecholamines and leucocyte redistribution during and following exercise. *Sports Med.* 33:261–284.

Snow, D.H., Ricketts, S.W., and Mason, D.K. (1983). Hematological responses to racing and training exercise in thoroughbred horses, with particular reference to the leukocyte response. *Equine Vet. J.* 15:149–154.

Spain, D.M., and Thalhimer, W. (1951). Temporary accumulation of eosinophilic leucocytes in spleen on mice following administration of cortisone. *Proc. Soc. Exp. Biol. Med.* 76:320–322.

Spencer, R.L., Young, E.A., Choo, P.H., and McEwen, B.S. (1990). Adrenal steroid type I and type II receptor binding: Estimates of in vivo receptor number, occupancy, and activation with varying level of steroid. *Brain Res.* 514:37–48.

Spencer, R.L., Miller, A.H., Stein, M., and McEwen, B.S. (1991). Corticosterone regulation of type I and type II adrenal steroid receptors in brain, pituitary, and immune tissue. *Brain Res.* 549:236–246.

Spencer, R.L., Miller, A.H., Moday, H., Stein, M., and McEwen, B.S. (1993). Diurnal differences in basal and acute stress levels of type I and type II adrenal steroid receptor activation in neural and immune tissues. *Endocrinology* 133:1941–1949.

Sprent, J., and Tough, D.F. (1994). Lymphocyte life-span and memory. *Science* 265: 1395–1400.

Spry, C.J.F. (1972). Inhibition of lymphocyte recirculation by stress and corticotropin. *Cell Immunol.* 4:86–92.

Stefanski, V. (2003). Social stress affects migration of blood T cells into lymphoid organs. *J. Neuroimmunol.* 138:17–24.

Stein, M., Ronzoni, E., and Gildea, E.F. (1951). Physiological responses to heat stress and ACTH of normal and schizophrenic subjects. *Am. J. Psychiatry* 6: 450–455.

Toft, P., Svendsen, P., Tonnesen, E., Rasmussen, J.W., and Christensen, N.J. (1993). Redistribution of lymphocytes after major surgical stress. *Acta. Anesthesiol. Scand.* 37:245–249.

Tonnesen, E., Christensen, N.J., and Brinklov, M.M. (1987). Natural killer cell activity during cortisol and adrenaline infusion in healthy volunteers. *Eur. J. Clin. Invest.* 17:497–503.

Ulich, T.R., Keys, M., Ni, R.X., del Castillo, J., and Dakay, E.B. (1988). The contributions of adrenal hormones, hemodynamic factors, and the endotoxin-related stress reaction to stable prostaglandin analog-induced peripheral lymphopenia and neutrophilia. *J. Leukoc. Biol.* 43:5–10.

Van Den Broek, A.A., Keuning, F.J., Soeharto, R., and Prop, N. (1983). Immune suppression and histophysiology of the immune response I. Cortisone acetate and lymphoid cell migration. *Virchows Arch. [Cell Pathol.].* 43:43–54.

Viswanathan, K., and Dhabhar, F.S. (2005). Stress-induced enhancement of leuko-cyte trafficking into sites of surgery or immune activation. *Proc. Natl. Acad. Sci. U.S.A.* 102:5808–5813.

Viswanathan, K., Daugherty, C., and Dhabhar, F.S. (2005). Stress as an endogenous adjuvant: augmentation of the immunization phase of cell-mediated immunity. *Int. Immunol.* 17:1059–1069.

Walzer, P.D., LaBine, M., Redington, T.J., and Cushion, M.T. (1984). Lymphocyte changes during chronic administration of and withdrawal from corticosteroids: Relation to *pneumocystis carinii* pneumonia. *J. Immunol.* 133:2502–2508.

Zalcman, S., and Anisman, H. (1993). Acute and chronic stressor effects on the anti-body response to sheep red blood cells. *Pharmacol. Biochem. Behav.* 46:445–452.

Zalcman, S., Henderson, N., Richter, M., and Anisman, H. (1991). Age-related enhancement and suppression of a T-cell-dependent antibody response following stressor exposure. *Behav. Neurosci.* 105:669–676.

Zatz, M.M. (1975). Effects of cortisone on lymphocyte homing. *Isr. J. Med. Sci.* 11:1368–1372.

3
Stress-induced Sympathetic Nervous System Activation Contributes to Both Suppressed Acquired Immunity and Potentiated Innate Immunity: The Role of Splenic NE Depletion and Extracellular Hsp72

Monika Fleshner

1. Introduction

The following chapter will review the evidence that exposure to the same intense acute stressor can produce dichotomous effects on immune function and host defense that are dependent on the type of immune response tested and will develop the hypothesis that a common neuroendocrine mechanism exists for both outcomes, that is, activation of the sympathetic nervous system (SNS). First, I will describe research supporting the hypothesis that excessive SNS activation can produce suppression of a measure of acquired immunity, the *in vivo* antibody response to a benign protein. Importantly, only "excessive" SNS activation can produce the immunosuppression. Frequently, stressor exposure stimulates the SNS response but does not excessively drive the response to produce a state of catecholamine depletion in innervated tissues. We predict in the absence of tissue norepinephrine (NE) depletion, stimulation of the SNS will not suppress acquired immunity. Second, I will review the research supporting the hypothesis that stress-induced activation of SNS facilitates innate immune host defense. The role of the SNS in stress-induced potentiation of innate immunity contrasts with that of SNS effects on acquired immunity in that it does not depend on excessive SNS drive or tissue catecholamine depletion. In our studies, innate immune function was assessed by measuring the host's initial response to an *in vivo* bacterial challenge that depends on innate and not acquired immune cells (Campisi *et al.*, 2005). Finally, I will introduce a previously unrecognized feature of the stress response, the extracellular release of an endogenous stress protein (eHsp72), which is stimulated by SNS activation and functions as a "danger signal" for the immune system contributing to facilitated host defense.

1.1. Excessive Sympathetic Nervous System Output Is Detrimental to Health

Stimulation of the SNS is a hallmark of the acute stress response (Goldstein, 1987). SNS activation has many physiological consequences that work in concert to promote the fight-or-flight response (Jansen et al., 1995; Goldstein, 1996) and facilitate features of innate immunity (Campisi and Fleshner, 2003; Johnson et al., 2005). SNS activation is a powerful feature of the acute stress response that is adaptive when the response is acute and constrained. If, however, SNS activation is frequent or excessive, it can produce negative health consequences (Seals and Dinenno, 2004). For example, chronically elevated SNS responses are believed to mechanistically contribute to the etiology of "metabolic syndrome," a key antecedent to clinical atherosclerotic diseases that includes visceral adiposity, glucose intolerance, insulin resistance, dyslipidemia, and hypertension (Baron, 1990; Julius et al., 1992; Lind and Lithell, 1993). In addition, it has been reported in both the human and animal literatures that chronic or excessive SNS activation can lead to arterial wall thickening (Pauletto et al., 1991; Chen et al., 1995; Xin et al., 1997), hypertension (Lind and Lithell, 1993), α- and β-adrenergic receptor desensitization (Abrass, 1986; Xiao and Lakatta, 1992; Dinenno et al., 2002), and immunosuppression (Irwin, 1993; Kennedy et al., 2005b). The negative consequences of frequent and/or excessive SNS activity have been convincingly demonstrated in transgenic mice lacking functional α_{2A} adrenergic receptor (ADR) autoinhibition in the midbrain. Due to the lack of normal α_{2A}ADR central nervous constraint on SNS drive, these mice have chronically activated peripheral SNS responses and rapidly develop cardiac dysfunction (Baum et al., 2002). Unfortunately, neither tissue NE nor immune function was assessed in this study.

1.2. Stress Modulates Immune Function: Acquired versus Innate

Exposure to physical and/or psychological stress modulates the immune response (Adell et al., 1988; Plotnikoff, 1991; Laudenslager, 1994; Maier et al., 1994). Stress is neither globally immunosuppressive nor immunopotentiating but rather immunomodulatory. Factors that impact the effect of stress on the immune response include the following: the duration and intensity of stressor exposure (Monjan, 1976); the perceived controllability of the stressor (Laudenslager, 1983); the physiological state of the organism (e.g., young vs. old, anxious vs. calm, healthy vs. ill, and physically active vs. sedentary) (Brown, 1988; Ader, 1991; Dishman, 1995; Bonneau, 1997; Moraska and Fleshner, 2001; Fleshner et al., 2002); and the timing and measure of the immune response (e.g., days vs. hours, acquired vs. innate) (Fleshner et al., 1998; Deak et al., 1999).

1.3. Animal Model of Acute Stress

Our laboratory has been studying the behavioral and physiological consequences of exposure to a well-characterized animal model of stress: inescapable tailshock. This model of stress involves exposing rats to random, intermittent (average intertrial interval 60 s), inescapable tailshocks (100, 1.6 mA, 5 s), administered when the rats are lightly restrained in Plexiglas tubes. The use of this stressor is important for several reasons. First, a great deal is known about the behavioral, neural, endocrine, and immunological consequences of exposure to this acute stressor (Watkins, 1990; Fleshner, 1993; Laudenslager, 1994; Maier *et al.*, 1994; Brennan, 1995; Fleshner *et al.*, 1995a, 1995b, 1995c; Brennan, 1996; Deak, 1997; Deak *et al.*, 1997; Fleshner *et al.*, 1998; Maier, 1998; Milligan *et al.*, 1998; Nguyen *et al.*, 1998a, 1998b; Deak, 1999; O'Conner, 1999; Nguyen *et al.*, 2000; Moraska and Fleshner, 2001; Campisi *et al.*, 2002; Fleshner *et al.*, 2002; Moraska *et al.*, 2002; Campisi and Fleshner, 2003; Campisi *et al.*, 2003b, 2003c; Gazda *et al.*, 2003; Greenwood *et al.*, 2003a, 2003b; Day *et al.*, 2004). Second, the effects of acute stressor exposure on immune function are stressor dependent (Ader, 1991; Plotnikoff, 1991), therefore the use of a consistent stressor is necessary to advance our understanding of the mechanism responsible for stress-induced immunomodulation. Third, tailshock stress allows the administration of a discrete, consistent, and quantifiable stressor that *does not* produce physical injury.

2. Stress Suppresses Acquired Immunity

2.1. In Vivo *Generation of Antibody Against KLH Is a Measure of Acquired Immunity*

Acquired immunity is characterized by two primary features: exquisite antigen specificity and immunological memory. The effector cells of the acquired immune response include T cells and B cells. Our assessment of acquired immune function has been the generation of an immunoglobulin response to keyhole limpet hemocyanin (αKLH Ig). This measure of immune function has both experimental advantages and clinical relevance that include the following: (1) the cells involved with the generation of this response remain in the hormonal milieu of the organism; (2) the kinetics of the developing response can be easily monitored; (3) use of a benign protein does not produce the behavioral confounds associated with the generation of sickness; (4) antibody reflects a functionally important end product of the immune system; (5) measurement of the antigen specific antibody response more accurately reflects the function of acquired immunity; (6) measurement of αKLH Ig is quantifiable making the results directly comparable across studies; (7) the cells involved with this response are T cells and B cells, two primary players in acquired immune responses; (8) the antibody response generated against KLH is similar to the immunological response generated after vaccination to tetanus toxoid; (9) a reduction in specific antibodies to bacteria, virus, or soluble toxin could render the organism more susceptible to disease caused by these pathogens; (10) KLH is clinically relevant because it is used as a immunotherapeutic in the treatment of cancer (Lamm, 1993; Jurincic-Winkler, 1995; Livingston, 1995; Gilewski, 1996; Jurincic-Winkler, 1996), and stress-induced modulation of the antibody response to KLH could affect the efficacy of this type of vaccination and immunotherapy; (11) a final advantage to measuring αKLH responses is that the results we find in animals can easily be tested in humans (Smith et al., 2004a, 2004b).

2.2. The Spleen Is Site for Stress-induced KLH Antibody Suppression

Rats that are immunized with KLH and exposed to a single session of inescapable tailshock have a long-term (+21 days) reduction in serum levels of αKLH IgM, IgG, and IgG2a (Laudenslager et al., 1988; Fleshner et al., 1995d, 1998; Gazda et al., 2003). We know that the final site of stress-induced immunomodulation is the spleen because if we remove the spleen from adult male rats prior to intraperitoneal immunization with KLH and stressor exposure, we eliminate the stress-induced reduction in αKLH Ig (Fleshner, 2005). Importantly, the stress-associated suppressive effect is specific to the generation of antibody to the antigen. Total serum IgM and IgG is not reduced (Fleshner et al., 1992; Smith et al., 2004b).

2.3. Cellular Mechanisms of Stress-induced KLH Antibody Suppression

The generation of an antibody response to a T cell–dependent soluble protein, such as KLH, involves the interaction of antigen presenting cells (APCs; B cells and/or dendritic cells), T helper cells (Th), and B cells. After intraperitoneal injection of KLH, antigen is transported to the draining lymph nodes and spleen. B cells expressing the B-cell receptor (BCR) that bind KLH must receive T-cell help from the KLH-specific T helper cells in the form of costimulation and cytokines. The Th "help" facilitates B-cell proliferation, B-cell differentiation into antibody-secreting cells (Clark and Ledbetter, 1994; Foy et al., 1996), and Ig isotype switching (IgM to IgG or IgG2a (Stevens et al., 1988)). The proliferation of KLH-specific Th and B cells is greatest in the draining lymph nodes and spleen 4–7 days after KLH (Fleshner et al., 1995d, 1998; Gazda et al., 2003). Using flow cytometric analysis (Fleshner et al., 1995d, 1998), ELISPOT (Laudenslager, 1994), and antigen-specific proliferative assays (Gazda et al., 2003), we have determined that the suppression in αKLH Ig is likely due to a failure of the stressed rats to increase KLH-specific T helper cell numbers (Fleshner *et al.*, 1995d, 1998). With fewer αKLH T helper cells, there is less T-cell help and fewer KLH-specific B in the spleen (Laudenslager, 1994). Fewer KLH-specific B cells leads to a reduction in serum αKLH Ig. Thus, tailshock-induced suppression of αKLH Ig is a well-characterized animal model of stress-induced immunosuppression.

2.4. Excessive Sympathetic Nervous System Response Suppresses Acquired Immunity

Although the specific mechanism responsible for stress-induced suppression of αKLH Ig remains under investigation, excessive SNS output likely plays a role. Most primary and secondary lymphoid tissues (including the spleen) receive dense SNS innervation (Felten, 1987; Felten and Olschowka, 1987; Meltzer, 1997), and Th cells (Sanders, 1997; Kohm and Sanders, 2000, 2001; Swanson *et al.*, 2001), B cells (Kasprowicz *et al.*, 2000; Kohm *et al.*, 2002; Podojil and Sanders, 2003; Podojil *et al.*, 2004), and mono-cytes-macrophages-dendritic cells (Takahashi *et al.*, 2004) express adrener-gic receptors β_2ADR. If we focus on the role of the SNS in stress-induced immunomodulation, there is evidence that SNS contributes to stress-induced suppression of specifically the αKLH Ig response (Irwin, 1993). Although earlier work suggested that high concentrations of NE could suppress various aspects of immunity, more recent data support the hypothesis that splenic NE depletion, not circulating or splenic NE elevation, may be responsible for stress-induced suppression of *in vivo* αKLH Ig responses.

There are several lines of evidence to support this shift in dogma from "too much NE" to "too little NE." First, if one examines the past literature demonstrating that high levels of NE are immunosuppressive, many studies were done *in vitro*, examined mitogen-stimulated proliferative or cytokine responses, and tested pharmacological concentrations of NE (Ramer-Quinn, 1997; Malarkey *et al.*, 2002). Under these circumstances, NE suppresses immune function and can be fatal to immune cells (Del Rey *et al.*, 2003). Second, activation status of the immune cells was rarely considered in these earlier studies. For example, it was recently reported that modulation of dendritic cell function after NE exposure occurred only in the early phases of dendritic cell activation (Maestroni, 2002), and β_2ADR are differentially expressed on naïve versus stimulated B cells (Sanders *et al.*, 2003). Thus past research supporting a simple view that too much NE is responsible for stress-induced suppression of *in vivo* immune responses has limitations.

Recent evidence is consistent with the dogmatic shift that too little NE may be responsible for stress-induced suppression of *in vivo* antibody responses and that dynamic interactions between SNS and immune cells occur to produce optimal Ig responses. For example, during the generation

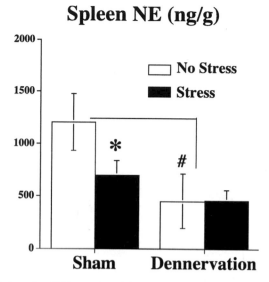

Spleen NE (ng/g)

FIGURE 3.1. Adult male F344 rats (n = 10 per group) were deeply anesthetized and, using a dissection scope, the incoming splenic nerve was transected under halothane anesthesia. Sham-operated controls were also surgical manipulated, but the splenic nerve remained intact. Rats were allowed 3 weeks to recover from surgery. After recovery, rats were either exposed to tailshock stress (100, 5s, 1.6mA) or no stress and sacrificed immediately after stressor termination. Spleens were removed and NE content was measured using HPLC. The NE depletion produced by dennervation was equal to that produced by tailshock stress. Because splenic dennervation prevents SNS drive from brain to spleen, tailshock stress did not deplete the NE in spleen further in dennervated rats. (*p < 0.05 sham stressed is different from sham not stressed control; #p < 0.05 dennervation not stressed is different from sham not stressed control.)

of an *in vivo* antibody response to KLH, NE is released from peripheral nerves innervating the spleen (Kohm *et al.*, 2000). NE binding to the B-cell β_2ADR stimulates the expression of costimulatory molecules (Kohm *et al.*, 2002), Ig production (Kasprowicz *et al.*, 2000), and splenic germinal center formation (Kohm, 1999). As depicted in Figure 3.1 and Figure 3.2, depletion of splenic NE content to an equal level as that produced by tailshock stress by cutting the splenic nerve or dennervating the spleen (Fig. 3.1) prior to *in vivo* KLH immunization reduces αKLH Ig (Fig. 3.2). In addition, splenic NE depletion produced by pharmacological lesion [6-OHDA (Kohm, 1999)] or pharmacological competition [α-methyl-*p*-tyrosine (Kennedy *et al.*, 2005b)] prior to immunization with KLH has also been reported to reduce the antibody response. Thus, splenic NE depletion in the absence of stress is sufficient for suppression αKLH Ig. We also have evidence that stress-induced suppression of αKLH Ig requires splenic NE depletion and not circulating NE elevation (Kennedy *et al.*, 2005b). Rats

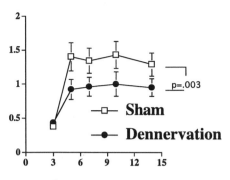
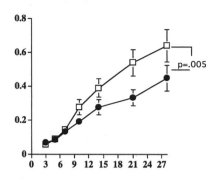

FIGURE 3.2. Adult male F344 rats were deeply anesthetized and, using a dissection scope, the incoming splenic nerve was transected. Sham-operated controls were also surgical manipulated, but the splenic nerve remained intact. Rats were allowed 3 weeks to recover from surgery and were immunized with KLH and returned to their home cages (n = 15; denervation vs. n = 15; sham). Blood samples were collected across days. Splenic dennervation suppressed αKLH IgM and IgG. Data are reported as relative optical density units (OD). (p values reported for surgery × time interactions.)

treated with a substrate for NE synthesis (tyrosine) prior to stressor exposure are protected from stress-induced splenic NE depletion and αKLH Ig suppression. Importantly, blood concentrations of NE in the tyrosine-treated stressed rats were equal to saline-injected stressed rats, yet tyrosine completely prevented the suppression in αKLH Ig. Tyrosine is a precursor for the synthesis of NE (and dopamine (DA)) and during times of intense SNS drive can be rate limiting (Gibson and Wurtman, 1977; Milner and Wurtman, 1987; Acworth et al., 1988). Furthermore, central activation of the SNS in the absence of stressor exposure with an α_{2A}ADR antagonist (Mirtazapine, Mirt) that acts in the brain to release the SNS from α_{2A}ADR-mediated inhibition (Dazzi et al., 2002) elevates blood NE for longer duration and to a higher level than that produced by stress. Yet, in spite of high blood concentrations of NE at the time of immunization, Mirt produces neither splenic NE depletion nor αKLH Ig suppression (Kennedy et al., 2005b). Blood NE is derived from spillover of NE released by nerve terminals in sympathetically innervated tissues. We speculate that the lack of splenic NE depletion in spite of equal or greater blood concentration on NE after Mirt injection may be due to a more global, whole-body activation of the SNS; whereas tailshock stress may activate more selective central SNS circuits (Greenwood et al., 2003b) perhaps excessively driving SNS output to select tissues such as the spleen.

2.5. Summary

These results suggest that excessive stress-induced activation of the SNS response is detrimental to acquired immunity. Why would our normally adaptive acute stress response produce immunosuppression? It would seem unlikely that this feature of stress response would have survived evolutionary selection. Although speculative, I can offer several possible explanations. First, our data suggest that organisms are vulnerable to antibody suppression only if antigen (KLH) is administered in the presence of splenic NE depletion. Our stress model produces splenic NE depletion and a window of susceptibility that is ~5–6 h in duration (Kennedy et al., 2005b). Clearly, exposure to intense acute stressors such as combat stress during war or intense physical activity (marathon) does not always lead to illness. This may due in part to the narrow window of susceptibility described in our work. Perhaps intense acute stressor exposure and pathogen exposure do not always align. This would be fortuitous for the host, making it less likely that the host will suffer the negative consequences of intense acute stressor exposure on acquired immunity and host defense, although, unfortunately, chronic or enduring mental or physical stressors may produce longer-lasting splenic NE depletions that would likely result in an extended window of stress susceptibility. In addition, we propose that SNS activation during acute stressor exposure is in most circumstances adaptive to host defense. The second half of this chapter will present evidence that SNS activation facilitates a different branch of the immunity, the innate immune response.

3. Stress Facilitates Innate Immunity

3.1. Introduction

The innate immune response is commonly referred to as the first line of defense. This more primitive type of immunity differs from acquired immunity in that innate immune cells recognize molecular patterns, rather than exquisitely specific amino acid sequences, and innate immunity lacks immunological memory. The primary effector cells of the innate immune system are macrophages and neutrophils. Several measures of innate immunity are optimized after exposure to an acute stressor in a healthy organism and include the following: the acute-phase response (APR), neutrophil and macrophage cellular function, local *in vivo* inflammatory responses, and recovery from bacterial challenge. The APR is a constellation of physiological changes initiated at a site of infection or trauma that collectively act to help clear the pathogen in a nonselective manner. Several components of the APR are triggered by exposure to acute stress. For example, uncontrollable tailshock stress increases levels of acute-phase proteins such as haptoglobulin (Deak et al., 1997) and $\alpha_1$acid glycoprotein (Deak et al.,

1997). In addition, circulating IL6 (Takaki *et al.*, 1994), complement function (Coe *et al.*, 1988; Fleshner *et al.*, 2002), and circulating neutrophils (Fleshner *et al.*, 2002) are increased after exposure to acute tailshock stress. Each of these substances can facilitate the development and resolution of the local inflammatory response (Baumann and Gauldie, 1994). In fact, Noursadighi *et al.* (2002) recently reported that prior pharmacological stimulation of the acute-phase response protected animals from lethal *Escherichia coli* infection, and this protection was due to enhanced early bacterial clearance, phagocytosis, and neutrophil oxidative burst. At the cellular level, acute stressor exposure increases the neutrophil oxidative burst (Smith and Pyne, 1997) and phagocytosis activity (Harmsen and Turney, 1985; Lyte *et al.*, 1990). In addition, lipopolysaccharide (LPS)-stimulated leukocyte nitric oxide (NO) (Fleshner *et al.*, 1998) and IL1β (Moraska *et al.*, 2002) responses are potentiated after tailshock exposure. Similar effects on antigen-stimulated NO production after exposure to a variety of acute stressors has also been reported (Coussons-Read *et al.*, 1994; Fecho *et al.*, 1994; Lysle *et al.*, 1995). Neutrophil oxidative burst, phagocytosis, NO production, TNF-α, and IL-1β production each play an important role in local inflammation (Ianaro *et al.*, 1994; Bellingan *et al.*, 1996; Ali *et al.*, 1997; Mac Micking, 1997; Dinarello, 2000). Many of the cellular products secreted by activated macrophages/neutrophils are potent, nonspecific suppressors of pathogen growth (Zidel, 1998). For example, NO, prostaglandin (PGE_2), and superoxide radicals are all toxic to cells and function in a nonspecific fashion to kill pathogens (Tomioka, 1992).

3.2. Subcutaneous Bacterial Challenge Is a Measure of Innate Immunity

To assess the effects of acute stress on innate immunity, we have completed a series of studies examining the effect of acute tailshock stress on the development and recovery from a subcutaneous challenge with live *E. coli*. The strain of bacteria and route of challenge were chosen for the following reasons: (1) the effect of subcutaneous *E. coli* in nonstressed rats has been thoroughly characterized. We (Campisi *et al.*, 2003a) have previously completed dose-response experiments testing the effect of *E. coli* on measures of innate immune activation (peripheral and local cytokines) and sickness responses (fever and spontaneous activity). (2) Previous work (Deak *et al.*, 1999) indicated that a subcutaneous injection of *E. coli*, rather than *Staphylococcus aureus*, stimulated a more consistent and robust *in vivo* inflammatory response. (3) *E. coli* challenge is known to stimulate an immediate innate inflammatory response (Campisi *et al.*, 2002), and macrophages and neutrophils are primarily responsible for host defense to this pathogen (Ali *et al.*, 1997; Fleshner *et al.*, 2002). (4) *E. coli* is a common pathogen found in nature as well as in the laboratory. (5) *E. coli* is not normally present in subcutaneous tissue of nonchallenged animals. Thus, we can be confident that any *E. coli* retrieved and cultured for assessment of colony-forming units (cfu) is from the experimental challenge and not from endogenous bacteria. (6) The robust inflammatory response induced by subcutaneous injection of *E. coli* allows *in vivo* measurements of various features of the sickness response, such as the diameter and grade of inflammation (Deak *et al.*, 1999; Campisi *et al.*, 2002; Fleshner *et al.*, 2002), fever (Campisi *et al.*, 2003a), circulating and local cytokines (Campisi *et al.*, 2003a), local NO (Campisi *et al.*, 2003b), body weight loss (Campisi *et al.*, 2002), and activity (Campisi *et al.*, 2003a). (7) A subcutaneous injection allows site-specific analyses of the *in vivo* immune response using an *ex vivo* approach. That is, after exposure to *in vivo* stress and bacterial challenge, we can remove the inflammatory site, place it in culture, and assess the ongoing response by sampling culture supernatants. This is especially advantageous because assessment of peripheral circulating leukocytes and circulating cytokines has limited relevance to the local immune response generated in response to challenge.

3.3. Stress Facilitates Recovery from Subcutaneous E. coli Challenge: A Role for NO

Using subcutaneous *E. coli*, we tested if exposure to acute tailshock prior to bacterial challenge would improve host defense due to potentiation of innate immunity (Campisi *et al.*, 2002; Campisi and Fleshner, 2003; Campisi *et al.*, 2003b). Rats exposed to tailshock stress and subcutaneously injected immediately after stressor termination with 2.5×10^9 CFU of freshly grown *E. coli* resolve their inflammation 10–14 days faster (Campisi *et al.*, 2002; Campisi and Fleshner, 2003; Campisi *et al.*, 2003b), experience less bacterial-induced body weight loss (Campisi *et al.*, 2002), and release 300% more nitric oxide (NO) at the inflammatory site compared with bacterially injected nonstressed controls (Campisi *et al.*, 2003b). It has been demonstrated that stress-induced potentiation of NO is important in host defense against bacteria because inhibition of NO at the inflammatory site with L-NIO (NOS inhibitor) reduces the effect of stress on facilitation of recovery from bacterial challenge (Campisi *et al.*, 2003b). NO contributes to almost every stage of inflammation by affecting leukocyte migration, adherence, antimicrobial activities, and phagocytic ability, and in fact can act to restrict the development of inflammation (Ali *et al.*, 1997). One beneficial effect of stress on recovery from bacterial challenge could be greater NO-mediated bacterial killing. Consistent with this idea is that rats challenged with *E. coli* after stress have fewer *E. coli* cfu retrieved from the inflammatory site 2, 4, and 6h after challenge compared with nonstressed *E. coli*–challenged controls (Campisi *et al.*, 2003b). The enhanced release of NO appears to be an important mediator; however, the mechanisms involved in stress-induced facilitation of NO and recovery from bacterial inflammation remain unknown. We propose that an endogenous stress protein, heat shock protein 72, may be responsible for this effect (Campisi and Fleshner, 2003).

4. Stress and Extracellular Heat Shock Proteins

4.1. Introduction

4.1.1. Intracellular Heat Shock Proteins

Heat shock proteins (Hsp) consist of several families of highly conserved proteins that play a role in a number of important cellular functions
(Morimoto, 1994). Induction of *intracellular* heat shock proteins were first
reported in 1962 (Ritossa, 1962) and the term *heat shock protein* was first
coined in 1974 (Tissieres *et al.*, 1974). The focus of the current chapter is on
one member of the 70-kDa Hsp (Hsp70) family of proteins, Hsp72. The
Hsp70 family of proteins includes the constituitive 73-kDa protein (HSC73)
and a highly stress-inducible 72-kDa protein (Hsp72) (Morimoto, 1994;
Hartl, 1996). *Intracellular* Hsp72 is found in nearly every cell of the body
and can be upregulated after exposure to a variety of cellular and organis-
mic stressors (Morimoto, 1994; Hartl, 1996). Although basal concentrations
of Hsp72 are low in most tissues, high concentrations of *intracellular* Hsp72
can be found in the absense of stressors in some tissues such as the frontal
cortex (Heneka *et al.*, 2003), pituitary (Campisi *et al.*, 2003c), adrenal
(Campisi *et al.*, 2003c), and brown fat (Matz *et al.*, 1996a, 1996b). The cellular functions of *intracellular* Hsp72 include limiting protein aggregation,
facilitating protein refolding, and chaperoning proteins (Morimoto, 1994;
Hartl, 1996) and function *en masse* to improve cell survival in the face of a
broad array of cellular stressors (Morimoto, 1994; Hartl, 1996). Induction
of Hsp72 is not simply a consequence of cellular stress but rather improves
resistance to death after cellular insult.

4.1.2. Extracellular Heat Shock Proteins

Clearly, a great deal is already understood about the function of intracellular Hsp72; however, the focus of the current review is on stress-induced
release of *extracellular* Hsp72. We have chosen to focus on *extracellular*
Hsp72 (eHsp72) because stress-induced release of eHsp72 into the blood
has only recently been documented and we are only now recognizing its
powerful immunological functions. In fact, we propose that the function
of *in vivo* endogenous eHsp72 is likely context dependent, such that in
a normal physiological state eHsp72 facilitates innate immune responses
to acute pathogenic challenge, whereas in a pathological state eHsp72
may exacerbate chronic inflammatory diseases (e.g., atherosclerosis,
Alzheimer disease, Crohn disease). The first reports that eHsp72 is
detectable in the circulation of humans were published by Pockely and collegues in 2000. This group reported that people suffering from a variety of
disease states such as renal disease (Wright *et al.*, 2000), hyptertension
(Pockley *et al.*, 2002), and atherosclerosis (Pockley *et al.*, 2003) have chronically elevated basal levels of eHsp72 relative to healthy age-matched

controls. In addition to elevated basal eHsp72 associated with disease pathology, Dybdahl *et al.* (2002) reported patients with coronary artery disease have an acute increase in eHsp72 in response to the stress of coronary bypass surgery. Not long after these reports, we (Campisi and Fleshner, 2003; Campisi *et al.*, 2003b; Fleshner *et al.*, 2003) and Febbraio *et al.*, (Walsh *et al.*, 2001; Febbraio *et al.*, 2002) reported that organisms in the **absence of clinical disease states** also rapidly increase the concentration of eHsp72 in blood after exposure to acute stressors. These papers were the first to demonstrate that an increase of eHsp72 in the blood occurs in healthy organisms after exposure to acute stressors and led us to suggest that stress-induced eHsp72 release may be a previously unrecognized feature of the normal stress response.

4.2. Stress-induced Sympathetic Nervous System Output Stimulates eHsp72 Release

4.2.1. Necrosis- versus Exocytosis-Mediated Release

There are currently two potential mechanisms for *extracellular* Hsp72 release. The first is that eHsp72 is released from the *intracellular* pool after necrotic or lytic cell death. The second is that eHsp72 is released via a receptor-mediated exocytosis mechanism. Gallucci, Loema, and Matzinger (Gallucci *et al.*, 1999) first suggested that Hsp72 is released only in pathological circumstances such as those that result in necrotic/lytic death and not after apoptosis or programmed cell death. More recently, Basu *et al.* (Basu *et al.*, 2000), Sauter *et al.* (Sauter *et al.*, 2000), and Berwin *et al.* (Berwin *et al.*, 2001) supported these ideas and demonstrated that indeed Hsp72 was released after necrotic/lytic but *not* apoptotic cell death. In these studies, cellular necrosis was induced *in vitro* by either repeated freeze/thaw exposures (Basu *et al.*, 2000; Sauter *et al.*, 2000; Berwin *et al.*, 2001), hypotonic lysis (Sauter *et al.*, 2000), or viral lysis (Berwin *et al.*, 2001). Apoptosis was induced by exposure to UV (Basu *et al.*, 2000; Sauter *et al.*, 2000) or serum-depleted culture media (Berwin *et al.*, 2001) and verified using flow cytometric assessment of AV and PI staining. Necrotic (lytic or messy) death versus apoptosis (controlled programmed death) was verified via cytometric assessment of annexin V (AV) + propridium iodide (PI) staining (Del Bino *et al.*, 1999; Hammill *et al.*, 1999; Honda *et al.*, 2000; Lecoeur *et al.*, 2001); necrotic (AV+PI+), apoptotic (AV+PI–), or viable (AV–PI–).

In contrast with necrotic/lytic release of eHsp72, we propose that eHsp72 released after exposure to a psychological and/or physical stressor most likely occurs via an exocytosis pathway in the absence of necrosis. This hypothesis is based on the following observations. First, there is precedent for an exocytotic eHsp72 releasing mechanism. In the brain, for example, glial cells may exocytotically release Hsp72 (Guzhova *et al.*, 2001; Tytell, 2005). In addition, there is recent evidence that suggests that eHsp72 released during times of stress are in exosomes (Lancaster and Febbraio, 2005), small membrane vesicles secreted by various cell types including antigen-presenting cells, B cells, and T cells of the immune system (Chaput et al., 2004). Exosomes contain numerous costimulatory and antigen-presenting molecules including Hsp70, and such release does not appear to depend on the classical secretory pathway (Lancaster and Febbraio, 2005). Second, eHsp72 is elevated in the blood within 10–25 min of tailshock or restraint stressor onset (Fleshner and Johnson, 2005). The rapidity of the response suggests the classic protein induction/necrosis release pathway is not likely. Third, eHsp72 is increased in the blood after exposure to psychological stressors such as conditioned contextual fear and predatory stress (Fleshner et al., 2004), stressors that are not likely to induce necrosis. Fourth, Febbraio and colleagues (Walsh et al., 2001; Febbraio et al., 2002) reported that intense exercise (~65% VO_{2max}) increases eHsp72 in blood within 30 min of exercise onset and that this occurs in the absence of cellular necrosis (Febbraio et al., 2002). Fifth, the increases in concentrations of eHsp72 released into the blood are two to six fold above baseline (pre-stress) levels. If necrotic/lytic cell death were the source of eHsp72 in the blood, it would require a large number of cells to simultaneously die a necrotic/lytic death. Nonetheless, it is still not possible to rule out necrotic release at this time because we only tested splenic necrosis and some other currently unidentified tissue in body may demonstrate greater necrosis than the spleen. In addition, stress may produce a low level of necrosis in a large number of tissues in the body, and that in combination with this global stress-induced cellular necrosis produces increases in circulation eHsp72.

Thus, we propose that release of eHsp72 via necrotic cell death does occur after exposure to some stressors; however, it likely results in a local, restricted, and tissue-specific increase in eHsp72 at the site of server tissue damage or injury. In this local fashion, eHsp72 released from necrotic cells may indeed function to facilitate local innate immune responses. In contrast with local release, we hypothesize that the observed large increases in eHsp72 in the blood after exposure to a whole-organism stressor is due to a receptor-mediated exocytosis releasing mechanism and is not dependent on necrotic/lytic cell death.

4.2.2. eHsp72 Release Involves Norepinephrine and α_1ADRs

We have recently reported the role of the sympathetic nervous system, and specifically norepinephrine (NE), in the stress-induced release of eHsp72 (Johnson et al., 2005). Using pharmacological blockade and stimulation of adrenergic receptors, we completed a series of studies that tested the effect of labetalol (α_1ADR and β_1ADR antagonist), propranolol (βADR antagonist), and prazosin (α_1ADR antagonist) on tailshock-induced release of eHsp72. The results of these studies were that labetalol and prazosin but *not* propranolol blocked the effect of tailshock on eHsp72. We also tested the effect of phenylephrine (α_1ADR agonist) and isoproterenol (β_1ADR agonist) in the absence of stress to release eHsp72. We found that phenylephrine but not isoproterenol released eHsp72 (Johnson *et al.*, 2005). Interestingly, as stated above there is accumulating evidence that Hsp72 is released in exosomes and is released in a calcium-dependent fashion upon stimulation of the cell (Savina *et al.*, 2003). Because activation of α_1ADR results in a rise in intracellular calcium (Schwietert *et al.*, 1992), the release of exosomes is one potential mechanism by which catecholamines trigger the rise in eHsp72. These data support our hypothesis that eHsp72 is being induced and/or released via an α_1ADR-mediated mechanism. We hypothesize that NE released from sympathetic nerve terminals binds to α_1ADR and stimulates Hsp72 release. We propose that NE and not epinephrine (E) is responsible because NE binds with a higher affinity to α_1ADR than does E (Hardman and Limbird, 2001), and adrenalectomy depletes ~95–99% of E (Hessman *et al.*, 1976; Vollmer *et al.*, 1995) yet has no effect on eHsp72 release after tailshock stress (Johnson *et al.*, 2005).

4.3. *Immunostimulatory Effect of Extracellular Hsp72*

Extracelluar Hsp72 can robustly stimulate inflammatory cytokine production and other innate immune responses (Multhoff *et al.*, 1999; Asea *et al.*, 2000a; Breloer *et al.*, 2001). We and others have reported that eHsp72 *in vitro* stimulates inducible NO synthase (Panjwani *et al.*, 2002), NO (Campisi *et al.*, 2003b), TNF-α (Asea *et al.*, 2000b; Campisi *et al.*, 2003b), IL1-β (Asea *et al.*, 2000b; Campisi *et al.*, 2003b), and IL6 (Asea *et al.*, 2000b; Campisi *et al.*, 2003b) production from macrophages and neutrophils. Importantly, these studies carefully demonstrated that stimulation of inflammatory cytokines and NO by eHsp72 *in vitro* was specific to eHsp72 and was ***not due*** to nonspecific effects of endotoxin contamination in the recombinant Hsp72 protein (Asea *et al.*, 2000b; Panjwani *et al.*, 2002; Campisi *et al.*, 2003b). Furthermore, eHsp72 has also been reported to stimulate the human complement pathway independent of antibodies (Prohaszka *et al.*, 2002). Thus, eHsp72 is immunostimulatory in its own right and acts as an adjuvant (Asea *et al.*, 2000a; Srivastava, 2002; van Eden *et al.*, 2005; Wang *et al.*, 2005). Although the immunological function of eHsp72 has been explored *in vitro*, little is known about what function eHsp72 may serve *in vivo*.

4.4. *Extracellular Hsp72 and the Danger Theory*

Extracellular Hsp72 is released during times of stress and can stimulate innate immunity. In fact it has been proposed that eHsp72 released during stress may function as a "messenger of stress" or "danger signal" to the immune system. Matzinger (Matzinger, 1994, 1998) first proposed the hypothesis that the body may release endogenous danger signals capable of stimulating immunity. In brief, the danger theory states that immune activation involves danger/nondanger molecular recognition schemas. The danger theory postulates that innate immune cells are activated by danger/alarm signals that are derived from stressed or damaged self. Although the danger theory is controversial when viewed as exclusionary, the ideas suggested are intriguing when viewed as complementary to other schemas. Clearly, innate immunity has evolved several strategies of activation. Consequently, it is reasonable to propose that innate immune cells can be activated by both exogenous antigens (i.e., lipopolysaccharide) binding to a limited number of germ-line encoded receptors [i.e., CD14 (Janeway and Medzhitov, 2002)] and by endogenous molecules that are released during times of cellular stress or danger.

One important and unresolved issue for the danger theory is what molecules serve as danger signals to the immune system. We (Fleshner et al., 2002; Campisi and Fleshner, 2003) and others (Colaco, 1998; Moseley, 1998; Chen et al., 1999; Asea et al., 2000b; Ohashi et al., 2000; Todryk et al., 2000; Breloer et al., 2001; Bethke et al., 2002; Habich et al., 2002; Vabulas et al., 2002) have suggested that eHsps may serve this function. Although Hsps fit the theoretical framework proposed by Matzinger, there is currently little supporting in vivo experimental evidence. It has been reported that humans who experienced trauma had increased serum levels of eHsp72 and that higher levels of eHsp72 correlated with improved survival (Pittet et al., 2002). Based on the danger theory, it would follow that if a danger signal serves to facilitate or target immune function, and eHsp72 acts as a danger signal, then organisms with increased eHsp72 should have improved immune responses and facilitated host defense to some types of pathogenic challenges.

4.5. Innate Immune Cell Activation: Toll-like Receptors Bind eHsp72

The search for the eHsp72 receptor is a topic of intense investigation. There is evidence of a cell-surface receptor for Hsp70 on macrophages/neutrophils (Asea et al., 2000b; Reed and Nicchitta, 2000; Sondermann et al., 2000; Asea et al., 2002), B cells (Arnold-Schild et al., 1999), and NK cells (Multhoff et al., 2001; Gross et al., 2003). A number of cell-surface binding proteins for eHsp have been implicated. Most research to date, however, suggests that eHsp72 transduces an inflammatory signal to innate immune cells (macrophages/dendritic/neutrophils) by binding to either Toll-like receptor-2 (TLR-2) and/or TLR-4 in a CD14-dependent fashion (Asea et al., 2000b; Visintin et al., 2001; Asea et al., 2002; Vabulas et al., 2002). Mammalian Toll-like receptors are transmembrane proteins that are evolutionarily conserved between very primitive organisms (such as insects) and humans (Akira et al., 2001). It has been suggested that just as released eHsps may function as danger signals or messenger of stress to the immune system, the TLRs may function as surveillance receptors for those signals (Johnson *et al.*, 2003a). In addition, exposure to prior injury stress was recently reported to produce a long-term (1–7 days) potentiatation of TLR-2– and TLR-4–induced IL-1β, IL-6, and TNF-α production by spleen cells (Paterson *et al.*, 2003), and chronic social stress (Avitsur *et al.*, 2003) modulates TLR-4–mediated responses. These data support the hypothesis that stress-induced modulation of innate immune function may involve TLR-2 and TLR-4. *Extracellular* Hsp72 exerts its effects on innate immune cells by stimulating the inflammatory MyD88/IRAK/NF-kappa-B signal transduction pathway (Vabulas *et al.*, 2002). A rapid intracellular Ca^{2+} flux ensues within 10s of eHsp72 binding with high affinity to monocytes or macrophages (Asea *et al.*, 2000b). This is important because it distinguishes eHsp72 signaling from LPS signaling, which does not induce Ca^{2+} flux (McLeish *et al.*, 1989). Based on work by Asea and colleagues (Asea *et al.*, 2000a, 2000b, 2002), eHsp72-induction of NF-kappa-B and inflammatory cytokines requires the expression of CD14, in addition to TLR-2 and TLR-4. Asea and colleagues have proposed that CD14 could function as a coreceptor for eHsp72 (Asea *et al.*, 2000b).

One implication of these results is that eHsp72 released into the blood after exposure to psychological and/or physical stressors may result in optimal stimulation of the inflammatory cascade only in the presence of CD14 activation. Interestingly, binding CD14 plus either TLR-2 and/or TLR-4 with selective receptor agonists (Pam3Cys binds TLR-2 or Taxol binds TLR-4) resulted in synergistic increases in NF-kappa-B (Asea *et al.*, 2002). In addition, we have preliminary data that low doses of LPS plus eHsp72 produced synergistic stimulation of NO from peritoneal macrophages. Thus, facilitation of innate immune responses by eHsp72 after exposure to stress may be restricted to cells that express CD14 and/or are binding bacteria or LPS via CD14. We hypothesize that acute stress–induced release of eHsp72 acts as a danger signal, preparing the immune system for possible subsequent pathogenic challenge. If no pathogenic challenge ensues, then eHsp72 has minimal impact on innate immune cell production of NO and/or inflammatory cytokines. If, however, the host is exposed to a pathogen, such as bacteria, eHsp72 via TLR-2 and/or TLR-4 could stimulate a potentiated NO and/or cytokine response, resulting in facilitated bacterial killing. One extension of this idea is that if, in contrast, the host suffers from chronic inflammatory disease (e.g., atherosclerosis, Alzheimer disease, Crohn's disease), then stress-induced eHsp72 may exacerbate the inflammatory disease states.

4.6. Stress Facilitates Recovery from Subcutaneous E. coli Challenge: A Role for eHsp72

We have completed a series of studies that lend support to the hypothesis that stress-induced increases in eHsp72 functions to facilitate innate immunity in the presence of pathogenic challenge (*E. coli*). First, rats exposed to tailshock stress and challenged with subcutaneous *E. coli* have an increase in eHsp72 at the site of inflammation (Campisi *et al.*, 2003b). Second, eHsp72 administered to the site of inflammation in the absence of stress improved recovery from bacterial challenge (Campisi *et al.*, 2003b). Third, prazosin *in vivo* blocked the tailshock-induced increase of eHsp72 in the blood (Johnson *et al.*, 2005), and preliminary data suggests that prazosin also blocks the increase eHsp72 at the inflammatory site and prevented the stress-induced reduction in bacterial load at the inflammatory site. Fourth, *in vivo* immunoneutralization of eHsp72 by anti-Hsp70-Ab46 at the site of inflammation attenuated the facilitory effect of tailshock stress on bacterial inflammation development and resolution. Importantly, anti-Hsp70-Ab46 (generously provided by Dr. Asea) blocked eHsp72 but not LPS-stimulated NO release from macrophages, tested *in vitro*. Finally, preliminary data suggest that low doses of LPS + eHsp72 *in vitro* results in synergistic NO response from macrophages.

4.7. Summary

As depicted in Figure 3.3 and previously discussed (Fleshner and Laudenslager, 2004; Fleshner and Johnson, 2005), we propose that exposure to a stressor activates the sympathetic nervous system leading to the release

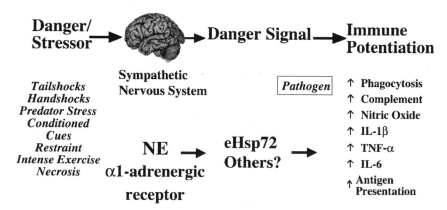

FIGURE 3.3. Depicted is our current hypothesis of how exposure to acute stress can lead to potentiated innate immunity. After exposure to a variety of stressors, the body responds by activating the sympathetic nervous system to release norepinephrine. Norepinephrine then binds to alpha$_1$-adrenergic receptor and stimulates the release of extracellular heat shock protein 72 (eHsp72) into the blood. Furthermore, we suggest that in presence of a pathogen and perhaps an inflammatory site, the circulating eHsp72 can extravasate into tissues and interact with innate immune cells to facilitate their responses.

of eHsp72 into the blood via an α_1ADR-medicated mechanism. If the animals are challenged with *E. coli*, eHsp72 extravasates from the blood into the subcutaneous space due to bacterial stimulated release of other inflammatory mediators that render the blood vessel leaky (PGE$_2$, BK, etc.) (Ali *et al.*, 1997). This is supported by recent evidence that the blockade of vascular leaking at the inflammatory cite prevents the local accumulation, but not elevated circulating levels, of eHsp72 after stress (Sharkey *et al.*, 2005). Extracellular Hsp72 at the inflammatory site binds to TLR-2 and TLR-4 on macrophage and/or neutrophils. Macrophages and/or neutrophils that have received a stimulatory signal via CD14 binding to LPS will mount potentiated innate immune responses (i.e., NO, TNF, IL-1, IL-6) that result in optimal bacterial killing. The release of eHsp72 in response to a global stressor such as uncontrollable tailshock, therefore, facilitates innate immune function only in the presence of pathogenic challenge. This is consistent with previous literature on the priming effects of stress on innate immunity (Johnson *et al.*, 2002a, 2002b, 2003b).

5. General Conclusions

These data support the hypothesis that exposure to intense acute stressor activates a cascade of physiological responses that work together to promote host survival. Here we suggest that in addition to classically associated consequences of activation of the SNS (i.e., pupil dilation, increased heart rate, increased respiration, increased muscular blood flow, etc.), the release of endogenous danger signals that prime immunity also occur. Thus we propose that SNS-induced release of eHsp72 should be considered a normal and adaptive feature of the stress response. If, however, SNS activation is chronic or excessive, then the response is maladaptive contributing to a plethora of negative effects such as "immunosuppression" and "metabolic syndrome," a key antecedent to clinical atherosclerotic diseases and immunosuppression (Irwin, 1993; Kennedy *et al.*, 2005b).

Based on our results, it follows that to prevent the negative consequences of activation of the acute stress response, one would need an intervention that can constrain excessive SNS output and prevent splenic NE depletion. Such an intervention should constrain but not eliminate SNS responses, so as to allow the host to reap the positive, while minimizing the negative, effects of SNS activation. We have evidence that exercise is such an intervention.

We have conducted a series of studies investigating the impact of tailshock on various aspects of the stress response including SNS activation, splenic NE depletion, αKLH Ig suppression, eHsp72 release, and increased host defense against a bacterial challenge. Physical active status was varied in these studies by housing animals with either mobile or locked running wheels. In these conditions, male F344 rats will run an average distance of

15 km/wk (Campisi *et al.*, 2003c; Greenwood *et al.*, 2003a). Nearly 100% of their running occurs during the dark part of their circadian cycle (Solberg, 1999). This level of activity produces physiological changes that are indicative of "metabolic fitness." In some rat strains, wheel running *reduces* body weight gain (Noble *et al.*, 1999), body fatness (Podolin, 1999), triglycerides concentrations (Suzuki, 1995) and *increases* lipid metabolism (Podolin, 1999), HDL/LDL ratio (Kennedy *et al.*, 2005a), muscular hypertrophy [triceps and plantaris (Ishihara *et al.*, 1998)], red blood cell hemoglobin content (Kennedy *et al.*, 2005a), and endurance.

What we found was that animals that lived a sedentary lifestyle with locked running wheels and were exposed to tailshock stress had excessive SNS responses leading to splenic NE depletion and αKLH Ig suppression (Kennedy *et al.*, 2005b). Sedentary rats exposed to stress, however, also have eHsp72 release and increased host defense against a bacterial challenge (Campisi *et al.*, 2002; Campisi *et al.*, 2003b). In contrast, rats that were physically active for 6 weeks prior to exposure to tailshock stress had constrained SNS responses such that tailshock elevated blood levels of NE but did not drive the response excessively, did not lead to splenic NE depletion (Greenwood *et al.*, 2003b), and did not produce αKLH Ig suppression (Moraska and Fleshner, 2001). Thus, physical activity prevented the negative effects of acute stress on acquired immunity by constraining SNS drive (Fleshner, 2005). Importantly, physically active rats exposed to stress still reap the immunopotentiating effects of stress. Rats that lived with a mobile running wheel for 4–6 weeks prior to exposure to tailshock stress have increases in circulating eHsp72 and potentiated host defense against bacterial challenge equal to or better than that produced by stress in sedentary rats (Fleshner *et al.*, 2002).

In conclusion, our work reveals the immunomodulatory effect of activation of the acute stress response and supports the hypothesis that a common neuroendocrine mechanism, that is, activation of the sympathetic nervous system (SNS), can be responsible for both suppression of acquired and facilitation of innate immune responses. Future work should strive to develop interventions, such as exercise, that allow us to reap the positive physiological and immunological effects, while minimizing the maladaptive consequences, of activation of the acute stress responses.

References

Abrass, I.B. (1986). Catecholamine levels and adrenergic responsiveness in aging. In M.J. Horan, G.M. Steinberg, J.B. Dunbar, and E.C. Hadley (eds.), *NIH Blood Pressure Regulation and Aging*. New York: Biomedical Information, pp. 123–130.

Acworth, I.N., During, M.J., and Wurtman, R.J. (1988). Tyrosine: Effects on catecholamine release. *Brain Res. Bull.* 21(3):473–477.

Adell, A., Garcia-Marquez, C., Armario, A., and Gelpi, E. (1988). Chronic stress increases serotonin and noradrenaline in rat brain and sensitizes their responses to a further acute stress. *J. Neurochem.* 50(6):1678–1681.

Ader, R., Felten, D.L., and Cohen, N. (1991). *Psychoneuroimmunology*. New York: Academic Press.

Akira, S., Takeda, K., and Kaisho, T. (2001). Toll-like receptors: Critical proteins linking innate and acquired immunity. *Nat. Immunol.* 2(8):675–680.

Ali, H., Haribabu, B., Richardson, R.M., and Snyderman, R. (1997). Mechanisms of inflammation and leukocyte activation. *Med. Clin. North Am.* 81(1):1–28.

Arnold-Schild, D., Hanau, D., Spehner, D., Schmid, C., Rammensee, H.G., de la Salle, H., and Schild, H. (1999). Cutting edge: Receptor-mediated endocytosis of heat shock proteins by professional antigen-presenting cells. *J. Immunol.* 162(7): 3757–3760.

Asea, A., Kabingu, E., Stevenson, M.A., and Calderwood, S.K. (2000a). HSP70 peptide-bearing and peptide-negative preparations act as chaperokines. *Cell Stress Chaperones* 5(5):425–431.

Asea, A., Kraeft, S.K., Kurt-Jones, E.A., Stevenson, M.A., Chen, L.B., Finberg, R.W., Koo, G.C., and Calderwood, S.K. (2000b). HSP70 stimulates cytokine production through a CD14-dependant pathway, demonstrating its dual role as a chaperone and cytokine. *Nat. Med.* 6(4):435–442.

Asea, A., Rehli, M., Kabingu, E., Boch, J.A., Bare, O., Auron, P.E., Stevenson, M.A., and Calderwood, S.K. (2002). Novel signal transduction pathway utilized by extracellular HSP70: Role of toll-like receptor (TLR) 2 and TLR4. *J. Biol. Chem.* 277(17):15028–15034.

Avitsur, R., Padgett, D.A., Dhabhar, F.S., Stark, J.L., Kramer, K.A., Engler, H., and Sheridan, J.F. (2003). Expression of glucocorticoid resistance following social stress requires a second signal. *J. Leukoc. Biol.* 74(4):507–513.

Baron, A., Laakso, M., Brechtel, G., Hoit, B., Watt, C., and Edelman, S. (1990). Reduced postprandial skeltal muscle blood flow contributes to glucose intolerance in human obesity. *J. Clin. Endocrinol. Metab.* 70:1525–1533.

Basu, S., Binder, R.J., Suto, R., Anderson, K.M., and Srivastava, P.K. (2000). Necrotic but not apoptotic cell death releases heat shock proteins, which deliver a partial maturation signal to dendritic cells and activate the NF-kappa B pathway. *Int. Immunol.* 12(11):1539–1546.

Baum, P.C., Kosek, J., Patterson, A., Bernstein, D., and Kobilka, B. (2002). Abnormal cardiac function associated with sympathetic nervous system hyperactivity in mice. *Am. J. Physiol.* 283:H1838–H1845.

Baumann, H., and Gauldie, J. (1994). The acute phase response. *Immunol. Today* 15(2):74–80.

Bellingan, G.J., Caldwell, H., Howie, S.E., Dransfield, I., and Haslett, C. (1996). In vivo fate of the inflammatory macrophage during the resolution of inflammation: Inflammatory macrophages do not die locally, but emigrate to the draining lymph nodes. *J. Immunol.* 157(6):2577–2585.

Berwin, B., Reed, R.C., and Nicchitta, C.V. (2001). Virally induced lytic cell death elicits the release of immunogenic GRP94/gp96. *J. Biol. Chem.* 276(24):21083–21088.

Bethke, K., Staib, F., Distler, M., Schmitt, U., Jonuleit, H., Enk, A.H., Galle, P.R., and Heike, M. (2002). Different efficiency of heat shock proteins (HSP) to activate human monocytes and dendritic cells: Superiority of HSP60. *J. Immunol.* 169(11):6141–6148.

Bonneau, R.H., Brehm, M.A., and Kern, A.M. (1997). The impact of psychological stress on the efficacy of anti-viral adoptive immunotherapy in an immunocomprimised host. *J. Neuroimmunol.* 78:19–33.

Breloer, M., Dorner, B., More, S.H., Roderian, T., Fleischer, B., and von Bonin, A. (2001). Heat shock proteins as "danger signals": eukaryotic Hsp60 enhances and accelerates antigen-specific IFN-gamma production in T cells. *Eur. J. Immunol.* 31(7):2051–2559.

Brennan, F.X., Fleshner, M., Watkins, L.R., and Maier, S.F. (1996). Macrophage stimulation reduces the cholesterol levels of stressed and unstressed rats. *Life Sci.* 58:1771–1776.

Brown, J.D., and Siegal, J.M. (1988). Exercise as a buffer of life stress: A prospective study of adolescent health. *Health Psychol.* 7(4):341–353.

Campisi, J., and Fleshner, M. (2003). Role of extracellular HSP72 in acute stress-induced potentiation of innate immunity in active rats. *J. Appl. Physiol.* 94(1): 43–52.

Campisi, J., Leem, T.H., and Fleshner, M. (2002). Acute stress decreases inflammation at the site of infection. A role for nitric oxide. *Physiol. Behav.* 77(2–3): 291–299.

Campisi, J., Hansen, M.K., O'Connor, K.A., Biedenkapp, J.C., Watkins, L.R., Maier, S.F., and Fleshner, M. (2003a). Effects of *E. coli* on core body temperature, activity, and cytokine levels in the circulation, brain & tissue. *J. Appl. Physiol.* 95:1973–1882.

Campisi, J., Leem, T.H., and Fleshner, M. (2003b). Stress-induced extracellular Hsp72 is a functionally significant danger signal to the immune system. *Cell Stress Chaperones* 8(3):272–286.

Campisi, J., Leem, T.H., Greenwood, B.N., Hansen, M.K., Moraska, A., Higgins, K., Smith, T.P., and Fleshner, M. (2003c). Habitual physical activity facilitates stress-induced HSP72 induction in brain, peripheral, and immune tissues. *Am. J. Physiol. Regul. Integr. Comp. Physiol.* 284(2):R520–530.

Campisi, J., Johnson, J.D., West, J., Cho, S., Sharkey, C., and Fleshner, M. (2006). Stress-induced facilitation of host defense against bacterial challenge is neither opsonization-phagocytosis nor T cell dependent. *Am. J. Physiol.* (submitted).

Chaput, N., Taieb, J., Schartz, N.E., Andre, F., Angevin, E., and Zitvogel, L. (2004). Exosome-based immunotherapy. *Cancer Immunol. Immunother.* 53(3):234–239.

Chen, L., Xin, X., Eckhart, A.D., Yang, N., and Faber, J.E. (1995). Regulation of vascular smooth muscle growth by alpha 1-adrenoreceptor subtypes in vitro and in vivo. *J. Biol. Chem.* 270:30980–30988.

Chen, W., Syldath, U., Bellmann, K., Burkart, V., and Kolb, H. (1999). Human 60-kDa heat-shock protein: A danger signal to the innate immune system. *J. Immunol.* 162(6):3212–3219.

Clark, E.A., and Ledbetter, J.A. (1994). How B and T cells talk to each other. *Nature* 367(6462):425–428.

Coe, C.L., Rosenberg, L.T., and Levine, S. (1988). Effect of maternal separation on the complement system and antibody responses in infant primates. *Int. J. Neurosci.* 40(3–4):289–302.

Colaco, C.A. (1998). Towards a unified theory of immunity: dendritic cells, stress proteins and antigen capture. *Cellular and molecular biology (Noisy-le-Grand, France).* 44(6):883–890.

Coussons-Read, M.E., Maslonek, K.A., Fecho, K., Perez, L., and Lysle, D.T. (1994). Evidence for the involvement of macrophage-derived nitric oxide in the modulation of immune status by a conditioned aversive stimulus. *J. Neuroimmunol.* 50(1):51–58.

Day, H.E., Greenwood, B.N., Hammack, S.E., Watkins, L.R., Fleshner, M., Maier, S.F., and Campeau, S. (2004). Differential expression of 5HT-1A, alpha 1b adrenergic, CRF-R1, and CRF-R2 receptor mRNA in serotonergic, gamma-aminobutyric acidergic, and catecholaminergic cells of the rat dorsal raphe nucleus. *J. Comp. Neurol.* 474(3):364–378.

Dazzi, L., Ladu, S., Spiga, F., Vacca, G., Rivano, A., Pira, L., and Biggio, G. (2002). Chronic treatment with imipramine or mirtazapine antagonizes stress- and FG7142-induced increase in cortical norepinephrine output in freely moving rats. *Synapse* 43(1):70–77.

Deak, T., Meriwether, J.L., Fleshner, M., Spencer, R.L., Abouhamze, A., Moldawer, L.L., Grahn, R.E., Watkins, L.R., and Maier, S.F. (1997). Evidence that brief stress may induce the acute phase response in rats. *Am. J. Physiol.* 273(6 Pt 2): R1998–2004.

Deak, T., Nguyen, K.T., Fleshner, M., Watkins, L.R., and Maier, S.F. (1999). Acute stress may facilitate recovery from a subcutaneous bacterial challenge. *Neuroimmunomodulation* 6(5):344–354.

Del Bino, G., Darzynkiewicz, Z., Degraef, C., Mosselmans, R., Fokan, D., and Galand, P. (1999). Comparison of methods based on annexin-V binding, DNA content or TUNEL for evaluating cell death in HL-60 and adherent MCF-7 cells. *Cell Prolif.* 32(1):25–37.

Del Rey, A., Kabiersch, A., Petzoldt, S., and Besedovsky, H.O. (2003). Sympathetic abnormalities during autoimmune processes: Potential relevance of noradrenaline-induced apoptosis. *Ann. N. Y. Acad. Sci.* 992:158–167.

Dinarello, C.A. (2000). Proinflammatory cytokines. *Chest* 118(2):503–508.

Dinenno, F.A., Dietz, N.M., and Joyner, M.J. (2002). Aging and forearm postjunctional alpha-adrenergic vasoconstriction in healthy men. *Circulation* 106: 1349–1354.

Dishman, R.K., Warren, J.M., Youngstedt, S.D., Yoo, H., Bunnell, B.N., and Mougey, E.H. (1995). Activity-wheel running attenuates suppression of natural killer cell activity after footshock. *J. Appl. Physiol.* 78(4):1547–1554.

Febbraio, M.A., Ott, P., Nielsen, H.B., Steensberg, A., Keller, C., Krustrup, P., Secher, N.H., and Pedersen, B.K. (2002). Exercise induces hepatosplanchnic release of heat shock protein 72 in humans. *J. Physiol.* 544(Pt 3):957–962.

Fecho, K., Maslonek, K.A., Coussons-Read, M.E., Dykstra, L.A., and Lysle, D.T. (1994). Macrophage-derived nitric oxide is involved in the depressed concanavalin A responsiveness of splenic lymphocytes from rats administered morphine in vivo. *J. Immunol.* 152(12):5845–5852.

Felten, S.Y., and Olschowka, J. (1987). Noradrenergic sympathetic innervation of the spleen II. Tyrosine hydroxylase (TH)- posititive nerve terminals from synaptic-like contacts on lymphocytes in splenic white pulp. *J. Neurosci. Res.* 18:37–42.

Felten, D.L., Ackerman, K.D., Wiegand, S.J., and Felten, S.Y. (1987). Noradrenergic sympathetic innervation of the spleen I. Nerve fibers associate with lymphocytes and macrophages in specific compartments of the splenic white pulp. *J. Neurosci. Res.* 18:28–36.

Fleshner, M. (2005). Physical activity & stress resistance: Sympathetic nervous system adaptations prevent stress-induced immunosuppression. *Exerc. Sports Sci. Rev.* 33:120–126.

Fleshner, M., and Johnson, J.D. (2005). Endogenous extra-cellular heat shock protein 72: Releasing signals and function. *Int. J. Hypertherm.* 33:120–126.

Fleshner, M., and Laudenslager, M.L. (2004). Psychoneuroimmunology: Then and now. *Behav. Cogn. Neurosci. Rev.* 3(2):114–130.

Fleshner, M., Watkins, L.R., Lockwood, L.L., Bellgrau, D., Laudenslager, M.L., and Maier, S.F. (1992). Specific changes in lymphocyte subpopulations: A potential mechanism for stress-induced immunomodulation. *J. Neuroimmunol.* 41(2):131–142.

Fleshner, M., Watkins, L.R., Lockwood, L.L., Brossard, C., Laudenslager, M.L., and Maier, S.F. (1993). Blockade of the hypothalamic-pituitary-adrenal response to stress by intraventricular injection of dexamethasone: A method for studying the stress-induced peripheral effects of Glucocorticoids. *Psychoneuroendocrinology* 18(4):251–263.

Fleshner, M., Bellgrau, D., Watkins, L.R., Laudenslager, M.L., and Maier, S.F. (1995a). Stress-induced reduction in the rat mixed lymphocyte reaction is due to macrophages and not to changes in T cell phenotypes. *J. Neuroimmunol.* 56(1):45–52.

Fleshner, M., Deak, T., Spencer, R.L., Laudenslauger, M.L., Watkins, L.R., and Maier, S.F. (1995b). A long term increase in basal levels of corticosterone and a decrease in corticosteroid-binding globulin after acute stressor exposure. *Endocrinology* 136(12):5336–5342.

Fleshner, M., Goehler, L.E., Hermann, J., Relton, J.K., Maier, S.F., and Watkins, L.R. (1995c). Subdiaphramatic vagotomy blocks hypothalamic NE depletion and attenuates serum corticosterone elevation produced by IL-1. *Brain Res. Bull.* 37: 605–610.

Fleshner, M., Hermann, J., Lockwood, L.L., Laudenslager, M.L., Watkins, L.R., and Maier, S.F. (1995d). Stressed rats fail to expand the CD45RC+CD4+ (Th1-like) T cell subset in response to KLH: Possible involvement of IFN-gamma. *Brain Behav. Immun.* 9(2):101–112.

Fleshner, M., Nguyen, K.T., Cotter, C.S., Watkins, L.R., and Maier, S.F. (1998). Acute stressor exposure both suppresses acquired immunity and potentiates innate immunity. *Am. J. Physiol.* 275(3 Pt 2):R870–878.

Fleshner, M., Campisi, J., Deak, T., Greenwood, B.N., Kintzel, J.A., Leem, T.H., Smith, T.P., and Sorensen, B. (2002). Acute stressor exposure facilitates innate immunity more in physically active than in sedentary rats. *Am. J. Physiol. Regul. Integr. Comp. Physiol.* 282(6):R1680–1686.

Fleshner, M., Campisi, J., and Johnson, J.D. (2003). Can exercise stress facilitate innate immunity? A functional role for stress-induced extracellular Hsp72. *Exerc. Immunol. Rev.* 9:6–24.

Fleshner, M., Campisi, J., Amiri, L., and Diamond, D.M. (2004). Cat exposure induces both intra- and extracellular Hsp72: The role of adrenal hormones. *Psychoneuroendocrinology* 29(9):1142–1152.

Foy, T.M., Aruffo, A., Bajorath, J., Buhlmann, J.E., and Noelle, R.J. (1996). Immune regulation by CD40 and its ligand GP39. *Annu. Rev. Immunol.* 14:591–617.

Gallucci, S., Lolkema, M., and Matzinger, P. (1999). Natural adjuvants: endogenous activators of dendritic cells. *Nat. Med.* 5(11):1249–1255.

Gazda, L.S., Smith, T., Watkins, L.R., Maier, S.F., and Fleshner, M. (2003). Stressor exposure produces long-term reductions in antigen-specific T and B cell responses. *Stress* 6(4):259–267.

Gibson, C.J., and Wurtman, R.J. (1977). Physiological control of brain catechol synthesis by brain tyrosine concentration. *Biochem. Pharmacol.* 26(12):1137–1142.

Gilewski, T., Adluri, R., Zhang, S., Houghton, A., Norton, L., and Livingston, P. (1996). Preliminary results: Vaccination of breast cancer patients (pts) lacking identifiable disease (NED) with MUC-1-keyhole limpet hemocyanin (KLH) conjugate and QS21. *Proceedings of the Annual Meeting of the American Society of Clinical Oncology* 15:A1807.

Goldstein, D.S. (1987). Stress-induced activation of the sympathetic nervous system. *Baillieres Clin. Endocrinol. Metab.* 1(2):253–278.

Goldstein, D.S. (1996). The sympathetic nervous system and the "fight-or-flight" response: Outmoded ideas? *Mol. Psychiatry* 1(2):95–97.

Greenwood, B.N., Foley, T.E., Day, H.E., Campisi, J., Hammack, S.H., Campeau, S., Maier, S.F., and Fleshner, M. (2003a). Freewheel running prevents learned helplessness/behavioral depression: role of dorsal raphe serotonergic neurons. *J. Neurosci.* 23(7):2889–2898.

Greenwood, B.N., Kennedy, S., Smith, T.P., Campeau, S., Day, H.E., and Fleshner, M. (2003b). Voluntary freewheel running selectively modulates catecholamine content in peripheral tissue and c-Fos expression in the central sympathetic circuit following exposure to uncontrollable stress in rats. *Neuroscience* 120(1):269–281.

Gross, C., Hansch, D., Gastpar, R., and Multhoff, G. (2003). Interaction of heat shock protein 70 peptide with NK cells involves the NK receptor CD94. *Biol. Chem.* 384(2):267–279.

Guzhova, I., Kislyakova, K., Moskaliova, O., Fridlanskaya, I., Tytell, M., Cheetham, M., and Margulis, B. (2001). In vitro studies show that Hsp70 can be released by glia and that exogenous Hsp70 can enhance neuronal stress tolerance. *Brain Res.* 914(1–2):66–73.

Habich, C., Baumgart, K., Kolb, H., and Burkart, V. (2002). The receptor for heat shock protein 60 on macrophages is saturable, specific, and distinct from receptors for other heat shock proteins. *J. Immunol.* 168(2):569–756.

Hammill, A.K., Uhr, J.W., and Scheuermann, R.H. (1999). Annexin V staining due to loss of membrane asymmetry can be reversible and precede commitment to apoptotic death. *Exp. Cell. Res.* 251(1):16–21.

Hardman, J.G., and Limbird, L.E., eds. (2001). *Goodman & Gilman's: The Pharmacological Basis of Therapeutics*, 9th edition. New York: McGraw-Hill.

Harmsen, A.G., and Turney, T.H. (1985). Inhibition of in vivo neutrophil accumulation by stress. Possible role of neutrophil adherence. *Inflammation* 9(1):9–20.

Hartl, F.U. (1996). Molecular chaperones in cellular protein folding. *Nature* 381(6583):571–579.

Heneka, M.T., Gavrilyuk, V., Landreth, G.E., O'Banion, M.K., Weinberg, G., and Feinstein, D.L. (2003). Noradrenergic depletion increases inflammatory responses in brain: Effects on IkappaB and HSP70 expression. *J. Neurochem.* 85(2):387–398.

Hessman, Y., Rentzhog, L., and Ekbohm, G. (1976). Effect of adrenal demedullation on urinary excretion of catecholamines in thermal trauma in rats. *Acta Chir. Scand.* 142(4):291–295.

Honda, O., Kuroda, M., Joja, I., Asaumi, J., Takeda, Y., Akaki, S., Togami, I., Kanazawa, S., Kawasaki, S., and Hiraki, Y. (2000). Assessment of secondary necrosis of Jurkat cells using a new microscopic system and double staining method with annexin V and propidium iodide. *Int. J. Oncol.* 16(2):283–288.

Ianaro, A., O'Donnell, C.A., Di Rosa, M., and Liew, F.Y. (1994). A nitric oxide synthase inhibitor reduces inflammation, down-regulates inflammatory cytokines and enhances interleukin-10 production in carrageenin-induced oedema in mice. *Immunology* 82(3):370–375.

Irwin, M. (1993). Brain corticotropin-releasing hormone and interleukin-1 beta-induced suppression of specific antibody production. *Endocrinology* 133:1352–1360.

Ishihara, A., Roy, R.R., Ohira, Y., Ibata, Y., and Edgerton, V.R. (1998). Hypertrophy of rat plantaris muscle fibers after voluntary running with increasing loads. *J. Appl. Physiol.* 84(6):2183–2189.

Janeway, C.A., Jr., and Medzhitov, R. (2002). Innate immune recognition. *Annu. Rev. Immunol.* 20:197–216.

Jansen, A.S., Nguyen, X.V., Karpitskiy, V., Mettenleiter, T.C., and Loewy, A.D. (1995). Central command neurons of the sympathetic nervous system: basis of the fight-or-flight response. *Science* 270(5236):644–646.

Johnson, J.D., O'Connor, K.A., Deak, T., Spencer, R.L., Watkins, L.R., and Maier, S.F. (2002a). Prior stressor exposure primes the HPA axis. *Psychoneuroendocrinology* 27(3):353–365.

Johnson, J.D., O'Connor, K.A., Deak, T., Stark, M., Watkins, L.R., and Maier, S.F. (2002b). Prior stressor exposure sensitizes LPS-induced cytokine production. *Brain Behav. Immun.* 16(4):461–476.

Johnson, G.B., Brunn, G.J., Tang, A.H., and Platt, J.L. (2003a). Evolutionary clues to the functions of the Toll-like family as surveillance receptors. *Trends Immunol.* 24(1):19–24.

Johnson, J.D., O'Connor, K.A., Hansen, M.K., Watkins, L.R., and Maier, S.F. (2003b). Effects of prior stress on LPS-induced cytokine and sickness responses. *Am. J. Physiol. Regul. Integr. Comp. Physiol.* 284:R422–432.

Johnson, J.D., Campisi, J., Sharkey, C.M., Kennedy, S., Nickerson, M., and Fleshner, M. (2005). Adrenergic receptors mediate stress-induced elevation in endogenous extracellular Hsp72. *Journal of Applied Physiology.* Nov; 99(5):1789–1795.

Julius, S., Gudbrandsson, T., Jamerson, K., and Andersson, O. (1992). The interconnection between sympathetics, microcirculation, and insulin resistance in hypertension. *Blood Pressure* 1:9–19.

Jurincic-Winkler, C., Metz, K.A., Beuth, J., Sippel, J., and Klippel, K.F. (1995). Effect of keyhole limpet hemocyanin (KLH) and bacillus Calmette-Guerin (BCG) instillation on carcinoma in situ of the urinary bladder. *Anticancer Res.* 15(6B): 2771–2776.

Jurincic-Winkler, C.D., von der Kammer, H., Beuth, J., Scheit, K.H., and Klippel, K.F. (1996). Antibody response to keyhole limpet hemocyanin (KLH) treatment in patients with superficial bladder carcinoma. *Anticancer Res.* 16(4A):2105–2110.

Kasprowicz, D.J., Kohm, A.P., Berton, M.T., Chruscinski, A.J., Sharpe, A., and Sanders, V.M. (2000). Stimulation of the B cell receptor, CD86 (B7-2), and the beta 2-adrenergic receptor intrinsically modulates the level of IgG1 and IgE produced per B cell. *J. Immunol.* 165(2):680–690.

Kennedy, S., Smith, T.P., and Fleshner, M. (2005a). Resting cellular and physiological effects of free wheel running. *Med. Sci. Sports Exerc.* 2005 Jan; 37(1): 79–83.

Kennedy, S.L., Nickerson, M., Campisi, J., Johnson, J.D., Smith, T.P., and Fleshner, M. (2005b). Splenic norepinephrine depletion and suppression of in vivo antibody. *J. Neuroimmunol.* 15:150–160.

Kohm, A.P., and Sanders, V.M. (1999). Suppression of antigen-specific Th2 cell-dependent IgM and IgG1 production following norepinephrine depletion in vivo. *J. Immunol.* 162:5299–5308.

Kohm, A.P., and Sanders, V.M. (2000). Norepinephrine: a messenger from the brain to the immune system. *Immunol. Today* 21(11):539–542.

Kohm, A.P., and Sanders, V.M. (2001). Norepinephrine and beta 2-adrenergic receptor stimulation regulate CD4+ T and B lymphocyte function in vitro and in vivo. *Pharmacol. Rev.* 53(4):487–525.

Kohm, A.P., Tang, Y., Sanders, V.M., and Jones, S.B. (2000). Activation of antigen-specific CD4+ Th2 cells and B cells in vivo increases norepinephrine release in the spleen and bone marrow. *J. Immunol.* 165(2):725–733.

Kohm, A.P., Mozaffarian, A., and Sanders, V.M. (2002). B cell receptor- and beta 2-adrenergic receptor-induced regulation of B7-2 (CD86) expression in B cells. *J. Immunol.* 168(12):6314–6322.

Lamm, D.L., DeHaven, J.I., Riggs, D.R., and Ebert, R.F. (1993). Immunotherapy of murine bladder cancer with keyhole limpet hemocyanin. *J. Urol.* 149(3):648–652.

Lancaster, G.I., and Febbraio, M.A. (2005). Exosome-dependent trafficking of HSP70: a novel secretory pathway for cellular stress proteins. *J. Biol. Chem.* 280(24):23349–23355.

Laudenslager, M.L., and Fleshner, M. (1994). Stress and Immunity: Of mice, monkeys, models, and mechanisms. In: R. Glaser, and J. Kiecolt-Glaser (eds.), *The Handbook of Human Stress and Immunity*. San Diego, California, Academic Press, pp. 161–181.

Laudenslager, M.L., Ryan, S.M., Drugen, R.L., Hyson, R.L., and Maier, S.F. (1983). Coping and immunosuppression: Inescapable but not escapable shock suppresses lymphocyte proliferation. *Science* 221:568–570.

Laudenslager, M.L., Fleshner, M., Hofstadter, P., Held, P.E., Simons, L., and Maier, S.F. (1988). Suppression of specific antibody production by inescapable shock: Stability under varying conditions. *Brain Behav. Immun.* 2(2):92–101.

Lecoeur, H., Prevost, M.C., and Gougeon, M.L. (2001). Oncosis is associated with exposure of phosphatidylserine residues on the outside layer of the plasma membrane: A reconsideration of the specificity of the annexin V/propidium iodide assay. *Cytometry* 44(1):65–72.

Lind, L., and Lithell, H. (1993). Decreased peripheral blood flow in the pathogenisis of the metabolic syndrome comprising hypertension, hyperlimidemia and hyperinsulinemia. *Am. Heart J.* 125:1474–1497.

Livingston, P.O. (1995). Approaches to augmenting the immunogenicity of melanoma gangliosides: From whole melanoma cells to ganglioside-KLH conjugate vaccines. *Immunol. Rev.* 145:147–166.

Lysle, D.T., Fecho, K., Maslonek, K.A., and Dykstra, L.A. (1995). Evidence for the involvement of macrophage-derived nitric oxide in the immunomodulatory effect of morphine and aversive Pavlovian conditioning. *Adv. Exp. Med. Biol.* 373:141–147.

Lyte, M., Nelson, S.G., and Thompson, M.L. (1990). Innate and adaptive immune responses in a social conflict paradigm. *Clin. Immunol. Immunopathol.* 57(1):137–147.

Mac Micking, J., Xie, Q.W., and Nathan, C. (1997). Nitric oxide and macrophage function. *Annu. Rev. Immunol.* 15:323–350.

Maestroni, G.J. (2002). Short exposure of maturing, bone marrow-derived dendritic cells to norepinephrine: Impact on kinetics of cytokine production and Th development. *J. Neuroimmunol.* 129(1–2):106–114.

Maier, S.F., Watkins, L.R., and Fleshner, M. (1994). Psychoneuroimmunology. The interface between behavior, brain, and immunity. *Am. Psychol.* 49(12): 1004–1017.

Maier, S.F., Fleshner, M., and Watkins, L.R. (1998). Neural, endocrine, and immune mechanisms of stress-induced immunomodulation. *New Frontiers in Stress Research: Modulation of Brain Function.* Aharon Levy, Ettic Orauer, David Ben-Nathan, E. Ronald De Kloet, Amsterdam, The Netherlands Harwood Academic Publishers, pp. 117–126.

Malarkey, W.B., Wang, J., Cheney, C., Glaser, R., and Nagaraja, H. (2002). Human lymphocyte growth hormone stimulates interferon gamma production and is inhibited by cortisol and norepinephrine. *J. Neuroimmunol.* 123(1–2):180–187.

Matz, J.M., LaVoi, K.P., and Blake, M.J. (1996a). Adrenergic regulation of the heat shock response in brown adipose tissue. *J. Pharmacol. Exp. Ther.* 277(3):1751–1758.

Matz, J.M., LaVoi, K.P., Moen, R.J., and Blake, M.J. (1996b). Cold-induced heat shock protein expression in rat aorta and brown adipose tissue. *Physiol. Behav.* 60(5):1369–1374.

Matzinger, P. (1994). Tolerance, danger, and the extended family. *Annu. Rev. Immunol.* 12:991–1045.

Matzinger, P. (1998). An innate sense of danger. *Semin. Immunol.* 10(5):399–415.

McLeish, K.R., Dean, W.L., Wellhausen, S.R., and Stelzer, G.T. (1989). Role of intracellular calcium in priming of human peripheral blood monocytes by bacterial lipopolysaccharide. *Inflammation* 13(6):681–692.

Meltzer, J.C., Grimm, P.C., Greenberg, A.H., and Nance, D.M. (1997). Enhanced immunohistochemical detection of autonomic nerve fibers, cytokines and inducible nitric oxide synthase by light and fluorescent microscopy in rat spleen. *Histochem. Cytochem.* 45:599–610.

Milligan, E.D., Nguyen, K.T., Deak, T., Hinde, J.L., Fleshner, M., Watkins, L.R., and Maier, S.F. (1998). The long term acute phase-like responses that follow acute stressor exposure are blocked by alpha-melanocyte stimulating hormone. *Brain Res.* 810(1–2):48–58.

Milner, J.D., and Wurtman, R.J. (1987). Tyrosine availability: A presynaptic factor controlling catecholamine release. *Adv. Exp. Med. Biol.* 221:211–221.

Monjan, A., and Collector, J. (1976). Stress-induced modulation of the immune response. *Science* 196:307–308.

Moraska, A., and Fleshner, M. (2001). Voluntary physical activity prevents stress-induced behavioral depression and anti-KLH antibody suppression. *Am. J. Physiol. Regul. Integr. Comp. Physiol.* 281(2):R484–489.

Moraska, A., Campisi, J., Nguyen, K.T., Maier, S.F., Watkins, L.R., and Fleshner, M. (2002). Elevated IL-1beta contributes to antibody suppression produced by stress. *J. Appl. Physiol.* 93(1):207–215.

Morimoto, R.I. (1994). *The Biology of Heat Shock Proteins and Molecular Chaperones.* Cold Spring Harbor, NY: Cold Spring Harbor Laboratory.

Moseley, P.L. (1998). Heat shock proteins and the inflammatory response. *Ann. N. Y. Acad. Sci.* 856:206–213.

Multhoff, G., Mizzen, L., Winchester, C.C., Milner, C.M., Wenk, S., Eissner, G., Kampinga, H.H., Laumbacher, B., and Johnson, J. (1999). Heat shock protein 70 (Hsp70) stimulates proliferation and cytolytic activity of natural killer cells. *Exp. Hematol.* 27(11):1627–1636.

Multhoff, G., Pfister, K., Gehrmann, M., Hantschel, M., Gross, C., Hafner, M., and Hiddemann, W. (2001). A 14-mer Hsp70 peptide stimulates natural killer (NK) cell activity. *Cell Stress Chaperones* 6(4):337–344.

Nguyen, K.T., Deak, T., Owens, S.M., Kohno, T., Fleshner, M., Watkins, L.R., and Maier, S.F. (1998). Exposure to acute stress induces brain interleukin-1beta protein in the rat. *J. Neurosci.* 18(6):2239–2246.

Nguyen, K.T., Deak, T., Will, M.J., Hansen, M.K., Hunsaker, B.N., Fleshner, M., Watkins, L.R., and Maier, S.F. (2000). Timecourse and corticosterone sensitivity of the brain, pituitary, and serum interleukin-1beta protein response to acute stress. *Brain Res.* 859(2):193–201.

Noble, E.G., Moraska, A., Mazzeo, R.S., Roth, D.A., Olsson, M.C., Moore, R.L., and Fleshner, M. (1999). Differential expression of stress proteins in rat myocardium after free wheel or treadmill run training. *J. Appl. Physiol.* 86(5):1696–1701.

Noursadeghi, M., Bickerstaff, M.C., Herbert, J., Moyes, D., Cohen, J., and Pepys, M.B. (2002). Production of granulocyte colony-stimulating factor in the nonspecific acute phase response enhances host resistance to bacterial infection. *J. Immunol.* 169(2):913–919.

O'Conner, K.A., Johnson, J.D., Nagel, T., Fleshner, M., Watkins, L.R., and Maier, S.F. (1999). Role of hormones in stress-induced nitric oxide production. *Society for Neuroscience Abstract.*

Ohashi, K., Burkart, V., Flohe, S., and Kolb, H. (2000). Cutting edge: Heat shock protein 60 is a putative endogenous ligand of the toll-like receptor-4 complex. *J. Immunol.* 164(2):558–561.

Panjwani, N.N., Popova, L., and Srivastava, P.K. (2002). Heat shock proteins gp96 and hsp70 activate the release of nitric oxide by APCs. *J. Immunol.* 168(6):2997–3003.

Paterson, H.M., Murphy, T.J., Purcell, E.J., Shelley, O., Kriynovich, S.J., Lien, E., Mannick, J.A., and Lederer, J.A. (2003). Injury primes the innate immune system for enhanced Toll-like receptor reactivity. *J. Immunol.* 171(3):1473–1483.

Pauletto, P., Scannapieco, G., and Pessina, A. (1991). Sympathetic drive and vascular damage in hypertension and atherosclerosis. *Hypertension* 17(Suppl): III75–III81.

Pittet, J.F., Lee, H., Morabito, D., Howard, M.B., Welch, W.J., and Mackersie, R.C. (2002). Serum levels of Hsp 72 measured early after trauma correlate with survival. *J. Trauma* 52(4):611–617.

Plotnikoff, N., Murgo, A., Faith, R., and Wybran, J., Ed. (1991). *Stress and Immunity.* Boca Raton, FL: CRC Press.

Pockley, A.G., De Faire, U., Kiessling, R., Lemne, C., Thulin, T., and Frostegard, J. (2002). Circulating heat shock protein and heat shock protein antibody levels in established hypertension. *J. Hypertens.* 20(9):1815–2180.

Pockley, A.G., Georgiades, A., Thulin, T., de Faire, U., and Frostegard, J. (2003). Serum heat shock protein 70 levels predict the development of atherosclerosis in subjects with established hypertension. *Hypertension* 42(3):235–238.

Podojil, J.R., and Sanders, V.M. (2003). Selective regulation of mature IgG1 transcription by CD86 and beta 2 adrenergic receptor stimulation. *J. Immunol.* 170(10):5143–5151.

Podojil, J.R., Kin, N.W., and Sanders, V.M. (2004). CD86 and beta2-adrenergic receptor signaling pathways, respectively, increase Oct-2 and OCA-B expression and binding to the 3'-IgH enhancer in B cells. *J. Biol. Chem.* 279(22):23394–23404.

Podolin, D.A., Wie, Y., and Paglizssotti, J. (1999). Effects of a high-fat diet and voluntary wheel running on gluconeogenesis and lipolysis in rats. *J. Appl. Phys.* 86:1374–1380.

Prohaszka, Z., Singh, M., Nagy, K., Kiss, E., Lakos, G., Duba, J., and Fust, G. (2002). Heat shock protein 70 is a potent activator of the human complement system. *Cell Stress Chaperones* 7(1):17–22.

Ramer-Quinn, D.S., Baker, R.A., and Sanders, V.M. (1997). Activated T helper 1 and T helper 2 cells differentially express the beta-2-adrenergic receptor. *J. Immunol.* 159:4857–4867.

Reed, R.C., and Nicchitta, C.V. (2000). Chaperone-mediated cross-priming: A hitchhiker's guide to vesicle transport (review). *Int. J. Mol. Med.* 6(3):259–264.

Ritossa, F. (1962). A new puffing pattern induced by temperature shock and DNP in *Drosophila*. *Experientia* 15:571–573.

Sanders, V.M., Baker, R.A., Ramer-Quinn, D.S., Kasprowicz, D.J., Fuchs, B.A., and Street, N.E. (1997). Differential expression of the beta2-adrenergic receptor by Th1 and Th2 clones. *J. Immunol.* 158:4200–4210.

Sanders, V.M., Kasprowicz, D.J., Swanson-Mungerson, M.A., Podojil, J.R., and Kohm, A.P. (2003). Adaptive immunity in mice lacking the beta(2)-adrenergic receptor. *Brain Behav. Immun.* 17(1):55–67.

Sauter, B., Albert, M.L., Francisco, L., Larsson, M., Somersan, S., and Bhardwaj, N. (2000). Consequences of cell death: Exposure to necrotic tumor cells, but not primary tissue cells or apoptotic cells, induces the maturation of immunostimulatory dendritic cells. *J. Exp. Med.* 191(3):423–434.

Savina, A., Furlan, M., Vidal, M., and Colombo, M.I. (2003). Exosome release is regulated by a calcium-dependent mechanism in K562 cells. *J. Biol. Chem.* 278(22):20083–20090.

Schwietert, H.R., Mathy, M.J., Wilhelm, D., Wilffert, B., Pfaffendorf, M., and Van Zwieten, P.A. (1992). Alpha 1-adrenoceptor-mediated Ca(2+)-entry from the extracellular fluid and Ca(2+)-release from intracellular stores: No role for alpha 1A,B-adrenoceptor subtypes in the pithed rat. *J. Auton. Pharmacol.* 12:125–136.

Seals, D., and Dinenno, F. (2004). Collateral damage: Cadiovascular consequences of chronic sympathetic activation with human aging. *Am. J. Physiol.* 287: 1895–1905.

Sharkey, C., Asea, A., and Fleshner, M. (2005). Extracellular Hsp72 extravasates via BK-leaking into bacterial inflammatory sites, facilitates innate immunity and host defense. *Experimental Biology Abstract*.

Smith, J.A., and Pyne, D.B. (1997). Exercise, training, and neutrophil function. *Exerc. Immunol. Rev.* 3:96–116.

Smith, T.P., Kennedy, S.L., and Fleshner, M. (2004a). Influence of age and physical activity on the primary in vivo antibody and T cell-mediated responses in men. *J. Appl. Physiol.* 97(2):491–498.

Smith, A., Vollmer-Conna, U., Bennett, B., Wakefield, D., Hickie, I., and Lloyd, A. (2004b). The relationship between distress and the development of a primary immune response to a novel antigen. *Brain Behav. Immun.* 18(1):65–75.

Solberg, L.C., Horton, T.H., and Turek, F.W. (1999). Circadian rhythms and depression: effect of exercise in an animal model. *Am. J. Physiol.* 276:R152–R161.

Sondermann, H., Becker, T., Mayhew, M., Wieland, F., and Hartl, F.U. (2000). Characterization of a receptor for heat shock protein 70 on macrophages and monocytes. *Biol. Chem.* 381(12):1165–1174.

Srivastava, P. (2002). Interaction of heat shock proteins with peptides and antigen presenting cells: Chaperoning of the innate and adaptive immune responses. *Annu. Rev. Immunol.* 20:395–425.

Stevens, T.L., Bossie, A., Sanders, V.M., Fernandez-Botran, R., Coffman, R.L., Mosmann, T.R., and Vitetta, E.S. (1988). Regulation of antibody isotype secretion by subsets of antigen-specific helper T cells. *Nature* 334(6179):255–258.

Suzuki, K., and Machida, K. (1995). Effectiveness of lower-level voluntary exercise in disease prevention of mature rats: Cardiovascular risk factor modification. *Eur. J. Appl. Physiol.* 71:240–244.

Swanson, M.A., Lee, W.T., and Sanders, V.M. (2001). IFN-gamma production by Th1 cells generated from naive CD4+ T cells exposed to norepinephrine. *J. Immunol.* 166(1):232–240.

Takahashi, H.K., Iwagaki, H., Mori, S., Yoshino, T., Tanaka, N., and Nishibori, M. (2004). Beta 2-adrenergic receptor agonist induces IL-18 production without IL-12 production. *J. Neuroimmunol.* 151(1–2):137–147.

Takaki, A., Huang, Q.H., Somogyvari-Vigh, A., and Arimura, A. (1994). Immobilization stress may increase plasma interleukin-6 via central and peripheral catecholamines. *Neuroimmunomodulation* 1(6):335–342.

Tissieres, A., Mitchell, H.K., and Tracy, U. (1974). Protein synthesis in salivary glands of Drosophila melanogaster: Relation to chromosome puffs. *J. Mol. Biol.* 84:389–398.

Todryk, S.M., Melcher, A.A., Dalgleish, A.G., and Vile, R.G. (2000). Heat shock proteins refine the danger theory. *Immunology* 99(3):334–337.

Tomioka, H., and Saito, H. (1992). Characterization of immunosuppressive funcitons of murine peritoneal macrophages induced with various agents. *J. Leukoc. Biol.* 51:24–31.

Tytell, M. (2005). Release of heat shock proteins (Hsps) and the effects of extracellular Hsps on neural cells and tissues. *Int. J. Hypertherm.* 2005 Aug; 21(5):445–455.

Vabulas, R.M., Ahmad-Nejad, P., Ghose, S., Kirschning, C.J., Issels, R.D., and Wagner, H. (2002). HSP70 as endogenous stimulus of the Toll/interleukin-1 receptor signal pathway. *J. Biol. Chem.* 277(17):15107–15112.

van Eden, W., van der Zee, R., and Prakken, B. (2005). Heat-shock proteins induce T-cell regulation of chronic inflammation. *Nat. Rev. Immunol.* 5(4):318–330.

Visintin, A., Mazzoni, A., Spitzer, J.H., Wyllie, D.H., Dower, S.K., and Segal, D.M. (2001). Regulation of Toll-like receptors in human monocytes and dendritic cells. *J. Immunol.* 166(1):249–255.

Vollmer, R.R., Meyers-Schoy, S.A., Kolibal-Pegher, S.S., and Edwards, D.J. (1995). The role of the adrenal medulla in neural control of blood pressure in rats. *Clin. Exp. Hypertens.* 17(4):649–667.

Walsh, R.C., Koukoulas, I., Garnham, A., Moseley, P.L., Hargreaves, M., and Febbraio, M.A. (2001). Exercise increases serum Hsp72 in humans. *Cell Stress Chaperones* 6(4):386–393.

Wang, Y., Whittall, T., McGowan, E., Younson, J., Kelly, C., Bergmeier, L.A., Singh, M., and Lehner, T. (2005). Identification of stimulating and inhibitory epitopes within the heat shock protein 70 molecule that modulate cytokine production and maturation of dendritic cells. *J. Immunol.* 174(6):3306–3316.

Watkins, L.R., Thurston, C.L., and Fleshner, M. (1990). Phenlyephrine-induced antinociception: Investigations of potential neural and endocrine bases. *Brain Res.* 528:273–284.

Wright, B.H., Corton, J.M., El-Nahas, A.M., Wood, R.F., and Pockley, A.G. (2000). Elevated levels of circulating heat shock protein 70 (Hsp70) in peripheral and renal vascular disease. *Heart Vessels* 15(1):18–22.

Xiao, R.P., and Lakatta, E.G. (1992). Deterioration of beta-adrenergic modulation of cardiovascular function with aging. *Ann. N. Y. Acad. Sci.* 673:293–310.

Xin, X., Yang, N., Eckhart, A.D., and Faber, J.E. (1997). alpha1D-adrenergic receptors and mitogen-activated protein kinase mediate increased protein synthesis by arterial smooth muscle. *Mol. Pharmacol.* 51:764–775.

Zidel, Z., and Masek, K. (1998). Erratic behavior of nitric oxide within the immune system: Illustrative review of conflicting data and their immunopharmacological aspects. *Int. J. Immunopharmacol.* 20:319–343.

4
Anthrax Lethal Factor Represses Glucocorticoid and Progesterone Receptor Activity

Jeanette I. Webster, Mahtab Moayeri, and Esther M. Sternberg

1. Introduction

Death from anthrax has been reported to occur from systemic shock. The lethal toxin (LeTx) is the major effector of anthrax mortality. Although the mechanism of entry of this toxin into cells is well understood, its actions once inside the cell are not as well understood. LeTx is known to cleave and inactivate mitogen activated protein kinase kinases (MAPKKs). We have recently shown that LeTx represses the glucocorticoid receptor both *in vitro* and *in vivo*. This repression is partial and specific, showing some receptor specificity and some promoter specificity. This toxin does not affect glucocorticoid receptor (GR) ligand binding or DNA binding in an *in vitro* electrophoretic mobility shift assay using a DNA probe. However, in chromatin immunoprecipitation assays, LeTx prevents GR binding to chromatin. We have suggested that LeTx may function by removing/inactivating one or more of the many cofactors and/or accessory proteins involved in nuclear hormone receptor signaling. Although the precise involvement of this nuclear hormone receptor repression in LeTx toxicity is unknown, examples of blunted hypothalamic-pituitary-adrenal (HPA) axis and glucocorticoid signaling in numerous autoimmune/inflammatory diseases suggest that such repression of critically important receptors could have deleterious effects on health. In addition, removal of endogenous glucocorticoids and treatment with glucocorticoids (in LT-resistant mice) increases susceptibility to LeTx, suggesting that a precise balance of glucocorticoid levels is required for LeTx survival.

2. Hypothalamic-Pituitary-Adrenal Axis and Glucocorticoid Responses

A balance within the body between the brain and immune systems is maintained and regulated by the hypothalamic-pituitary-adrenal (HPA) axis and the resultant immunomodulatory hormones, glucocorticoids (Webster

57

et al., 2002). The endogenous glucocorticoid in man is cortisol, whereas in rodents it is corticosterone. The expression of corticotrophin releasing hormone (CRH) in the hypothalamic region of the brain is activated by inflammatory or other stimuli. In turn, CRH stimulates the release of adrenocorticotropin hormone (ACTH) into the bloodstream from the anterior pituitary gland. ACTH then stimulates the synthesis and release of glucocorticoids from the adrenal glands. In order to maintain regulation of this axis, glucocorticoids feed back and downregulate the HPA axis at the level of the hypothalamus and pituitary. In addition to regulation of the immune system, glucocorticoids are also essential for the regulation of several homeostatic systems in the body, including the central nervous system, cardiovascular system, and metabolic homeostasis. Glucocorticoid regulation of the immune system will not be discussed in detail here but has been the subject of another recent review (Webster *et al.*, 2002).

The many functions of glucocorticoids are elicited through the glucocorticoid receptor (GR), a cytosolic receptor. This receptor, along with receptors such as the thyroid hormone, mineralocorticoid, estrogen and progesterone receptors, is a member of the nuclear hormone receptor superfamily (Evans, 1988). For GR, the receptor is located in the cytoplasm in a protein complex, which includes Hsp90 and Hsp70, in the absence of ligand. When the ligand binds, GR is released from the protein complex, dimerizes, and translocates to the nucleus. Once in the nucleus, GR regulates gene expression by binding to specific DNA sequences called glucocorticoid response elements (GREs) (Aranda and Pascual, 2001; Schoneveld *et al.*, 2004). GR is able to upregulate gene expression, such as for the gluconeogenic enzyme tyrosine animotransferase (TAT) (Jantzen *et al.*, 1987), through direct DNA binding. However, it can also repress gene activation, such as the POMC gene, by direct binding to DNA sequences called negative GREs (nGREs) (Drouin *et al.*, 1989). GR can also negatively regulate gene expression without direct binding to DNA. In this case, GR interferes with the action of other signaling pathways, such as NF-κB and AP-1. It is through such interference with other signaling pathways that glucocorticoids exert many of their anti-inflammatory actions (McKay and Cidlowski, 1999; Adcock, 2000; De Bosscher *et al.*, 2003; Smoak and Cidlowski, 2004). GR is essential for life, and mice lacking GR die shortly after birth due to defects in lung maturation (Cole *et al.*, 1995). However, mice with a point mutation that inhibits GR dimerization (GR$^{dim/dim}$) are viable. In these mice, GR functions that require dimerization, such as GREmediated gene activation, are prevented, but GR functions that do not require dimerization, such as interactions with NF-κB and AP-1, are still possible. This suggests that the anti-inflammatory actions of GR mediated through protein-protein interactions rather than direct DNA binding are essential for life (Reichardt *et al.*, 1998).

2.1. Protective Features of an Intact HPA Axis and Glucocorticoid Response

An intact HPA axis and resultant glucocorticoid responses are critical in maintaining body homeostasis and protecting against insults of a variety of sources. The importance of this has been shown in several systems but most compelling is the fact that lack of GR is incompatible with life (Cole *et al.*, 1995). However, endogenous glucocorticoids can be removed by adrenalectomy, and adrenalectomized animals can survive provided the appropriate hormones are replaced exogenously. In addition, strain differences in glucocorticoid responsiveness are associated with differential inflammatory disease susceptibility.

Blockade of the HPA axis, either by adrenalecomy or hypophysectomy, or removal of functional GR by the antagonist RU486, has been shown to exacerbate disease course even to the extent of death in response to numerous bacterial or viral infections. Conversely, corticosterone replacement promotes survival and disease remittance. This was shown first in mice where adrenalectomy significantly reduced the lethal amount (LD_{50}) of *Escherichia coli* serotype O111:B4 endotoxin (McCallum and Stith, 1982). In F334/N rats, treatment with RU486 and streptococcal cell walls (SCWs) resulted in higher mortality rates than SCW or RU486 alone (Sternberg *et al.*, 1989a). Similarly, in LEW/N rats, myelin basic protein (MBP)-induced experimental allergic encephalomyelitis (EAE) was exacerbated by adrenalectomy. Replacement of corticosterone, depending on dose used, either returned disease status to that of control animals or completely alleviated symptoms (MacPhee *et al.*, 1989). Furthermore, intervention of the HPA axis by hypophysectomy resulted in increased mortality rates from *Salmonella* (Edwards *et al.*, 1991), and adrenalectomy increased mortality rates after murine cytomegalovirus (MCMV) virus infection, which was reversed by dexamethasone treatment (Ruzek *et al.*, 1999).

In addition, *Clostridium difficile* toxin A–induced fluid secretion and inflammation, (Castagliuolo *et al.*, 2001; Mykoniatis *et al.*, 2003) and mortality from Shiga toxin in BALB/c mice (Gómez *et al.*, 1998) were enhanced by RU486 and adrenalectomy. Again glucocorticoid treatment reversed these effects, increasing survival rates from Shiga toxin (Gómez *et al.*, 1998; Palermo *et al.*, 2000) and reversing the inflammatory responses to *Clostridium difficile* toxin A (Castagliuolo *et al.*, 2001; Mykoniatis *et al.*, 2003) in adrenalectomized animals. In fact, replacement of a physiological corticosterone dose resulted in an inflammatory response equivalent to sham-operated animals, whereas replacement of a high corticosterone dose resulted in a reduction of the inflammatory response (Castagliuolo *et al.*, 2001).

2.2. Diminished HPA Axis and GR Responses in Disease

A blunted HPA axis response (i.e., blunted glucocorticoid secretion in response to stimuli) has been associated with numerous autoimmune/ inflammatory diseases in both animal models and humans. In animals, a blunted HPA axis has been associated with autoimmune thyroiditis in chickens (Wick *et al.*, 1998), lupus in mice (Hu *et al.*, 1993; Lechner *et al.*, 1996), and numerous autoimmune/inflammatory diseases in rats (Wilder *et al.*, 1982; Sternberg *et al.*, 1989a, 1989b).

The inflammatory resistant Fischer (F334/N) rats have a hyper-HPA axis with hypersecretion of CRH, ACTH, and corticosterone in response to a stimuli thereby allowing corticosterone to suppress the immune system. Conversely, inflammatory-prone Lewis (LEW/N) rats have a blunted HPA axis with minimal production of CRH, ACTH, and corticosterone in response to stimuli. If the stimulus is a proinflammatory or antigenic molecule, then the immune system will not be downregulated due to the lack of corticosterone, and these animals will be susceptible to development of a variety of autoimmune/inflammatory diseases (Wilder *et al.*, 1982; Sternberg *et al.*, 1989a, 1989b; Moncek *et al.*, 2001). Interestingly, BALB/c mice also have a hyper-HPA axis and C57/BJ mice a relative blunted HPA axis response similar to the F334/N and LEW/N rats (Shanks *et al.*, 1990).

In humans, a blunted HPA axis response has been associated with rheumatoid arthritis (Cash *et al.*, 1992; Chikanza *et al.*, 1992; Crofford *et al.*, 1997; Cutolo *et al.*, 1999; Eijsbouts and Murphy, 1999; Gutierrez *et al.*, 1999), system lupus erythematosus (SLE) (Gutierrez *et al.*, 1998; Crofford, 2002), Sjögren syndrome (Johnson *et al.*, 1998; Valtysdottir *et al.*, 2001; Crofford, 2002), allergic asthma and atopic skin disease (Rupprecht *et al.*, 1995; Buske-Kirschbaum *et al.*, 1997, 1998, 2003; Buske-Kirschbaum and Hellhammer, 2003), chronic fatigue syndrome (Demitrack *et al.*, 1991; Demitrack and Crofford, 1998; Neeck and Crofford, 2000; Racciatti *et al.*, 2001; Gaab *et al.*, 2002, 2004; Crofford *et al.*, 2004; Roberts *et al.*, 2004), fibromyalgia (Demitrack and Crofford, 1998; Crofford *et al.*, 1994, 2004; Neeck and Crofford, 2000; Calis *et al.*, 2004), and multiple sclerosis (Michelson *et al.*, 1994; Wei and Lightman, 1997; Huitinga *et al.*, 2003).

Glucocorticoid resistance and/or diminished function of GR has also been associated with numerous diseases. Familial glucocorticoid resistance is a hereditary disease usually involving mutations in the GR. To date, three point mutations and a deletion in the ligand binding domain of GR and a point mutation in the hinge region have been identified in families with familiar glucocorticoid resistance (Kino and Chrousos, 2001). Polymorphisms in GR-α could also cause changes in glucocorticoid sensitivity, as has been shown for a polymorphism in codon 363 (Huizenga *et al.*, 1998). However, no relationship between this and other polymorphisms and glucocorticoid resistance was seen in a normal population (Koper *et al.*, 1997). For a recent update on all known GR mutations and polymorphisms in both patients and cell lines, see the recent review by Bray and Cotton (Bray and Cotton, 2003). GR-β, a splice variant of the glucocorticoid receptor, has been suggested to function as a dominant negative repressor of GR (Oakley *et al.*, 1996, 1999; Vottero and Chrousos, 1999), although other investigators failed to find such a repressor function (Hecht *et al.*, 1997; Brogan *et al.*, 1999; Carlstedt-Duke, 1999). However, an increased expression of GR-β relative to GR-α has been shown in several autoimmune/inflammatory diseases including glucocorticoid resistant asthma (Leung *et al.*, 1997; Hamid *et al.*, 1999; Sousa *et al.*, 2000; Strickland *et al.*, 2001), ulcerative colitis (Honda *et al.*, 2000; Orii *et al.*, 2002), chronic lymphocytic leukemia (Shahidi *et al.*, 1999), nasal polyposis disease (Hamilos *et al.*, 2001; Pujols *et al.*, 2003), rhinosinusitis (Fakhri *et al.*, 2003, 2004), interstitial lung disease (Pujols *et al.*, 2004), and rheumatoid arthritis (DeRijk *et al.*, 2001; Chikanza, 2002). In childhood leukemia, glucocorticoid resistance was not associated with over-expression of GR-β but is possibly due to overexpression of another splice variant, GR-γ (Haarman *et al.*, 2004). In addition, decreased GR numbers have been associated with various diseases, including Crohn's disease (Hori *et al.*, 2002) and rheumatoid arthritis (Schlaghecke *et al.*, 1992).

Glucocorticoid insensitivity need not result from a mutation in the glucocorticoid receptor itself. There are multiple steps in glucocorticoid signaling, and a problem/defect at any of these points could result in glucocorticoid insensitivity. These include transport of the hormone in the blood, availability of the hormone, entry of the hormone into the cell, dissociation from the heat shock protein complex, dimerization, translocation to the nucleus, and interaction with DNA, cofactors and the transcriptional machinery. In the blood, cortisol is bound to cortisol binding globulin (CBG) and only the free, non-protein-bound portion is active. Therefore, increased expression of CBG would decrease the bioavailability of cortisol. This has been suggested to cause the glucocorticoid resistance seen in patients with long-standing Crohn's disease (Mingrone *et al.*, 1999). The enzyme 11β-dehydroxysteroid dehydrogenase (11β-HSD) converts active glucocorticoids into an inactive state (Seckl and Walker, 2001). Therefore, changes in this enzyme would result in changes in the ratio of glucocorticoids in the active versus inactive state. Furthermore, dysregulation of this enzyme has been correlated with obesity and type 2 diabetes (Rask *et al.*, 2001; Lindsay *et al.*, 2003; Westerbacka *et al.*, 2003; Seckl *et al.*, 2004; Valsamakis *et al.*, 2004) and decreased 11β-HSD mRNA is seen in ulcerative colitis (Takahasi *et al.*, 1999). Multidrug resistance proteins (MDR) are members of the ATP-binding cassette (ABC) family of transporters, which have been shown to be capable of exporting glucocorticoids from cells and thereby regulating the intracellular hormonal concentration (Kralli and Yamamoto, 1996; Medh *et al.*, 1998; Webster and Carlstedt-Duke, 2002). Overexpression of MDR-1 has been shown in glucocorticoid-resistant inflammatory bowel disease/ulcerative colitis (Farrell *et al.*, 2000; Hirano *et al.*, 2004), rheumatoid arthritis (Llorente *et al.*, 2000), systemic lupus erythematosos (SLE) (Diaz-Borjon *et al.*, 2000), and myasthenia gravis (Richaud-Patin *et al.*, 2004). In addition, defects in the cofactors involved in the interaction between the glucocorticoid receptor and the transcriptional machinery may cause changes in glucocorticoid resistance. The HIV protein, virion-associated protein (Vpr), functions as a cofactor to enhance glucocorticoid responses resulting in the HIV-associated glucocorticoid hypersensitive state (Kino *et al.*, 1999). Conversely, a defect in a cofactor has been proposed to be the cause of resistance to multiple steroids in two sisters (New *et al.*, 1999, 2001).

In light of these facts and the importance of an intact HPA axis and functional GR for health, we suggested that another mechanism by which glucocorticoid resistance may occur is through effects of bacterial proteins on glucocorticoid receptor signaling. The observations that normally inflammatory-resistant F334/N rats die from a severe inflammatory response that results if they are simultaneously treated with RU486, a GR antagonist, and streptococcal cell walls (Sternberg *et al.*, 1989a) and that this strain of rats are highly susceptible to death from anthrax lethal toxin (Ezzell *et al.*, 1984) led to the hypothesis that LeTx may act as a GR antagonist and facilitate shock in a manner similar to RU486.

3. Anthrax Lethal Toxin

The spore-forming, Gram-positive bacteria *Bacillus anthracis* contains two plasmids [for a review on the bacterium, please refer to the review by Mock and Fouet (Mock and Fouet, 2001)]. The three proteins that comprise the two toxins are encoded on one of these plasmids, pXO1. Together these proteins form a variation of the classical A-B toxins. Protective antigen (PA) and edema factor (EF) constitute the edema toxin and PA and lethal factor (LF) the lethal toxin (LeTx) (Collier and Young, 2003). The involvement of the lethal toxin and anthrax infection is evident as strains lacking the pXO1 plasmid are attenuated in virulence, and the lethal toxin alone manifests distinct symptoms of *B. anthracis* infection such as pleural effusions (Klein *et al.*, 1962; Fish *et al.*, 1968; Pezard *et al.*, 1991; Moayeri *et al.*, 2003; Cui *et al.*, 2004). Therefore, much of anthrax research has focused on LeTx and its mechanism of action, and this toxin will also be the focus of this review.

LeTx entry into cells is now fairly well understood at the molecular level (Abrami *et al.*, 2003, 2004) and is summarized in Figure 4.1. Only macrophages from a limited group of inbred mice have been shown to be sensitive to rapid lysis by LeTx and thus have until recently been suggested to be the target of LeTx action (Friedlander, 1986; Singh *et al.*, 1989; Friedlander *et al.*, 1993; Roberts *et al.*, 1998). LeTx at sublytic concentrations induces macrophage apoptosis (Park *et al.*, 2002; Popov *et al.*, 2002b). Similar proapoptotic behavior is seen in human peripheral blood mononuclear cells (PBMC) but these cells are not lysed by LeTx (Popov *et al.*,

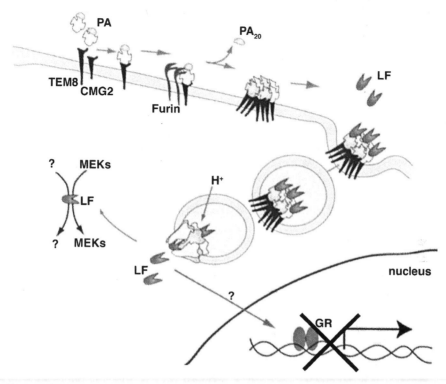

FIGURE 4.1. Mechanism of action of anthrax lethal toxin. PA binds to two different receptors, tumor endothelial marker 8 (TEM-8) and capillary morphogenesis protein 2 (CMG-2) (Bradley *et al.*, 2001; Scobie *et al.*, 2003), which seem to be ubiquitously expressed. PA is then cleaved by the enzyme furin (Klimpel *et al.*, 1992), heptamerizes, and then binds EF and/or LF (Singh *et al.*, 1999). This complex is then internalized by clathrin-dependent raft-mediated endocytosis (Abrami *et al.*, 2003), and the LF/EF are translocated across the endosomal membrane and into the cytosol via a pH- and voltage-dependent mechanism (Blaustein *et al.*, 1989; Zhao *et al.*, 1995; Wesche *et al.*, 1998). Once inside the cell, the mechanism of action of LF is less well understood. It is known to cleave and inactivate members of the MAPKK family, and we now show that it can inactivate the glucocorticoid receptor probably by directly or indirectly destabilizing or interfering with GR binding to chromatin.

2002a). Recent data has suggested that macrophage lysis is not essential for LeTx toxicity, however, it appears to potentially exacerbate the toxin's effects in mice (Moayeri *et al.*, 2003, 2005). Most species harbor LeTx-resistant macrophages. However, a more complicated involvement of macrophage sensitivity in the pathogenesis of anthrax is suggested by discoveries showing that resistant C57BL/6J macrophages or human macrophages can be made sensitive to LeTx by treatment with poly-D-glutamic acid (the major component of the *B. anthracis* capsule), peptidoglycan (a component of Gram-positive bacterial cell walls), lipopolysaccharide (LPS; a component of Gram-negative bacterial cell walls), or TNF-α (Park *et al.*, 2002; Popov *et al.*, 2002a; Kim *et al.*, 2003). Recent studies show that the differentiation state of human monocytic cells determines their LeTx sensitivity (Kassam *et al.*, 2005). Therefore, it is possible to imagine sensitization of macrophages during the course of infection in hosts normally harboring LT-resistant macrophages may play a role in *B. anthracis* pathogenesis.

LF, a zinc metalloprotease, is known to cleave and inactivate some members of the mitogen activated protein (MAP) kinase kinase (MAPKK/MEK) family (Duesbery *et al.*, 1998; Vitale *et al.*, 1998; Pellizzari *et al.*, 2000). The cleavage and inactivation of MAPKKs results in inhibition of downstream signaling pathways such as AP-1 and NFAT (Paccani *et al.*, 2005). However, LF cleavage of MAPKKs alone cannot account for macrophage lysis as LF internalization (Menard *et al.*, 1996; Roberts *et al.*, 1998; Singh *et al.*, 1989) and MAPKK cleavage in sensitive and resistant macrophages and cell lines (Pellizzari *et al.*, 1999, 2000; Watters *et al.*, 2001) are the same. Inhibition of the proteolytic function of LF prevents LeTx toxicity in sensitive cells (Klimpel *et al.*, 1994; Menard *et al.*, 1996; Duesbery *et al.*, 1998; Hammond and Hanna, 1998; Vitale *et al.*, 1998) suggesting that cleavage of MAPKKs or other potentially unidentified substrates is necessary for LeTx macrophage lysis. It is possible that the response to MAPKK cleavage in different cells may lead to a different cascade of events that in sensitive cells leads to cell lysis but not in resistant cells, thereby defining differential sensitivity to LeTx. The resistance response, in turn, can be overcome by pretreatment with LPS or *B. anthracis* cell wall products. One such potential factor involved in this differential response may be the kinesin Kif1C, which was identified as the macrophage sensitivity locus for LeTx (Watters *et al.*, 2001). The function of Kif1C is unknown, but it has been suggested to be involved in endoplasmic reticulum (ER) transport (Dorner *et al.*, 1998). Although a single locus is linked to macrophage sensitivity to LeTx, another group has shown contribution of two additional loci, Ltxs2 and Ltxs3, to mouse susceptibility to LeTx (McAllister *et al.*, 2003). The relative contribution of the different macrophage sensitivity loci to LeTx susceptibility has been shown to vary among different mouse strains (Moayeri *et al.*, 2004).

Despite recent reports on endothelial cell sensitivity to LeTx (Kirby, 2004; Pandey and Warburton, 2004), most studies on the mechanism of LeTx-mediated cell death use the macrophage as a model. However, the mechanism by which LeTx induces apoptosis (Park *et al.*, 2002) or necrosis (Kim *et al.*, 2003) in macrophages or other cell types is currently unknown. Recent studies have focused on gene array and proteome changes in response to toxin using dying LeTx-sensitive macrophages at late time points, yielding a variety of changes primarily associated with stress responses or energy production in cells (Bergman *et al.*, 2005; Chandra *et al.*, 2005; Comer *et al.*, 2005) and not providing too many clues to the early events involved in toxicity. Other studies on LeTx have reported on the toxin's ability to interfere with immune cell function through its inhibition of MAPKK function. For example, LeTx represses LPS-induced cytokine (TNF-α, IL-1α, IL-6, and IL-12) production in dendritic cells (Agrawal *et al.*, 2003) and anthrax cell wall–induced cytokine release in peripheral blood mononuclear cells (PBMCs) (Popov *et al.*, 2002a). Recently, the toxin has been shown to inhibit T-cell activation (Paccani *et al.*, 2005).

Death from anthrax lethal toxin has been reported to occur from systemic shock (Smith *et al.*, 1955), resembling cytokine-mediated LPS-induced shock (Hanna *et al.*, 1993). However, the involvement of cytokines in LeTx toxicity has been a matter of some controversy. In sensitive macrophages, LeTx (1 pg/ml for 16 h) alone did not induce TNF-α, IL-6, IL-1α, or IL-1β (Erwin *et al.*, 2001). However, others have shown that in a LeTx-sensitive macrophage cell line and in macrophages from ICR mice, sublytic LeTx concentrations (1 pg/ml for 6 h) alone was able to induce TNF-α and IL-1β (Hanna *et al.*, 1993; Shin *et al.*, 2000). Analysis of more than 40 cytokines and inflammatory mediators in BALB/cJ and C57BL/6J mice after LeTx administration showed an early transitory increase of numerous factors in BALB/cJ but not C57BL/6J mice. However, no inflammatory cascade was induced and no TNF-α induction was seen (Moayeri *et al.*, 2003) while animals from both strains were susceptible to LeTx. The transitory response seen in the Balb/c animals was directly linked to the macrophage lysis event (Moayeri *et al.*, 2004). Inflammatory responses were also absent in a rat infusion model of LeTx killing (Cui *et al.*, 2004). Knockout mice for the TNF or IL-1 receptors or iNOS did not differ in their susceptibility to anthrax infection compared with control mice (Kalns *et al.*, 2002). In fact, in both sensitive and resistant macrophage cell lines, LeTx inhibits LPS/IFN-γ stimulation of TNF-α and NO probably through inactivation of the MAPKK pathways (Pellizzari *et al.*, 1999, 2000; Erwin *et al.*, 2001) much in the manner LPS-mediated cytokine responses are shut down by LT in dendritic cells (Agrawal *et al.*, 2003). Although there is conflicting data regarding cytokine induction by LeTx alone, there seems to be consistency in that LeTx inhibits bacterial or LPS-induced cytokine release probably through inhibition of MAPKK. It is likely that an inflammatory cytokine release is not critical for LeTx lethality and induction of circulatory shock (Moayeri *et al.*, 2003; Cui *et al.*, 2004). However, cytokines are likely released in response to exposure to *B. anthracis* spores (Pickering and Merkel, 2004; Pickering *et al.*, 2004; Popov *et al.*, 2004) and may alter macrophage and other cell responses to LeTx in ways we currently cannot predict. In rats, LeTx-mediated killing can occur as rapidly as 45 min (Ezzell *et al.*, 1984) and clearly does not require transcription-translation events, and thus cytokine expression. The method by which LeTx induces circulatory shock in different animal models remains to be elucidated.

4. Anthrax Lethal Toxin Repression of GR

We have recently shown that LeTx does inhibit dexamethasone-induced GR transactivation in two cellular systems. In a transient transfection system, the ability of GR to activate a reporter gene was repressed 50% by LeTx, and in cells endogenously expressing GR the ability of dexamethasone to induce the tyrosine aminotransferase gene was also repressed 50% by LeTx. In addition, we also showed that the ability of dexamethasone to induce the liver tyrosine aminotransferase gene *in vivo* in BALB/c mice was also repressed 50% by LeTx (Figure 4.1) (Webster *et al.*, 2003). This LeTx-mediated repression of GR function is specific to gene activation as it has no effect on GR repression of NF-κB (Webster and Sternberg, 2005). LeTx does not function as a true GR antagonist, such as RU486, to repress GR gene activation as it does not prevent ligand binding to the receptor (Webster *et al.*, 2003).

LeTx not only represses GR but also represses the progesterone receptor B (PR-B) and estrogen receptor α (ER-α) to different extents. However, not all nuclear hormone receptors are repressed by LeTx, for example ER-β is not repressed by LeTx (Webster *et al.*, 2003). Upon further analysis, we have shown that LeTx shows receptor specificity but also some promoter specificity. For example, on a simple promoter, (GRE)$_2$ tk-luc, which contains 2 GRE sequences before the minimal tk promoter, LeTx represses GR and PR-B but has no effect on MR. LeTx also has no effect on the androgen receptor (AR) on its simple promoter, (ARE)$_4$ luc. However, on the complex promoter, mouse mammary tumor virus (MMTV), which contains 1000 base pairs of the MMTV promoter, LeTx is able to repress GR, PR-B, AR, and MR (Webster and Sternberg, 2005). Therefore, LeTx shows some receptor and some promoter specificity in the repression of nuclear hormone receptors.

The features that determine LeTx repression are a matter of current research. LeTx, unlike a true antagonist such as RU486, does not fully repress these receptors. This led to the suggestion that LeTx may be removing/inactivating one or more of the many cofactors involved in the interaction of nuclear hormone receptors and the transcriptional machinery (Webster *et al.*, 2003). It is feasible that removal of one or more of these multiple pathways may result in partial repression as some activity could still be afforded through the remaining intact pathways.

We had previously shown that LeTx does not interfere with the GR-DNA interaction, at least on a simple GRE sequence in an *in vitro* electrophoretic mobility shift assay (EMSA) experiment (Webster *et al.*, 2003). However, further analysis using chromatin immunoprecipitation (ChIP) assays showed that LeTx did, in fact, prevent GR binding to DNA in the context of native chromatin (Webster and Sternberg, 2005). The observation of this difference in assays employing an oligonucleotide versus native chromatin eludes toward the molecular mechanism of action of LeTx. As stated above, we had previously suggested that the site of action of LeTx is one of the many cofactors involved in nuclear hormone receptor gene activation (Webster et al., 2003), but this could now be extended to included the accessory proteins that interact with GR at the level of DNA binding.

Interestingly, arsenic has been shown to inhibit GR in a similar manner. Low levels of arsenic inhibit GR-mediated gene activation but not gene repression, and this appears to be also dependent on the DNA binding region of the receptor (Bodwell et al., 2004). This suggests that it is possible that other toxins (bacterial or environmental) may be able to interfere with GR signaling through a similar mechanism. Other bacterial proteins have previously been shown to affect the HPA axis and glucocorticoid signaling [reviewed by us (Webster and Sternberg, 2004)], but this mechanism of repression of nuclear hormone receptors by LeTx is novel. Recently, activation of some nuclear hormone receptors, namely RXR and LXR, has been shown to prevent cell death and apoptosis of macrophage cell lines infected by *Bacillus anthracis* (Valledor et al., 2004). These data suggest that nuclear hormone receptors may play a role in bacteria-mediated cell killing.

One unanswered question has been the role of HPA axis/glucocorticoid responses during exposure to anthrax lethal toxin. As we have previously described, an intact HPA axis response is essential for survival from a range of proinflammatory, bacterial, and viral insults. We have recently shown that adrenalectomy also enhances sensitivity to LeTx in all mouse strains tested including DBA/2J mice, which are completely resistant to LeTx when the adrenals are intact (Moayeri et al., 2005). However, unlike the studies described earlier, dexamethasone treatment does not prevent this increased mortality rate but rather sensitizes mice to LeTx (Moayeri et al., 2005). Although this data is contrary to what one might initially expect, there is data to demonstrate that the HPA axis needs to be carefully balanced for optimal host responses in bacterial exposures. Perturbations in either direction are detrimental to health. Indeed, prolonged treatment of septic shock with high doses of glucocorticoids has been shown to be detrimental to health, and increased cortisol levels have been shown in critically ill patients (Vermes and Beishuizen, 2001; Thompson, 2003; Hamrahian et al., 2004). These data suggest that a carefully balanced HPA axis is required for survival from LeTx. In agreement with this is the observation that blood cortisol levels increase in BALB/cJ mice, which are normally sensitive to LeTx, after LeTx administration, whereas there is no change in DBA/2J blood cortisol levels (Moayeri et al., 2005).

4.1. p38 MAPK and GR

Because LeTx has been shown to cleave and inactivate members of the MAPKK family (Duesbery et al., 1998; Vitale et al., 1998; Pellizzari et al., 2000), we tested the effect of other MAPKK inhibitors on dexamethasone induction of GR-mediated transactivation. We have shown that dexamethasone-induced GR transactivation was specifically repressed by inhibitors of the p38 MAPKK pathway but not by inhibitors of the other MAPKK pathways (Webster *et al.*, 2003). Whether repression of GR by LeTx is a direct effect of p38 MAPKK inactivation or if it is purely a correlative effect remains to be determined. Published literature shows that activation of MAPKK pathways, not repression of basal MAPKKs, inactivates GR (Lucibello *et al.*, 1990; Schule *et al.*, 1990; Rogatsky *et al.*, 1998; Krstic *et al.*, 1997; Herrlich, 2001; Karin and Chang, 2001; Lopez *et al.*, 2001; Irusen *et al.*, 2002; Szatmary *et al.*, 2004; Wang *et al.*, 2004). Therefore, if LeTx repression of GR is mediated through its inactivation of p38 MAPKK, then this is a novel mechanism of action. There are, however, a few studies emerging that suggest an interaction between p38 and GR. In osteoclasts, the p38 inhibitor SB203580 was shown to prevent dexamethasone induction of Hsp27 suggesting that p38 is involved in the dexamethasone induction of Hsp27 in these cells (Kozawa *et al.*, 2002). It is possible that the effect of p38 on GR (or other nuclear hormone receptors) is dependent on the presence of specific cofactors, such as peroxisome-proliferator-activated receptor γ co-activator 1 (PGC-1). In HeLa cells and in the absence of PGC-1, the p38 inhibitor had no effect on GR-mediated transactivation suggesting that p38 does not act directly on GR. However, in the presence of PGC-1, coexpression of constitutively active MKK6 enhanced GR transactivation, and this could be repressed by SB203580 (Knutti *et al.*, 2001). The role of p38 in LeTx repression of GR is currently under investigation.

4.2. Therapeutic Implications

Reduced glucocorticoid production as a result of a blunted HPA axis response or diminished glucocorticoid sensitivity has been associated with numerous autoimmune/inflammatory diseases, as described above. We have recently described a new mechanism of glucocorticoid resistance related to exposure to the bacterial toxin anthrax lethal toxin. Although we have yet to show the full extent of the involvement of LeTx repression of nuclear hormone receptors in LeTx toxicity, it is reasonable to predict that there may be potential clinical therapeutic implications of LeTx repression of GR. It is clear that a balanced HPA axis and glucocorticoid production is required for survival from LeTx. If other bacterial toxins also induce GR resistance, this could potentially also contribute to their toxicity. This data suggests that care should be taken when treating exposure to LeTx with glucocorticoids (and potentially the anthrax bacterium) as glucocorticoid treatment may not be in the patient's best interest. Identification of the precise mechanism by which LeTx represses GR and other nuclear hormone receptors could provide novel targets for therapy of anthrax and possibly other bacterial toxins.

References

Abrami, L., Liu, S., Cosson, P., Leppla, S.H., and van der Goot, F.G. (2003). Anthrax toxin triggers endocytosis of its receptor via a lipid raft-mediated clathrin-dependent process. *J. Cell. Biol.* 160:321–328.

Abrami, L., Lindsay, M., Parton, R.G., Leppla, S.H., and van der Goot, F.G. (2004). Membrane insertion of anthrax protective antigen and cytoplasmic delivery of lethal factor occur at different stages of the endocytic pathway. *J. Cell. Biol.* 166: 645–651.

Adcock, I.M. (2000). Molecular mechanisms of glucocorticosteroid actions. *Pulm. Pharmacol. Ther.* 13:115–126.

Agrawal, A., Lingappa, J., Leppla, S.H., Agrawal, S., Jabbar, A., Quinn, C., and Pulendran, B. (2003). Impairment of dendritic cells and adaptive immunity by anthrax lethal toxin. *Nature* 424:329–334.

Aranda, A., and Pascual, A. (2001). Nuclear hormone receptors and gene expression. *Physiol. Rev.* 81:1269–1304.

Bergman, N.H., Passalacqua, K.D., Gaspard, R., Shetron-Rama, L.M., Quackenbush, J., and Hanna, P.C. (2005). Murine macrophage transcriptional responses to *Bacillus anthracis* infection and intoxication. *Infect. Immun.* 73:1069–1080.

Blaustein, R.O., Koehler, T.M., Collier, R.J., and Finkelstein, A. (1989). Anthrax toxin: Channel-forming activity of protective antigen in planar phospholipid bilayers. *Proc. Natl. Acad. Sci. U.S.A.* 86:2209–2213.

Bodwell, J.E., Kingsley, L.A., and Hamilton, J.W. (2004). Arsenic at very low concentrations alters glucocorticoid receptor (GR)-mediated gene activation but not GR-mediated gene repression: Complex dose-response effects are closely correlated with levels of activated GR and require a functional GR DNA binding domain. *Chem. Res. Toxicol.* 17:1064–1076.

Bradley, K.A., Mogridge, J., Mourez, M., Collier, R.J., and Young, J.A. (2001). Identification of the cellular receptor for anthrax toxin. *Nature* 414:225–229.

Bray, P.J., and Cotton, R.G. (2003). Variations of the human glucocorticoid receptor gene (NR3C1): Pathological and *in vitro* mutations and polymorphisms. *Hum. Mutat.* 21:557–568.

Brogan, I.J., Murray, I.A., Cerillo, G., Needham, M., White, A., and Davis, J.R.E. (1999). Interaction of glucocorticoid receptor isoforms with transcription factors AP-1 and NF-κB: Lack of effect of glucocorticoid receptor β. *Mol. Cell. Endocrinol.* 157:95–104.

Buske-Kirschbaum, A., and Hellhammer, D.H. (2003). Endocrine and immune responses to stress in chronic inflammatory skin disorders. *Ann. N.Y. Acad. Sci.* 992:231–240.

Buske-Kirschbaum, A., Jobst, S., Psych, D., Wustman, A., Kirschbaum, C., Rauh, W., and Hellhammer, D. (1997). Attenuated free cortisol response to psychosocial stress in children with atopic dermatitis. *Psychosom. Med.* 59:419–426.

Buske-Kirschbaum, A., Jobst, S., and Hellhammer, D.H. (1998). Altered reactivity of the hypothalamus-pituitary-adrenal axis in patients with atopic dermatitis: Pathologic factor or symptom? *Ann. N. Y. Acad. Sci.* 840:747–754.

Buske-Kirschbaum, A., Von Auer, K., Krieger, S., Weis, S., Rauh, W., and Hellhammer, D. (2003). Blunted cortisol responses to psychosocial stress in asthmatic children: A general feature of atopic disease? *Psychosom. Med.* 65: 806–810.

Calis, M., Gokce, C., Ates, F., Ulker, S., Izgi, H.B., Demir, H., Kirnap, M., Sofuoglu, S., Durak, A.C., Tutus, A., and Kelestimur, F. (2004). Investigation of the hypo-thalamo-pituitary-adrenal axis (HPA) by 1 microg ACTH test and metyrapone test in patients with primary fibromyalgia syndrome. *J. Endocrinol. Invest.* 27: 42–46.

Carlstedt-Duke, J. (1999). Glucocorticoid receptor β: View II. *Trends Endocrinol. Metab.* 10:339–342.

Cash, J.M., Crofford, L.J., Gallucci, W.T., Sternberg, E.M., Gold, P.W., Chrousos, G.P., and Wilder, R.L. (1992). Pituitary-adrenal axis responsiveness to ovine corti-cotropin releasing hormone in patients with rheumatoid arthritis treated with low dose prednisone. *J. Rheumatol.* 19:1692–1696.

Castagliuolo, I., Karalis, K., Valenick, L., Pasha, A., Nikulasson, S., Wlk, M., and Pothoulakis, C. (2001). Endogenous corticosteroids modulate *Clostridium difficile* toxin A-induced enteritis in rats. *Am. J. Physiol.* 280:G539–545.

Chandra, H., Gupta, P.K., Sharma, K., Mattoo, A.R., Garg, S.K., Gade, W.N., Sirdeshmukh, R., Maithal, K., and Singh, Y. (2005). Proteome analysis of mouse macrophages treated with anthrax lethal toxin. *Biochim. Biophys. Acta* 1747: 151–159.

Chikanza, I.C. (2002). Mechanisms of corticosteroid resistance in rheumatoid arthri-tis: A putative role for the corticosteroid receptor beta isoform. *Ann. N.Y. Acad. Sci.* 966:39–48.

Chikanza, I.C., Petrou, P., Kingsley, G., Chrousos, G.P., and Panayi, G.S. (1992): Defective hypothalamic response to immune and inflammatory stimuli in patients with rheumatoid arthritis. *Arthritis Rheum.* 35:1281–1288.

Cole, T.J., Blendy, J.A., Monaghan, A.P., Krieglstein, K., Schmid, W., Aguzzi, A., Fantuzzi, G., Hummler, E., Unsicker, K., and Schutz, G. (1995). Targeted disrup-tion of the glucocorticoid receptor gene blocks adrenergic chromaffin cell devel-opment and severely retards lung maturation. *Genes Dev.* 9:1608–1621.

Collier, R.J., and Young, J.A. (2003). Anthrax toxin. *Annu. Rev. Cell. Dev. Biol.* 19: 45–70.

Comer, J.E., Galindo, C.L., Chopra, A.K., and Peterson, J.W. (2005). Genechip analy-ses of global transcriptional responses of murine macrophages to the lethal toxin of *Bacillus anthracis. Infect. Immun.* 73:1879–1885.

Crofford, L.J. (2002). The hypothalamic-pituitary-adrenal axis in the pathogenesis of rheumatic diseases. *Endocrinol. Metab. Clin. North Am.* 31:1–13.

Crofford, L.J., Pillemer, S.R., Kalogeras, K.T., Cash, J.M., Michelson, D., Kling, M.A., Sternberg, E.M., Gold, P.W., Chrousos, G.P., and Wilder, R.L. (1994): Hypothalamic-pituitary-adrenal axis perturbations in patients with fibromyalgia. *Arthitis Rheum.* 37:1583–1592.

Crofford, L.J., Kalogeras, K.T., Mastorakos, G., Magiakou, M.A., Wells, J., Kanik, K.S., Gold, P.W., Chrousos, G.P., and Wilder, R.L. (1997): Circadian relationships between interleukin (IL)-6 and hypothalamic-pituitary-adrenal axis hormones: Failure of IL-6 to cause sustained hypercortisolism in patients with early untreated rheumatoid arthritis. *J. Clin. Endocrinol. Metab.* 82:1279–1283.

Crofford, L.J., Young, E.A., Engleberg, N.C., Korszun, A., Brucksch, C.B., McClure, L.A., Brown, M.B., and Demitrack, M.A. (2004). Basal circadian and pulsatile ACTH and cortisol secretion in patients with fibromyalgia and/or chronic fatigue syndrome. *Brain Behav. Immun.* 18:314–325.

Cui, X., Moayeri, M., Li, Y., Li, X., Haley, M., Fitz, Y., Correa-Araujo, R., Banks, S.M., Leppla, S.H., and Eichacker, P.Q. (2004). Lethality during continuous anthrax lethal toxin infusion is associated with circulatory shock but not inflammatory cytokine or nitric oxide release in rats. *Am. J. Physiol.* 286:R699–709.

Cutolo, M., Foppiani, L., Perete, C., Ballarino, P., Sulli, A., Villaggio, B., Seriolo, B., Giusti, M., and Accardo, S. (1999): Hypothalamic-pituitary-adrenocortical axis function in premenopausal women with rheumatoid arthritis not treated with glucocorticoids. *J. Rheumatol.* 26:282–288.

De Bosscher, K., Vanden Berghe, W., and Haegeman, G. (2003). The interplay between the glucocorticoid receptor and nuclear factor-kappaB or activator protein-1: Molecular mechanisms for gene repression. *Endocr. Rev.* 24:488–522.

Demitrack, M.A., and Crofford, L.J. (1998). Evidence for and pathophysiologic implications of hypothalamic-pituitary-adrenal axis dysregulation in fibromyalgia and chronic fatigue syndrome. *Ann. N.Y. Acad. Sci.* 840:684–697.

Demitrack, M.A., Dale, J.K., Straus, S.E., Laue, L., Listwak, S.J., Kruesi, M.J., Chrousos, G.P., and Gold, P.W. (1991): Evidence for impaired activation of the hypothalamic-pituitary-adrenal axis in patients with chronic fatigue syndrome. *J. Clin. Endocrinol. Metab.* 73:1224–1234.

DeRijk, R.H., Schaaf, M.J., Turner, G., Datson, N.A., Vrcugdcnhil, E., Cidlowski, J., de Kloet, E.R., Emery, P., Sternberg, E.M., and Detera-Wadleigh, S.D. (2001). A human glucocorticoid receptor gene variant that increases the stability of the glucocorticoid receptor beta-isoform mRNA is associated with rheumatoid arthritis. *J. Rheumatol.* 28:2383–2388.

Diaz-Borjon, A., Richaud-Patin, Y., Alvarado de la Barrera, C., Jakez-Ocampo, J., Ruiz-Arguelles, A., and Llorente, L. (2000). Multidrug resistance-1 (MDR-1) in rheumatic autoimmune disorders. Part II: Increased P-glycoprotein activity in lymphocytes from systemic lupus erythematosus patients might affect steroid requirements for disease control. *Joint Bone Spine* 67:40–48.

Dorner, C., Ciossek, T., Muller, S., Moller, P.H., Ullrich, A., and Lammers, R. (1998): Characterization of *Kif1c*, a new kinesin-like protein involved in vesicle transport from the golgi apparatus to the endoplasmic reticulum. *J. Biol. Chem.* 273: 20267–20275.

Drouin, J., Sun, Y.L., and Nemer, M. (1989). Glucocorticoid repression of pro-opiomelanocortin gene transcription. *J. Steroid Biochem.* 34:63–69.

Duesbery, N.S., Webb, C.P., Leppla, S.H., Gordon, V.M., Klimpel, K.R., Copeland, T.D., Ahn, N.G., Oskarsson, M.K., Fukasawa, K., Paull, K.D., and Vande Woude, G.F. (1998). Proteolytic inactivation of map-kinase-kinase by anthrax lethal factor. *Science* 280:734–737.

Edwards, C.K.I., Yunger, L.M., Lorence, R.M., Dantzer, R., and Kelley, K.W. (1991). The pituitary gland is required for protection against lethal effects of *Salmonella typhimurium*. *Proc. Natl. Acad. Sci. U.S.A.* 88:2274–2277.

Eijsbouts, A.M., and Murphy, E.P. (1999). The role of the hypothalamic-pituitary-adrenal axis in rheumatoid arthritis. *Baillieres Best Pract. Res. Clin. Rheumatol.* 13:599–613.

Erwin, J.L., DaSilva, L.M., Bavari, S., Little, S.F., Friedlander, A.M., and Chanh, T.C. (2001). Macrophage-derived cell lines do not express proinflammatory cytokines after exposure to *Bacillus anthracis* lethal toxin. *Infect. Immun.* 69:1175–1177.

Evans, R.M. (1988). The steroid and thyroid hormone receptor superfamily. *Science* 240:889–895.

Ezzell, J.W., Ivins, B.E., and Leppla, S.H. (1984). Immunoelectrophoretic analysis, toxicity, and kinetics of *in vitro* production of the protective antigen and lethal factor components of *Bacillus anthracis* toxin. *Infect. Immun.* 45:761–767.

Fakhri, S., Christodoulopoulos, P., Tulic, M., Fukakusa, M., Frenkiel, S., Leung, D.Y., and Hamid, Q.A. (2003). Role of microbial toxins in the induction of glucocorticoid receptor beta expression in an explant model of rhinosinusitis. *J. Otolaryngol.* 32:388–393.

Fakhri, S., Tulic, M., Christodoulopoulos, P., Fukakusa, M., Frenkiel, S., Leung, D.Y., and Hamid, Q.A. (2004). Microbial superantigens induce glucocorticoid receptor beta and steroid resistance in a nasal explant model. *Laryngoscope* 114:887–892.

Farrell, R.J., Murphy, A., Long, A., Donnelly, S., Cherikuri, A., O'Toole, D., Mahmud, N., Keeling, P.W., Weir, D.G., and Kelleher, D. (2000). High multidrug resistance (P-glycoprotein 170) expression in inflammatory bowel disease patients who fail medical therapy. *Gastroenterology* 118:279–288.

Fish, D.C., Klein, F., Lincoln, R.E., Walker, J.S., and Dobbs, J.P. (1968). Pathophysiological changes in the rat associated with anthrax toxin. *J. Infect. Dis.* 118: 114–124.

Friedlander, A.M. (1986). Macrophages are sensitive to anthrax lethal toxin through an acid-dependent process. *J. Biol. Chem.* 261:7123–7126.

Friedlander, A.M., Bhatnagar, R., Leppla, S.H., Johnson, L., and Singh, Y. (1993). Characterization of macrophage sensitivity and resistance to anthrax lethal toxin. *Infect. Immun.* 61:245–252.

Gaab, J., Huster, D., Peisen, R., Engert, V., Heitz, V., Schad, T., Schurmeyer, T.H., and Ehlert, U. (2002). Hypothalamic-pituitary-adrenal axis reactivity in chronic fatigue syndrome and health under psychological, physiological, and pharmacological stimulation. *Psychosom. Med.* 64:951–962.

Gaab, J., Engert, V., Heitz, V., Schad, T., Schurmeyer, T.H., and Ehlert, U. (2004). Associations between neuroendocrine responses to the insulin tolerance test and patient characteristics in chronic fatigue syndrome. *J. Psychosom. Res.* 56:419–424.

Gutierrez, M.A., Garcia, M.E., Rodriguez, J.A., Rivero, S., and Jacobelli, S. (1998). Hypothalamic-pituitary-adrenal axis function and prolactin secretion in systemic lupus erythematosus. *Lupus* 7:404–408.

Gutierrez, M.A., Garcia, M.E., Rodriguez, J.A., Mardonez, G., Jacobelli, S., and Rivero, S. (1999). Hypothalamic-pituitary-adrenal axis function in patients with active rheumatoid arthritis: A controlled study using insulin hypoglycemia stress test and prolactin stimulation. *J. Rheumatol.* 26:277–281.

Gómez, F., de Kloet, E.R., and Armario, A. (1998). Glucocorticoid negative feedback on the HPA axis in five inbred rat strains. *Am. J. Physiol.* 274:R420–R427.

Haarman, E.G., Kaspers, G.J., Pieters, R., Rottier, M.M., and Veerman, A.J. (2004). Glucocorticoid receptor alpha, beta and gamma expression vs. *in vitro* glucocorticoid resistance in childhood leukemia. *Leukemia* 18:530–537.

Hamid, Q.A., Wenzel, S.E., Hauk, P.J., Tsicopoulos, A., Wallaert, B., Lafitte, J.-J., Chrousos, G.P., Szefler, S.J., and Leung, D.Y.M. (1999). Increased glucocorticoid receptor β in airway cells of glucocorticoid-insensitive asthma. *Am. J. Respir. Crit. Care Med.* 159:1600–1604.

Hamilos, D.L., Leung, D.Y., Muro, S., Kahn, A.M., Hamilos, S.S., Thawley, S.E., and Hamid, Q.A. (2001). GRbeta expression in nasal polyp inflammatory cells and its relationship to the anti-inflammatory effects of intranasal fluticasone. *J. Allergy Clin. Immunol.* 108:59–68.

Hammond, S.E., and Hanna, P.C. (1998). Lethal factor active-site mutations affect catalytic activity *in vitro*. *Infect. Immun.* 66:2374–2378.

Hamrahian, A.H., Oseni, T.S., and Arafah, B.M. (2004). Measurements of serum free cortisol in critically ill patients. *N. Engl. J. Med.* 350:1629–1638.

Hanna, P.C., Acosta, D., and Collier, R.J. (1993). On the role of macrophages in anthrax. *Proc. Natl. Acad. Sci. U.S.A.* 90:10198–10201.

Hecht, K., Carlstedt-Duke, J., Stierna, P., Gustafsson, J.-Å., Brönnegård, M., and Wikström, A.-C. (1997). Evidence that the β-isoform of the human glucocorticoid receptor does not act as a physiologically significant repressor. *J. Biol. Chem.* 272: 26659–26664.

Herrlich, P. (2001). Cross-talk between glucocorticoid receptor and AP-1. *Oncogene* 20:2465–2475.

Hirano, T., Onda, K., Toma, T., Miyaoka, M., Moriyasu, F., and Oka, K. (2004). MDR1 mRNA expressions in peripheral blood mononuclear cells of patients with ulcerative colitis in relation to glucocorticoid administration. *J. Clin. Pharmacol.* 44: 481–486.

Honda, M., Orii, F., Ayabe, T., Imai, S., Ashida, T., Obara, T., and Kohgo, Y. (2000). Expression of glucocorticoid receptor β in lymphocytes of patients with glucocorticoid-resistant ulcerative colitis. *Gastroenterology* 118:859 866.

Hori, T., Watanabe, K., Miyaoka, M., Moriyasu, F., Onda, K., Hirano, T., and Oka, K. (2002). Expression of mRNA for glucocorticoid receptors in peripheral blood mononuclear cells of patients with Crohn's disease. *J. Gastroenterol. Hepatol.* 17: 1070–1077.

Hu, Y., Dietrich, H., Herold, M., Heinrich, P.C., and Wick, G. (1993). Disturbed immuno-endocrine communication via the hypothalamo-pituitary-adrenal axis in autoimmune disease. *Int. Arch. Allergy Immunol.* 102:232–241.

Huitinga, I., Erkut, Z.A., van Beurden, D., and Swaab, D.F. (2003). The hypothalamo-pituitary-adrenal axis in multiple sclerosis. *Ann. N. Y. Acad. Sci.* 992:118–128.

Huizenga, N.A.T.M., Koper, J.W., De Lange, P., Pols, H.A.P., Stolk, R.P., Burger, H., Grobbee, D.E., Brinkmann, A.O., De Jong, F.H., and Lamberts, S.W.J. (1998). A polymorphism in the glucocorticoid receptor gene may be associated with an increased sensitivity to glucocorticoids *in vivo. J. Clin. Endocrinol. Metab.* 83: 144–151.

Irusen, E., Matthews, J.G., Takahashi, A., Barnes, P.J., Chung, K.F., and Adcock, I.M. (2002). p38 mitogen-activated protein kinase-induced glucocorticoid receptor phosphorylation reduces its activity: Role in steroid-insensitive asthma. *J. Allergy Clin. Immunol.* 109:649–657.

Jantzen, H.M., Strahle, U., Gloss, B., Stewart, F., Schmid, W., Boshart, M., Miksicek, R., and Schutz, G. (1987). Cooperativity of glucocorticoid response elements located far upstream of the tyrosine aminotransferase gene. *Cell* 49:29–38.

Johnson, E.O., Vlachoyiannopoulos, P.G., Skopouli, F.N., Tzioufas, A.G., and Moutsopoulos, H.M. (1998). Hypofunction of the stress axis in Sjogren's syndrome. *J. Rheumatol.* 25:1508–1514.

Kalns, J., Scruggs, J., Millenbaugh, N., Vivekananda, J., Shealy, D., Eggers, J., and Kiel, J. (2002). TNF receptor 1, IL-1 receptor, and iNOS genetic knockout mice are not protected from anthrax infection. *Biochem. Biophys. Res. Commun.* 292:41–44.

Karin, M., and Chang, L. (2001). AP-1-glucocorticoid receptor crosstalk taken to a higher level. *J. Endocrinol.* 169:447–451.

Kassam, A., Der, S.D., and Mogridge, J. (2005). Differentiation of human monocytic cell lines confers susceptibility to *Bacillus anthracis* lethal toxin. *Cell. Microbiol.* 7:281–292.

74   J.I. Webster *et al.*

Kim, S.O., Jing, Q., Hoebe, K., Beutler, B., Duesbery, N.S., and Han, J. (2003). Sensitizing anthrax lethal toxin-resistant macrophages to lethal toxin-induced killing by tumor necrosis factor-alpha. *J. Biol. Chem.* 278:7413–7421.

Kino, T., and Chrousos, G.P. (2001). Glucocorticoid and mineralocorticoid resistance/hypersensitivity syndromes. *J. Endocrinol.* 169:437–445.

Kino, T., Gragerov, A., Kopp. J.B., Stauber, R.H., Pavlakis, G.N., and Chrousos, G.P. (1999). The HIV-1 virion-associated protein Vpr is a coactivator of the human glucocorticoid receptor. *J. Exp. Med.* 189:51–61.

Kirby, J.E. (2004). Anthrax lethal toxin induces human endothelial cell apoptosis. *Infect. Immun.* 72:430–439.

Klein, F., Hodges, D.R., Mahlandt, B.G., Jones, W.I., Haines, B.W., and Lincoln, R.E. (1962). Anthrax toxin: Causative agent in the death of rhesus monkeys. *Science* 138:1331–1333.

Klimpel, K.R., Molloy, S.S., Thomas, G., and Leppla, S.H. (1992). Anthrax toxin protective antigen is activated by a cell surface protease with the sequence specificity and catalytic properties of furin. *Proc. Natl. Acad. Sci. U.S.A.* 89:10277–10281.

Klimpel, K.R., Arora, N., and Leppla, S.H. (1994). Anthrax toxin lethal factor contains a zinc metalloprotease consensus sequence which is required for lethal toxin activity. *Mol. Microbiol.* 13:1093–1100.

Knutti, D., Kressler, D., and Kralli, A. (2001). Regulation of the transcriptional coactivator PGC-1 via MAPK-sensitive interaction with a repressor. *Proc. Natl. Acad. Sci. U.S.A.* 98:9713–9718.

Koper, J.W., Stolk, R.P., De Lange, P., Huizenga, N.A.T.M., Molijn, G.-J., Pols, H.A.P., Grobbee, D.E., Karl, M., De Jong, F.H., Brinkmann, A.O., and Lamberts, S.W.J. (1997). Lack of association between five polymorphisms in the human glucocorticoid receptor gene and glucocorticoid resistance. *Hum. Genet.* 99:663–668.

Kozawa, O., Niwa, M., Hatakeyama, D., Tokuda, H., Oiso, Y., Matsuno, H., Kato, K., and Uematsu, T. (2002). Specific induction of heat shock protein 27 by glucocorticoid in osteoblasts. *J. Cell. Biochem.* 86:357–364.

Kralli, A.R., and Yamamoto. K.R. (1996). An FK506-sensitive transporter selectively decreases intracellular levels and potency of steroid hormones. *J. Biol. Chem.* 271: 17152–17156.

Krstic, M.D., Rogatsky, I., Yamamoto, K.R., and Garabedian, M.J. (1997). Mitogen-activated and cyclin-dependent protein kinases selectively and differentially modulate transcriptional enhancement by the glucocorticoid receptor. *Mol. Cell. Biol.* 17:3947–3954.

Lechner, O., Hu, Y., Jafarian Tehrani, M., Dietrich, H., Schwartz, S., Herold, M., Haour, F., and Wick, G. (1996). Disturbed immunoendocine communication via the hypothalamo-pituitary-adrenal axis in murine lupus. *Brain Behav. Immun.* 10: 337–350.

Leung, D.Y.M., Hamid, Q., Vottero, A., Szefler, S.J., Surs, W., Minshall, E., Chrousos, G.P., and Klemm, D.J. (1997). Association of glucocorticoid insensitivity with increased expression of glucocorticoid receptor β. *J. Exp. Med.* 186:1567–1574.

Lindsay, R.S., Wake, D.J., Nair, S., Bunt, J., Livingstone, D.E., Permana, P.A., Tataranni, P.A., and Walker, B.R. (2003). Subcutaneous adipose 11 beta-hydroxysteroid dehydrogenase type 1 activity and messenger ribonucleic acid levels are associated with adiposity and insulinemia in Pima Indians and Caucasians. *J. Clin. Endocrinol. Metab.* 88:2738–2744.

Llorente, L., Richaud-Patin, Y., Diaz-Borjon, A., Alvarado de la Barrera, C., Jakez-Ocampo, J., de la Fuente, H., Gonzalez-Amaro, R., and Diaz-Jouanen, E. (2000). Multidrug resistance-1 (MDR-1) in rheumatic autoimmune disorders. Part I: Increased P-glycoprotein activity in lymphocytes from rheumatoid arthritis patients might influence disease outcome. *Joint Bone Spine* 67:30–39.

Lopez, G.N., Turck, C.W., Schaufele, F., Stallcup, M.R., and Kushner, P.J. (2001). Growth factors signal to steroid receptors through mitogen-activated protein kinase regulation of p160 coactivator activity. *J. Biol. Chem.* 276:22177–22182.

Lucibello, F.C., Slater, E.P., Jooss, K.U., Beato, M., and Muller, R. (1990). Mutual transrepression of Fos and the glucocorticoid receptor: Involvement of a functional domain in Fos which is absent in FosB. *EMBO J.* 9:2827–2834.

MacPhee, I.A.M., Antoni, F.A., and Mason, D.W. (1989). Spontaneous recovery of rats from experimental allergic encephalomyelitis is dependent on regulation of the immune system by endogenous adrenal corticosteroids. *J. Exp. Med.* 169: 431–445.

McAllister, R.D., Singh, Y., Du Bois, W.D., Potter, M., Boehm, T., Meeker, N.D., Fillmore, P.D., Anderson, L.M., Poynter, M.E., and Teuscher, C. (2003). Susceptibility to anthrax lethal toxin is controlled by three linked quantitative trait loci. *Am. J. Pathol.* 163:1735–1741.

McCallum, R.E., and Stith, R.D. (1982). Endotoxin-induced inhibition of steroid binding by mouse liver cytosol. *Circ. Shock* 9:357–367.

McKay, L.I., and Cidlowski, J.A. (1999). Molecular control of immune/inflammatory responses: Interactions between nuclear factor-kappa B and steroid receptor-signaling pathways. *Endocr. Rev.* 20:435–459.

Medh, R.D., Lay, R.H., and Schmidt, T.J. (1998). Agonist-specific modulation of glucocorticoid receptor-mediated transcription by immunosuppressants. *Mol. Cell. Endocrinol.* 138:11–23.

Menard, A., Papini, E., Mock, M., and Montecucco, C. (1996). The cytotoxic activity of *Bacillus anthracis* lethal factor is inhibited by leukotriene A4 hydrolase and metallopeptidase inhibitors. *Biochem. J.* 320:687–691.

Michelson, D., Stone, L., Galliven, E., Magiakou, M.A., Chrousos, G.P., Sternberg, E.M., and Gold, P.W. (1994). Multiple sclerosis is associated with alterations in hypothalamic-pituitary-adrenal axis function. *J. Clin. Endocrinol. Metab.* 79:848–853.

Mingrone, G., DeGaetano, A., Pugeat, M., Capristo, E., Greco, A.V., and Gasbarrini, G. (1999). The steroid resistance of Crohn's disease. *J. Invest. Med.* 47:319–325.

Moayeri, M., Haines, D., Young, H.A., and Leppla, S.H. (2003): *Bacillus anthracis* lethal toxin induces TNF-alpha-independent hypoxia-mediated toxicity in mice. *J. Clin. Invest.* 112:670–682.

Moayeri, M., Martinez, N.W., Wiggins, J., Young, H.A., and Leppla, S.H. (2004). Mouse susceptibility to anthrax lethal toxin is influenced by genetic factors in addition to those controlling macrophage sensitivity. *Infect. Immun.* 72:4439–4447.

Moayeri, M., Webster, J.I., Wiggins, J.F., Leppla, S.H., and Sternberg, E.M. (2005). Endocrine perturbation increases susceptibility of mice to anthrax lethal toxin. *Infect. Immun.* 73:4238–4244.

Mock, M., and Fouet, A. (2001). Anthrax. *Annu. Rev. Microbiol.* 55:647–671.

Moncek, F., Kvetnansky, R., and Jezova, D. (2001). Differential responses to stress stimuli of Lewis and Fischer rats at the pituitary and adrenocortical level. *Endocr. Reg.* 35:35–41.

Mykoniatis, A., Anton, P.M., Wlk, M., Wang, C.C., Ungsunan, L., Bluher, S., Venihaki, M., Simeonidis, S., Zacks, J., Zhao, D., Sougioultzis, S., Karalis, K., Mantzoros, C., and Pothoulakis, C. (2003). Leptin mediates *Clostridium difficile* toxin A-induced enteritis in mice. *Gastroenterology* 124:683–691.

Neeck, G., and Crofford, L.J. (2000). Neuroendocrine perturbations in fibromyalgia and chronic fatigue syndrome. *Rheum. Dis. Clin. North Am.* 26:989–1002.

New, M.I., Nimkarn, S., Brandon, D.D., Cunningham-Rundles, S., Wilson, R.C., Newfield, R.S., Vandermeulen, J., Barron, N., Russo, C., Loriaux, D.L., and O'Malley, B. (1999). Resistance to several steroids in two sisters. *J. Clin. Endocrinol. Metab.* 84:4454–4464.

New, M.I., Nimkarn, S., Brandon, D.D., Cunningham-Rundles, S., Wilson, R.C., Newfield, R.S., Vandermeulen, J., Barron, N., Russo, C., Loriaux, D.L., and O'Malley, B. (2001). Resistance to multiple steroids in two sisters. *J. Steroid Biochem. Mol. Biol.* 76:161–166.

Oakley, R.H., Sar, M., and Cidlowski, J.A. (1996). The human glucocorticoid receptor b isoform. Expression, biochemical properties and putative function. *J. Biol. Chem.* 271:9550–9559.

Oakley, R.H., Jewell, C.M., Yudt, M.R., Bofetiado, D.M., and Cidlowski, J.A. (1999). The dominant negative activity of the human glucocorticoid receptor β isoform. Specificity and mechanisms of action. *J. Biol. Chem.* 274:27857–27866.

Orii, F., Ashida, T., Nomura, M., Maemoto, A., Fujiki, T., Ayabe, T., Imai, S., Saitoh, Y., and Kohgo, Y. (2002). Quantitative analysis for human glucocorticoid receptor alpha/beta mRNA in IBD. *Biochem. Biophys. Res. Commun.* 296:1286–1294.

Paccani, S.R., Tonello, F., Ghittoni, R., Natale, M., Muraro, L., D'Elios, M.M., Tang, W.J., Montecucco, C., and Baldari, C.T. (2005). Anthrax toxins suppress T lymphocyte activation by disrupting antigen receptor signaling. *J. Exp. Med.* 201: 325–331.

Palermo, M., Alves-Rosa, F., Rubel, C., Fernandez, G.C., Fernandez-Alonso, G., Alberto, F., Rivas, M., and Isturiz, M. (2000). Pretreatment of mice with lipopolysaccharide (LPS) or IL-1beta exerts dose-dependent opposite effects on Shiga toxin-2 lethality. *Clin. Exp. Immunol.* 119:77–83.

Pandey, J., and Warburton, D. (2004). Knock-on effect of anthrax lethal toxin on macrophages potentiates cytotoxicity to endothelial cells. *Microbes Infect.* 6: 835–843.

Park, J.M., Greten, F.R., Li, Z.W., and Karin, M. (2002). Macrophage apoptosis by anthrax lethal factor through p38 map kinase inhibition. *Science* 297:2048–2051.

Pellizzari, R., Guidi-Rontani, C., Vitale, G., Mock, M., and Montecucco, C. (1999). Anthrax lethal factor cleaves MKK3 in macrophages and inhibits the LPS/IFNγ-induced release of NO and TNFα. *FEBS Lett.* 462:199–204.

Pellizzari, R., Guidi-Rontani, C., Vitale, G., Mock, M., and Montecucco, C. (2000). Lethal factor of *Bacillus anthracis* cleaves the N-terminus of MAPKKs: Analysis of the intracellular consequences in macrophages. *Int. J. Med. Microbiol.* 290: 421–427.

Pezard, C., Berche, P., and Mock, M. (1991). Contribution of individual toxin components to virulence of *Bacillus anthracis*. *Infect. Immun.* 59:3472–3477.

Pickering, A.K., and Merkel, T.J. (2004). Macrophages release tumor necrosis factor alpha and interleukin-12 in response to intracellular *Bacillus anthracis* spores. *Infect. Immun.* 72:3069–3072.

Pickering, A.K., Osorio, M., Lee, G.M., Grippe, V.K., Bray, M., and Merkel, T.J. (2004). Cytokine response to infection with *Bacillus anthracis* spores. *Infect. Immun.* 72:6382–6389.

Popov, S.G., Villasmil, R., Bernardi, J., Grene, E., Cardwell, J., Popova, T., Wu, A., Alibek, D., Bailey, C., and Alibek, K. (2002a). Effect of *Bacillus anthracis* lethal toxin on human peripheral blood mononuclear cells. *FEBS Lett.* 527:211–215.

Popov, S.G., Villasmil, R., Bernardi, J., Grene, E., Cardwell, J., Wu, A., Alibek, D., Bailey, C., and Alibek, K. (2002b). Lethal toxin of *Bacillus anthracis* causes apoptosis of macrophages. *Biochem. Biophys. Res. Commun.* 293:349–355.

Popov, S.G., Popova, T.G., Grene, E., Klotz, F., Cardwell, J., Bradburne, C., Jama, Y., Maland, M., Wells, J., Nalca, A., Voss, T., Bailey, C., and Alibek, K. (2004). Systemic cytokine response in murine anthrax. *Cell. Microbiol.* 6:225–233.

Pujols, L., Mullol, J., Benitez, P., Torrego, A., Xaubet, A., de Haro, J., and Picado, C. (2003). Expression of the glucocorticoid receptor alpha and beta isoforms in human nasal mucosa and polyp epithelial cells. *Respir. Med.* 97:90–96.

Pujols, L., Xaubet, A., Ramirez, J., Mullol, J., Roca-Ferrer, J., Torrego, A., Cidlowski, J.A., and Picado, C. (2004). Expression of glucocorticoid receptors alpha and beta in steroid sensitive and steroid insensitive interstitial lung diseases. *Thorax* 59: 687–693.

Racciatti, D., Guagnano, M.T., Vecchiet, J., De Remigis, P.L., Pizzigallo, E., Della Vecchia, R., Di Sciascio, T., Merlitti, D., and Sensi, S. (2001). Chronic fatigue syndrome: Circadian rhythm and hypothalamic-pituitary-adrenal (HPA) axis impairment. *Int. J. Immunopathol. Pharmacol.* 14:11–15.

Rask, E., Olsson, T., Soderberg, S., Andrew, R., Livingstone, D.E., Johnson, O., and Walker, B.R. (2001). Tissue-specific dysregulation of cortisol metabolism in human obesity. *J. Clin. Endocrinol. Metab.* 86:1418–1421.

Reichardt, H.M., Kaestner, K.H., Tuckermann, J., Kretz, O., Wessely, O., Bock, R., Gass, P., Schmid, W., Herrlich, P., Angel, P., and Schütz, G. (1998). DNA binding of the glucocorticoid receptor is not essential for survival. *Cell* 93:531–541.

Richaud-Patin, Y., Vega-Boada, F., Vidaller, A., and Llorente, L. (2004). Multidrug resistance-1 (MDR-1) in autoimmune disorders IV. P-glycoprotein overfunction in lymphocytes from myasthenia gravis patients. *Biomed. Pharmacother.* 58:320–324.

Roberts, J.E., Watters, J.W., Ballard, J.D., and Dietrich, W.F. (1998). Ltx1, a mouse locus that influences the susceptibility of macrophages to cytolysis caused by intoxication with *Bacillus anthracis* lethal factor, maps to chromosome 11. *Mol. Microbiol.* 29:581–591.

Roberts, A.D., Wessely, S., Chalder, T., Papadopoulos, A., and Cleare, A.J. (2004). Salivary cortisol response to awakening in chronic fatigue syndrome. *Br. J. Psychiatry* 184:136–141.

Rogatsky, I., Logan, S.K., and Garabedian, M.J. (1998). Antagonism of glucocorticoid receptor transcriptional activation by the c-Jun N-terminal kinase. *Proc. Natl. Acad. Sci. U.S.A.* 95:2050–2055.

Rupprecht, M., Hornstein, O.P., Schluter, D., Schafers, H.J., Koch, H.U., Beck, G., and Rupprecht, R. (1995). Cortisol, corticotropin, and beta-endorphin responses to corticotropin-releasing hormone in patients with atopic eczema. *Psychoneuroendocrinology* 20:543–551.

Ruzek, M.C., Pearce, B.D., Miller, A.H., and Biron, C.A. (1999). Endogenous glucocorticoids protect against cytokine-mediated lethality during viral infection. *J. Immunol.* 162:3527–3533.

Schlaghecke, R., Kornely, E., Wollenhaupt, J., and Specker, C. (1992). Glucocorticoid receptors in rheumatoid arthritis. *Arthritis Rheum.* 35:740–744.

Schoneveld, O.J., Gaemers, I.C., and Lamers, W.H. (2004). Mechanisms of glucocorticoid signalling. *Biochim. Biophys. Acta* 1680:114–128.

Schule, R., Rangarajan, P., Kliewer, S., Ransone, L.J., Bolado, J., Yang, N., Verma, I.M., and Evans, R.M. (1990). Functional antagonism between oncoprotein c-Jun and the glucocorticoid receptor. *Cell* 62:1217–1226.

Scobie, H.M., Rainey, G.J., Bradley, K.A., and Young, J.A. (2003). Human capillary morphogenesis protein 2 functions as an anthrax toxin receptor. *Proc. Natl. Acad. Sci. U.S.A.* 100:5170–5174.

Seckl, J.R., and Walker, B.R. (2001). Minireview: 11β-hydroxysteroid dehydrogenase type 1—A tissue-specific amplifier of glucocorticoid action. *Endocrinology* 142: 1371–1376.

Seckl, J.R., Morton, N.M., Chapman, K.E., and Walker, B.R. (2004). Glucocorticoids and 11beta-hydroxysteroid dehydrogenase in adipose tissue. *Recent Prog. Horm. Res.* 59:359–393.

Shahidi, H., Vottero, A., Stratakis, C., Taymans, S.E., Karl, M., Longui, C.A., Chrousos, G.P., Daughaday, W.H., Gregory, S.A., and Plate, J.M.D. (1999). Imbalanced expression of the glucocorticoid receptor isoforms in cultured lymphocytes from a patient with systemic glucocorticoid resistance and chronic lymphocytic leukemia. *Biochem. Biophys. Res. Commun.* 254:559–565.

Shanks, N., Griffiths, J., Zalcman, S., Zacharko, R.M., and Anisman, H. (1990). Mouse strain differences in plasma corticosterone following uncontrollable footshock. *Pharmacol. Biochem. Behav.* 36:515–519.

Shin, S., Hur, G.H., Kim, Y.B., Yeon, G.B., Park, K.J., Park, Y.M., and Lee, W.S. (2000). Dehydroepiandrosterone and melatonin prevent *Bacillus anthracis* lethal toxin-induced TNF production in macrophages. *Cell. Biol. Toxicol.* 16:165–174.

Singh, Y., Leppla, S.H., Bhatnagar, R., and Friedlander, A.M. (1989). Internalization and processing of *Bacillus anthracis* lethal toxin by toxin-sensitive and -resistant cells. *J. Biol. Chem.* 264:11099–11102.

Singh, Y., Klimpel, K.R., Goel, S., Swain, P.K., and Leppla, S.H. (1999). Oligomerization of anthrax toxin protective antigen and binding of lethal factor during endocytic uptake into mammalian cells. *Infect. Immun.* 67:1853–1859.

Smith, H., Keppie. J., Stanley, J.L., and Harris-Smith, P.W. (1955). The chemical basis of the virulence of *Bacillus anthracis*. IV: Secondary shock as the major factor in death of guinea-pigs from anthrax. *Br. J. Exp. Pathol.* 36:323–335.

Smoak, K.A., and Cidlowski, J.A. (2004). Mechanisms of glucocorticoid receptor signaling during inflammation. *Mech. Ageing Dev.* 125:697–706.

Sousa, A.R., Lane, S.J., Cidlowski, J.A., Staynov, D.Z., and Lee, T.H. (2000). Glucocorticoid resistance in asthma is associated with elevated *in vivo* expression of the glucocorticoid receptor beta-isoform. *J. Allergy Clin. Immunol.* 105:943–950.

Sternberg, E.M., Hill, J.M., Chrousos, G.P., Kamilaris, T., Listwak, S.J., Gold, P.W., and Wilder, R.L. (1989a). Inflammatory mediator-induced hypothalamic-pituitary-adrenal axis activation is defective in streptococcal cell wall arthritis-susceptible Lewis rats. *Proc. Natl. Acad. Sci. U.S.A.* 86:2374–2378.

Sternberg, E.M., Young, W.S.d., Bernardini, R., Calogero, A.E., Chrousos, G.P., Gold, P.W., and Wilder, R.L. (1989b). A central nervous system defect in biosynthesis of corticotropin-releasing hormone is associated with susceptibility to streptococcal cell wall-induced arthritis in Lewis rats. *Proc. Natl. Acad. Sci. U.S.A.* 86:4771–4775.

Strickland, I., Kisich, K., Hauk, P.J., Vottero, A., Chrousos, G.P., Klemm, D.J., and Leung, D.Y.M. (2001). High constitutive glucocorticoid receptor β in human neutrophils enables them to reduce their spontaneous rate of cell death in response to corticosteroids. *J. Exp. Med.* 193:585–593.

Szatmary, Z., Garabedian, M.J., and Vilcek, J. (2004). Inhibition of glucocorticoid receptor-mediated transcriptional activation by p38 mitogen-activated protein (MAP) kinase. *J. Biol. Chem.* 279:43708–43715.

Takahasi, K.I., Fukushima, K., Sasano, H., Sasaki, I., Matsuno, S., Krozowski, Z.S., and Nagura, H. (1999). Type II 11beta-hydroxysteroid dehydrogenase expression in human colonic epithelial cells of inflammatory bowel disease. *Dig. Dis. Sci.* 44: 2516–2522.

Thompson, B.T. (2003). Glucocorticoids and acute lung injury. *Crit. Care Med.* 31: S253–257.

Valledor, A.F., Hsu, L.C., Ogawa, S., Sawka-Verhelle, D., Karin, M., and Glass, C.K. (2004). Activation of liver X receptors and retinoid X receptors prevents bacterial-induced macrophage apoptosis. *Proc. Natl. Acad. Sci. U.S.A.* 101:17813–17818.

Valsamakis, G., Anwar, A., Tomlinson, J.W., Shackleton, C.H., McTernan, P.G., Chetty, R., Wood, P.J., Banerjee, A.K., Holder, G., Barnett, A.H., Stewart, P.M., and Kumar, S. (2004). 11beta-hydroxysteroid dehydrogenase type 1 activity in lean and obese males with type 2 diabetes mellitus. *J. Clin. Endocrinol. Metab.* 89:4755–4761.

Valtysdottir, S.T., Wide, L., and Hallgren, R. (2001). Low serum dehydroepiandrosterone sulfate in women with primary Sjogren's syndrome as an isolated sign of impaired HPA axis function. *J. Rheumatol.* 28:1259–1265.

Vermes, I., and Beishuizen, A. (2001). The hypothalamic-pituitary-adrenal response to critical illness. *Best Pract. Res. Clin. Endocrinol. Metab.* 15:495–511.

Vitale, G., Pellizzari, R., Recchi, C., Napolitani, G., Mock, M., and Montecucco, C. (1998). Anthrax lethal factor cleaves the N-terminus of MAPKKs and induces tyrosine/threonine phosphorylation of MAPKs in cultured macrophages. *Biochem. Biophys. Res. Commun.* 248:706–711.

Vottero, A., and Chrousos, G.P. (1999). Glucocorticoid receptor β: View I. *Trends Endocrinol. Metab.* 10:333–338.

Wang, X., Wu, H., and Miller, A.H. (2004). Interleukin 1alpha (IL-1alpha) induced activation of p38 mitogen-activated protein kinase inhibits glucocorticoid receptor function. *Mol. Psychiatry* 9:65–75.

Watters, J.W., Dewar, K., Lehoczky, J., Boyartchuk, V., and Dietrich, W.F. (2001). *Kif1c*, a kinesin-like motor protein, mediates mouse macrophage resistance to anthrax lethal factor. *Curr. Biol.* 11:1503–1511.

Webster, J.I., and Carlstedt-Duke, J. (2002). Involvement of multidrug-resistance proteins (MDR) in the modulation of glucocorticoid response. *J. Steroid Biochem. Mol. Biol.* 82:277–288.

Webster, J.I., and Sternberg, E.M. (2004). Role of the hypothalamic-pituitary-adrenal axis, glucocorticoids and glucocorticoid receptors in toxic sequelae of exposure to bacterial and viral products. *J. Endocrinol.* 181:207–221.

Webster, J.I., and Sternberg, E.M. (2005). Anthrax lethal toxin represses glucocorticoid receptor (GR) transactivation by inhibiting GR-DNA binding *in vivo*. *Mol. Cell. Endocrinol.* 241:21–31.

Webster, J.I., Tonelli, L., and Sternberg, E.M. (2002). Neuroendocrine regulation of immunity. *Annu. Rev. Immunol.* 20:125–163.

Webster, J.I., Tonelli, L.H., Moayeri, M., Simons, S.S., Jr., Leppla, S.H., and Sternberg, E.M. (2003). Anthrax lethal factor represses glucocorticoid and progesterone receptor activity. *Proc. Natl. Acad. Sci. U.S.A.* 100:5706–5711.

Wei, T., and Lightman, S.L. (1997). The neuroendocrine axis in patients with multiple sclerosis. *Brain* 120:1067–1076.

Wesche, J., Elliott, J.L., Falnes, P.O., Olsnes, S., and Collier, R.J. (1998). Characterization of membrane translocation by anthrax protective antigen. *Biochemistry* 37: 15737–15746.

Westerbacka, J., Yki-Jarvinen, H., Vehkavaara, S., Hakkinen, A.M., Andrew, R., Wake, D.J., Seckl, J.R., and Walker, B.R. (2003). Body fat distribution and cortisol metabolism in healthy men: Enhanced 5beta-reductase and lower cortisol/ cortisone metabolite ratios in men with fatty liver. *J. Clin. Endocrinol. Metab.* 88:4924–4931.

Wick, G., Sgonc, R., and Lechner, O. (1998). Neuroendocrine-immune disturbances in animal models with spontaneous autoimmune diseases. *Ann. N.Y. Acad. Sci.* 840:591–598.

Wilder, R.L., Calandra, G.B., Garvin, A.J., Wright, K.D., and Hansen, C.T. (1982). Strain and sex variation in the susceptibility to streptococcal cell wall-induced polyarthritis in the rat. *Arthritis Rheum.* 25:1064–1072.

Zhao, J., Milne, J.C., and Collier, R.J. (1995). Effect of anthrax toxin's lethal factor on ion channels formed by the protective antigen. *J. Biol. Chem.* 270:18626–18630.

5
Adrenergic Regulation of Adaptive Immunity

VIRGINIA M. SANDERS

1. Introduction

The early hypothesis that the brain and immune system communicated with each other was first proposed from the results of a study on the effect of taste aversion conditioning of humoral immune responsiveness (Ader and Cohen, 1975). Many studies have since confirmed the existence of such a bidirectional regulation [reviewed in (Besedovsky and Del Rey, 1996; Ader, 2000; Kohm and Sanders, 2001)] and provide plausible mechanisms by which the immune system alerts the brain that it is responding to an antigen, as well as mechanisms by which the brain regulates the level of immune cell activity that develops (Figure 5.1). Four key discoveries indicate that mechanisms exist by which the brain is able to communicate with cells of the peripheral immune system. First, primary and secondary lymphoid organs are innervated with sympathetic nerve fibers, and mechanisms exist by which signals are sent from the activated immune system to the brain. Second, the sympathetic neurotransmitter norepinephrine (NE) is released from nerve terminals residing within the parenchyma of lymphoid tissues after antigen or cytokine administration. Third, lymphoid cells, except for Th2 cells, express the β_2-adrenergic receptor (β_2AR) that binds NE to transduce extracellular signals to the cell interior. And finally, NE regulates lymphocyte activity at the level of gene expression. Although NE appears to regulate immune system activity overall, we will focus this chapter to a discussion of the role NE plays in regulating CD4$^+$ T-cell and B-cell activity, with special emphasis placed on the role it plays in regulating the level of cytokine and antibody produced.

2. Sympathetic Innervation of Lymphoid Tissue and Norepinephrine Release

In general, the Sympathetic Nervous System (SNS) maintains homeostasis by regulating the activity of organ systems that are not under voluntary, conscious control, and is typically associated with the physiological

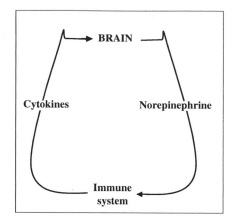

FIGURE 5,1. Pathways of communication between the central nervous and immune systems. The presence of sympathetic nerve fibers in lymphoid organs and the release of norepinephrine from nerve terminals located in the direct vicinity of immune cells provide a mechanism by which norepinephrine might influence immune cell function. The activity of the central nervous system (CNS) may be influenced by products of activated immune cells because circulating cytokines are either actively transported into the CNS or cytokine receptors expressed on the vagal nerve transmit signals to the CNS. Hormone production resulting from activation of the hypothalamic-pituitary-adrenal (HPA) axis may also influence a variety of systemic immune cell activities but is not shown in this diagram or discussed in this chapter.

flight-or-fight response, such as that involved in the regulation of cardio-vascular, respiratory, and metabolic function during times of critical need. A dense perivascular network of sympathetic nerve fibers are present in the splenic white pulp, thymus, and lymph nodes of mice and humans. Nerve endings that terminate in the parenchyma of the white pulp are especially numerous in the periarteriolar lymphoid sheath (PALS) and are adjacent to both CD4+ and CD8+ T cells, as well as macrophages [reviewed in (Felten et al., 1987a, 1998)].

Systemic infection decreased the level of norepinephrine (NE) in the spleen via the endotoxin released from bacterial cell walls, suggesting that the decrease was due to either decreased NE production, increased NE release, increased NE diffusion/metabolism, and/or decreased reuptake of NE back into the nerve terminal. Later studies showed that lipopolysaccharide (LPS)- or infection-induced activation of immune cell populations increased the rate of NE release in both the spleen and the heart during the first 12 h of exposure, indicating that the decrease in splenic NE levels induced by LPS exposure was due to an increase in the level of systemic sympathetic nerve activity [reviewed in detail in (Kohm and Sanders, 2001)]. Other types of immune stimuli also influenced the rate of NE release in lymphoid organs. One of the earliest studies showed that sheep erythrocyte (sRBC)-induced immune cell activation decreased the total lymphoid tissue content of NE by increasing the level of sympathetic nerve activity (Besedovsky et al., 1979) and the rate of NE release in the spleen without disrupting the homeostatic mechanisms responsible for maintaining constant levels of NE tissue content (Fuchs et al., 1988). Likewise, using [3H]-NE turnover analysis, immunization of scid mice reconstituted with antigen-specific B cells and Th2 cells showed that the rate of NE release was increased in an antigen-specific manner in the spleen and bone marrow 18–25h, but not 1–8h, after immunization. This increase in rate was only partially blocked by the ganglionic-blocker chlorisondamine, indicating that pre- and postganglionic mechanisms were involved in regulating NE release. Thus, sympathetic outflow is increased into the spleen and bone marrow during the early immune cell activation events that take place in response to antigen, suggesting that NE may serve as one mechanism by which the brain communicates with the immune system to regulate immune cell activity.

3. Initial Evidence That Norepinephrine Plays a Role in Regulating Adaptive Immune Cell Activity

After infection, immune cells of the adaptive immune system, CD4$^+$ and CD8$^+$ T cells and B cells, protect the host by specifically destroying the infectious pathogen and by providing pathogen-specific long-term protection. Thus, the adaptive immune system acts in an antigen-specific manner, generates long-term protection, and is divided into two major branches: the cell-mediated and humoral. Due to the ability of CD4$^+$ T cells to activate or "help" B cells to make antibody, these cells were called T helper (Th) cells. The ability of CD4$^+$ T cells to promote either a cell-mediated or a humoral response was first understood when Mosmann et al. (1986), and later Romagnani et al. (1991), showed that clones of CD4$^+$ T cells from mice and humans, respectively, could be divided into two distinct effector subsets, Th1 and Th2 cells, based on the cytokines that they produced. We now know that the two effector cell subsets are derived from a common precursor cell called the naive CD4$^+$ T cell [reviewed in (Swain et al., 1996)]. The B cell is responsible for generating the humoral antibody response and is provided help to do so by the CD4$^+$ effector cells. The following sections in this chapter will focus on the evidence that NE plays a role in regulating CD4$^+$ T-cell and B-cell activity.

Pharmacological evidence for involvement of the sympathetic nervous system in regulating the level of an immune response has been obtained using the chemical neurotoxin 6-hydroxydopamine (6-OHDA), a drug that depletes NE from peripheral sympathetic nerve terminals by first displacing NE from the terminal and then reversibly destroying the terminal of adult mice for 4–8 weeks. When the SNS of mice was deleted using 6-

OHDA, immune responses were found to be either enhanced, suppressed, or unaltered [reviewed in (Kohm and Sanders, 2001)], suggesting that NE either suppressed, enhanced, or had no effect, respectively, on an immune response when it was released within the vicinity of immune cells responding to antigen. However, differences in the dose and/or time of 6-OHDA administration and the type of immune response being measured make it difficult to conclude whether these disparate findings reflect different aspects of a related phenomenon.

With these caveats in mind, studies that used NE-depleted mice showed that Th1 cell–mediated immune responses were suppressed. For example, when mice were exposed to 6-OHDA before sensitization with the hapten Trinitrochlorobenzene (TNCB), the resulting Th1 cell–mediated contact hypersensitivity response was decreased in comparison with NE-intact controls (Madden et al., 1989). This finding suggested that NE was required for the generation of a Th1 cell–mediated immune response and may have affected the precursor cell from which the Th1 cell develops. Also, if mice were treated with 6-OHDA at least 3–5 days after sensitization with TNCB, a time when activated naïve CD4$^+$ T cells were committed to differentiate into Th1 cells, the resulting contact hypersensitivity response was also decreased (Madden et al., 1989). Thus, this study indicated that chemical sympathectomy was suppressive both during naive T-cell priming and at a time when the cells were committed to the Th1-cell phenotype, suggesting that NE was needed at both the naïve and effector stages for an optimal Th1 cell–mediated response to develop. However, one limitation to this experimental design is that the cells may have been exposed to the large bolus of NE released after 6-OHDA treatment, so that it is unclear whether this NE signal, or the lack of NE, was responsible for the effects measured.

To begin to address this possibility, mice that are genetically deficient for the enzyme dopamine β-hydroxylase, which is required to synthesize NE from dopamine, were used to determine if NE regulates the magnitude of a Th1 cell–driven response (Alaniz et al., 1999). These NE-deficient mice showed a significant decrease in the level of IFN-γ produced by CD4$^+$ T cells and in immunological protection when exposed to the Th1-promoting pathogens Listeria monocytogenes or Mycobacterium tuberculosis. These data show that the absence of NE results in a diminished Th1 cell–driven response in vivo, suggesting that NE plays a role in upregulating the magnitude of a Th1 cell–mediated immune response. However, one caution is that due to the deficiency of dopamine β-hydroxylase, these mice expressed an increase in the level of dopamine (Alaniz et al., 1999). Because dopamine has been shown to decrease CD4$^+$ T-cell function (Kauassi et al., 1987), the decrease in the NE-deficient mice may not be due to the absence of NE but rather due to the exposure of the T cells to concentrations of dopamine that are higher than in wild-type mice. Nonetheless, taken together, these studies are the first to indicate that NE may exert an enhancing effect on either early naïve CD4$^+$ T-cell development into a Th1 cell, the commitment to becoming a Th1 cell, and/or the amount of IFN-γ secreted by the Th1 cell.

The role that NE plays in a Th2 cell–driven response is less clear, and the role it plays in a Th1 cell–mediated response has been challenged. When two strains of mice, C57Bl/6J (Th1 cell–slanted strain) and Balb/c (Th2 cell–slanted strain), were depleted of NE and immunized 2 days later with the T cell–dependent antigen KLH, splenic cells from both strains of mice produced significantly higher levels of IL-2 and IL-4 after reactivation *in vitro* when compared with cells isolated from NE-intact controls (Kruszewska *et al.*, 1995). Although IFN-γ levels were not determined, the increase in serum IgG$_{2a}$ in these mice suggested that this cytokine was also increased when NE was depleted. These results refute the results described above indicating that NE may exert an enhancing effect on Th1 cell development and/or IFN-γ production. Thus, it remains unclear whether or not NE is needed to obtain an optimal Th1 cell–driven response *in vivo*.

The role played by NE in regulating the magnitude of an antibody response has also been studied in 6-OHDA–induced NE-depleted mice. Unfortunately, however, *in vivo* results show either a decrease (Kasahara *et al.*, 1977b; Hall *et al.*, 1982; Livnat *et al.*, 1985; Cross *et al.*, 1986; Fuchs *et al.*, 1988; Madden *et al.*, 1989; Ackerman *et al.*, 1991b) or increase (Besedovsky *et al.*, 1979; Miles *et al.*, 1981; Chelmicka-Schorr *et al.*, 1988) in Th cell–dependent antibody production. Both the hemagglutinin titer and number of plaque-forming cells that formed in response to primary immunization with sRBC were decreased in NE-depleted mice when compared with NE-intact mice, but the secondary response to antigen in these mice was unchanged (Kasahara *et al.*, 1977a; 1977b), suggesting that NE was needed for the development of an optimal primary antibody response but not the secondary response. However, the secondary antibody response was suppressed when 6-OHDA was administered at the same time as the secondary exposure to antigen, suggesting that the concentration of NE at the time of antigen exposure may be a determining factor in the development of an optimal primary or secondary antibody response. More recently, the level of serum antigen-specific antibody produced by dopamine β-hydroxylase–deficient mice was significantly lower than the level of antibody produced by B cells in wild-type mice (Alaniz *et al.*, 1999). Thus, these results suggested that NE was needed to produce an optimal level of antibody *in vivo* after immunization with antigen.

In contrast, NE depletion was also reported to increase the number of antibody-secreting cells after immunization with the T cell—dependent antigen sRBC (Besedovsky *et al.*, 1979). As well, NE depletion in mice increased the number of antibody-forming cells activated by T-independent antigens but in contrast did not alter the number of antibody-secreting cells activated by a T cell–dependent antigen (Miles *et al.*, 1981). Another study reported a strain-specific enhancement in antibody production against a T cell–dependent antigen in NE-depleted C57Bl/6J (Th1-slanted strain) and

Balb/c (Th2-slanted strain) mice (Kruszewska *et al.*, 1995). In this study, the serum levels of Keyhole Limpet Hemocyanin KLH-specific IgM, total IgG, IgG$_1$, and IgG$_{2a}$ were enhanced in NE-depleted C57Bl/6J mice 1–2 weeks postimmunization, whereas only IgG$_1$ was enhanced in NE-depleted Balb/c mice. To address the possibility that the 6-OHDA–displaced NE might affect immune cells before a state of NE depletion was established and before antigen was delivered, another model system had to be developed.

Antigen-specific Th2 and B cells were adoptively transferred into NE-depleted severe combined immunodeficient (scid) mice after the mice that were depleted of NE by 6-OHDA prior to cell transfer (Kohm and Sanders, 1999). Four weeks after the primary immunization, a significantly lower serum level of antigen-specific IgM and IgG$_1$ was measured in NE-depleted mice as compared with NE-intact mice, and this effect was prevented by the administration of a β2AR-selective agonist. Secondary immunization of these mice depleted of NE 9 weeks earlier induced serum levels of antigen-specific IgG$_1$ that were not only lower but also delayed in reaching a maximal level. NE-depletion did not alter T- and B-cell trafficking to the spleen but was found to decrease follicular expansion and germinal center formation in the spleen when compared with NE-intact controls. Taken together, these data suggest that stimulation of the β$_2$AR during the course of a T-dependent immune response is necessary to maintain an optimal level of antibody production during a primary and secondary response *in vivo*.

4. Receptors for Adrenergic Receptors on Adaptive Immune Cells

Norepineprhine, also known as noradrenaline, and epinephrine, also known as adrenaline, are catecholamines that bind to adrenergic receptors. Radioligand binding studies on immune cell subsets show binding sites that are saturable, reversible, high affinity, and almost exclusively of the β$_2$AR subtype [reviewed in (Sanders *et al.*, 2001)].

Radioligand binding analysis identified the βAR on both human and murine T cells, with most characterized as being of the β$_2$AR subtype [reviewed in (Sanders *et al.*, 2001)]. Thus far, no radioligand binding data have shown the presence of a high-affinity β$_1$AR or β$_3$AR on T cells. On average, there are approximately 400 binding sites per CD4$^+$ T cell. Although it is difficult to conduct a radioligand binding analysis on naïve CD4$^+$ T cells because of the low numbers of cells obtainable, RT-PCR analysis was used to show that naïve cells express the β$_2$AR (Swanson *et al.*, 2001). Resting clones of Th1 effector cells, but not clones of Th2 effector cells, express a detectable level of the β$_2$AR, as determined using both radioligand binding with iodopindolol and immunofluorescence staining with a polyclonal anti-β$_2$AR antibody directed against the cytoplasmic region of the β$_2$AR

(Sanders *et al.*, 1997). This finding was confirmed when a β_2AR-selective agonist induced an increase in the intracellular concentration of cAMP in clones of Th1 cells but not in clones of Th2 cells (Sanders *et al.*, 1997). Upon cell activation, with either Concanavalin A (Con A), phorbol-12 myristate 13-acetate ester (PMA)/calcium ionophore, contact sensitization activation, or an anti–T cell receptor antibody, the level of β_2AR expression on activated unfractionated splenic or lymph node T cells either increased (Sanders and Munson, 1985b; Westly and Kelley, 1987; Madden *et al.*, 1989; Radojcic *et al.*, 1991; Ramer-Quinn *et al.*, 1997) or decreased (Radojcic *et al.*, 1991; Cazaux *et al.*, 1995), but remained undetectable on Th2 clones (Ramer-Quinn *et al.*, 1997). Because the β_2AR was expressed by naïve CD4$^+$ T cells, the lack of β_2AR expression on Th2 cells suggests that the β_2AR gene is repressed as the naïve T cell differentiates into a Th2 cell (Fig. 5.2). Stimulation of the βAR on CD4$^+$ T cells increases both the intracellular concentration of cAMP and adenylate cyclase activity [reviewed in (Sanders *et al.*, 2001)].

Radioligand binding analysis revealed that B cells express almost twice the level of the βAR than CD4$^+$ T cells and, again, the receptor is of the β_2AR subtype [reviewed in (Sanders *et al.*, 2001)]. Expression of the β_2AR on naïve antigen-specific B cells was confirmed using both radioligand binding with iodopindolol, immunofluorescence staining with a polyclonal anti-β_2AR antibody directed against the cytoplasmic region of the β_2AR (Kohm and Sanders, 1999), and RT-PCR. No data show the presence of β_1AR or β_3AR on B cells. A few radioligand binding studies report the presence of the αAR on B cells (McPherson and Summers, 1982;

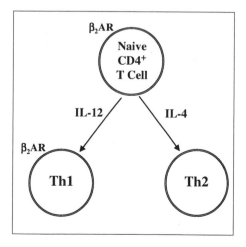

Figure 5.2. β_2AR expression by CD4$^+$ T-cell subsets. Th1 and Th2 cells are derived from a common precursor naïve CD4$^+$ T cell in the presence of IL-12 and IL-4, respectively. Expression of the β_2AR is restricted to the naïve and Th1 cell but is repressed as the naïve cell differentiates into a Th2 cell.

Titinchi and Clark, 1984; Goin *et al.*, 1991). However, the results from these studies may be misleading because αAR-expressing platelets were not removed from lymphocyte samples and, therefore, may have complicated the interpretation of binding results. Stimulation of the β_2AR on B cells enhances the intracellular accumulation of cAMP and adenylate cyclase activation [reviewed in (Sanders *et al.*, 2001)].

Taken together, these data suggest that CD4$^+$ naïve T cells, Th1 cells, and B cells, but not Th2 cells, express the β_2AR, and that stimulation of the receptor is capable of increasing both the intracellular concentration of cAMP and the activity of PKA.

5. Effect of Norepinephrine and β_2-Adrenergic Receptor Stimulation on Adaptive Immune Cell Activity

The findings summarized thus far in this chapter suggest strongly that NE serves as a messenger from the brain, which is translated by the β_2AR expressed on immune cells into an intracellular signal that regulates the magnitude of a specific immune cell activity.

The Th1 cell was defined by its ability to produce the cytokines interleukin-2 (IL-2), IL-3, interferon-γ (IFN-γ), tumor necrosis factor-α (TNF-α), and granulocyte/macrophage-colony stimulating factor (GMCSF). Through the production of these cytokines, Th1 cells augment a cell-mediated immune response through the activation of phagocytes, cytotoxic CD8$^+$ T cells, NK cells, and B cells [reviewed in (Mosmann and Coffman, 1989)]. The Th2 cell was defined by its ability to produce the cytokines IL-3, IL-4, IL-5, IL-6, IL-10, and IL-13. Th2 cells promote humoral immunity by activating B cells to increase expression of MHC class II, to proliferate, and to produce IgG$_1$ and IgE [reviewed in (Mosmann and Coffman, 1989)]. Later experiments showed that both Th1 and Th2 cells derived from a common precursor cell, called the naïve CD4$^+$ cell [reviewed in (Swain *et al.*, 1996)].

Because the naïve and Th1 cells expressed the β_2AR, it was possible that NE could send a message to a precursor naïve T cell to directly change the number of cells that differentiate into a Th1 cell and/or change the level of IFN-γ produced by the effector Th1 cell, thereby affecting any effector response dependent on IFN-γ. The literature records that an elevation of cAMP, the second messenger activated by β_2AR stimulation, inhibits proliferation of a polyclonally activated population of unfractionated CD4$^+$ T cells by decreasing IL-2 expression and secretion (Mary *et al.*, 1989; Wacholtz *et al.*, 1991; Chen and Rothenberg, 1994; Tamir and Isakov, 1994) and IL-2 receptor expression (Feldman *et al.*, 1987; Krause and Deutsch, 1991; Tamir and Isakov, 1994). Likewise, an increase in intracellular cAMP in Th1 cells inhibits the production of IL-2 (Gajewski *et al.*, 1990; Munoz *et al.*, 1990; Novak and Rothenberg, 1990) and IFN-γ (Gajewski *et al.*, 1990) by Th1 cells, but in contrast, either does not change (Pochet and Delespesse,

1983; Gajewski *et al.*, 1990; Novak and Rothenberg, 1990; Betz and Fox, 1991; Katamura *et al.*, 1995; Paliogianni and Boumpas, 1996; Yoshimura *et al.*, 1998), inhibits (Parker *et al.*, 1995; Borger *et al.*, 1996a), or enhances (Betz and Fox, 1991; Lacour *et al.*, 1994a, 1994b; Borger *et al.*, 1996b; Crocker *et al.*, 1996; Naito *et al.*, 1996; Wirth *et al.*, 1996) the production of IL-4 and IL-5 by Th2 cells.

However, if NE is to send a message to the cell, it must stimulate a receptor. Thus, it was found that although cAMP affects Th2 cytokine production, NE has no effect on a Th2 cell because the Th2 cell does not express the β_2AR (Sanders *et al.*, 1997). For the Th1 cell that expressed the β_2AR and is responsive to NE, it appears that the timing of β_2AR stimulation to cell activation plays a role in determining the effector response. For example, exposure of Th1 cells to a β_2AR-selective agonist before their activation by antigen-presenting B cells decreases both IL-2 and IFN-γ production (Sanders *et al.*, 1997). However, stimulation of the β_2AR on Th1 cells either at the time of or after cell activation induces a trend toward an increase in IFN-γ (Ramer-Quinn *et al.*, 1997). Thus, an elevation in intracellular cAMP within each CD4$^+$ T-cell subset appears to affect cell activity, but NE and β_2AR stimulation specifically affect naïve and Th1 cell activity only, with the effect depending on the time of β_2AR stimulation in relation to cell activation.

Because both NE and cytokines are present within the naïve CD4$^+$ T-cell microenvironment during cell activation and differentiation within lymphoid tissues, it is possible that NE and cytokines act together to regulate Th1 cell development and subsequent function as an effector cell. Data show that NE stimulates the β_2AR on a naïve CD4$^+$ T cell to generate Th1 cells that produce more IFN-γ per cell upon reactivation, without affecting the number of Th1 cells that develop, when compared with naïve T cells not exposed to NE (Swanson *et al.*, 2001). It was also found that IL-12 was essential for NE and β_2AR stimulation to exert this enhancing effect, suggesting that the IL-12R and β_2AR signaling pathways may affect each other. Thus, these findings may have relevance for vaccination protocols in which the goal might be to increase IFN-γ production by a limited number of Th1 cells that develop. However, the mechanism by which the NE message delivered to a naïve T cell translates into an enhancing effect on the level of Th1 cell IFN-γ production remains unknown.

Likewise, β_2AR stimulation and elevation of cAMP affect polyclonally activated B-cell proliferation (Diamantstein and Ulmer, 1975; Watson, 1975; Vischer, 1976; Johnson *et al.*, 1981; Muraguchi *et al.*, 1984; Blomhoff *et al.*, 1987; Holte *et al.*, 1988; Cohen and Rothstein, 1989; Muthusamy *et al.*, 1991; Whisler *et al.*, 1992) by either inhibiting early biochemical events and proliferation (Blomhoff *et al.*, 1987; Holte *et al.*, 1988; Muthusamy *et al.*, 1991) or enhancing B-cell proliferation (Cohen and Rothstein, 1989; Li *et al.*, 1989; Whisler *et al.*, 1992) and specific IgG isotype production in the presence of specific cytokines (Roper *et al.*, 1990, 2002; Stein and Phipps, 1991a, 1991b,

1992). A pharmacologic characterization of the enhancing effect of NE on the IgM response showed that either NE alone or NE in the presence of phentolamine produced an enhanced IgM response, suggesting that βAR activation was responsible for the enhancement (Sanders and Munson, 1984a). $β_2AR$ stimulation with the agonist terbutaline enhanced the antibody response with a similar magnitude and kinetics to that produced by NE, and the enhancement was blocked with propranolol (Sanders and Munson, 1984a), suggesting that $β_2AR$ activation was responsible for mediating the enhancing effect of NE. For a review of the early history of findings in this area, please refer to Sanders and Munson (1985a).

The enhancing effect induced by NE and $β_2AR$ stimulation also occurs for IgG_1 and IgE (Kasprowicz et al., 2000). The mechanism by which these two antibody isotypes are made is complicated, but a working knowledge of these mechanisms is essential to understand the mechanism by which NE and $β_2AR$ stimulation affect the level to which these isotypes are produced. Highly regulated events take place at the DNA level during B-cell differentiation. First, somatic hypermutation occurs when a germinal center forms to generate immunoglobulin diversity and high-affinity B-cell-receptor selection (Diaz and Casali, 2002). Second, class switch recombination occurs so that a B cell can differentiate into a cell that produces an antibody isotype other than IgM. The production of germline (GL) γ1 mRNA is necessary before a B-cell class switches to production of the mature form of IgG_1 that will be secreted from the cell. Once GLγ1 expression occurs, the switch recombination site at the 5′ end of the constant region of IgG_1 becomes accessible to the recombination transcriptional machinery (Stavnezer-Nordgren and Sirlin, 1986). Once a B cell has class-switched to IgG_1, the level of mature IgG_1 produced by a B cell is regulated by activity at the 3′-IgH enhancer region within the IgH locus in the mouse (Khamlichi et al., 2000). A number of octamer sequences are contained within the 3′-IgH enhancer region (Clerc et al., 1988), one of which binds the B cell—specific transcription factor Oct-2 that regulates activity of the 3′-IgH enhancer either alone or synergistically with its coactivator OCA-B (Tang and Sharp, 1999). Both Oct-2 and OCA-B are induced when CD40 is stimulated on the B cell (Pinaud et al., 2001; Stevens et al., 2000), and deletion of Oct-2 decreases serum IgG_1 (Corcoran and Karvelas, 1994; Humbert and Corcoran, 1997), while deletion of OCA-B decreases serum IgG_1 and germinal center formation (Kim et al., 1996; Nielsen et al., 1996; Schubart et al., 1996). Taken together, these data indicate that 3′-IgH enhancer activity is regulated by Oct-2 and OCA-B and that these proteins determine if a B cell will produce an optimal level of IgG_1.

Although the Th2 cell that helps a B cell to make IgG_1 or IgE is not affected by $β_2AR$ stimulation because it does not express the receptor, it was possible that the $β_2AR$-induced effect on B-cell activity could result in an increase in T-cell activity to increase the level of IL-4 produced by the Th2 cell, a cytokine that is needed for the isotype switch to IgG_1 or IgE.

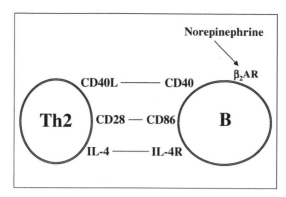

FIGURE 5.3. Critical B-cell surface molecules involved in activating a B cell to IgG1 production. Th2 cells do not express the β_2AR, while B cells do, suggesting that any effect mediated by norepinephrine on a Th2 cell–dependent antibody response, such as IgG$_1$, will occur through a direct effect on the B cell. The Th2-associated CD40L, IL-4, and CD28 serve as essential activating ligands for the B cell–associated CD40, IL-4R, and CD86 (B7-2). These essential signals can also be delivered to the B cell without the presence of an intact Th2 cell. In this model system, a resting B cell can be fully activated *in vitro* by the exogenous addition of CD40L-infected Sf9 cells, recombinant IL-4, and either an anti-CD86 antibody or a CD28 fusion protein.

Thus, to study the effect of β_2AR stimulation on the B cell alone and to eliminate any contribution of an indirect effect on Th2 cell activity, the Th2 cell was removed from subsequent experiments and was substituted with the two essential T cell–derived B-cell activation signals, CD40L and IL-4 (Fig. 5.3). Using this model system, data show that β_2AR stimulation increases the amount of IgG$_1$ produced per B cell without affecting the number of cells that are generated to produce IgG$_1$, as compared with B cells not exposed to β_2AR stimulation. The mechanism responsible for the β_2AR-induced increase in the amount of antibody produced per cell appears to involve both a direct and an indirect mechanism. The direct mechanism appears to involve a direct signaling cascade from CD86 to affect 3′-IgH-enhancer activity in the B cell. Via this mechanism, stimulation of the β_2AR increases the rate of mature IgG$_1$ transcription (Podojil and Sanders, 2003) without affecting mature IgG$_1$ transcript stability, the number of IgG$_1^+$ and IgG$_1$-secreting B cells generated, or the level of sterile germline γ1 transcript produced (Podojil and Sanders, 2003). These findings suggest that β_2AR signaling in a B cell affects the steady-state level of mature IgG$_1$ transcript produced without affecting class switch recombination.

The indirect mechanism appears to involve a β_2AR-induced increase in CD86. *In vivo* data show that antigen induces an increase in CD86 expression on a B cell and that NE depletion prevents this increase in CD86 from occurring. *In vitro*, β_2AR stimulation on a B cell increased CD86 expression (Kasprowicz *et al.*, 2000) but did not affect the level of expression for other B-cell surface molecules. Stimulation of CD86 alone on a CD40L/IL-4–activated B cell increases the level of IgG_1 produced (Jeannin *et al.*, 1997; Kasprowicz *et al.*, 2000) and differentially affects the level of antiapoptotic and proapoptotic molecules expressed (Suvas *et al.*, 2002). Taken together, these findings suggest that the enhancing effect of β_2AR stimulation on a B cell–mediated IgG_1 response is partially mediated indirectly via a direct effect on the level of CD86 expression, which, upon stimulation, signals for an increase in IgG_1 to occur to a level above that induced by CD40L and IL-4 alone.

A combination of β_2AR and CD86 stimulation produces an additive enhancing IgG_1 effect (Fig. 5.4). The mechanism responsible for mediating this additive increase in IgG_1 involves a CD86-induced increase in the expression of the transcription factor Oct-2 and a β_2AR-induced increase in the coactivator OCA-B, respectively (Podojil *et al.*, 2004). The increased

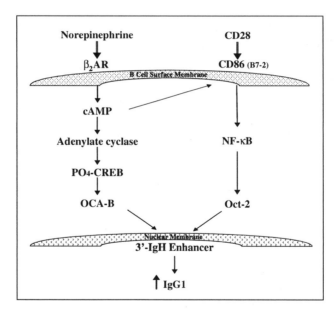

FIGURE 5.4. The proposed signaling pathways by which β_2AR and CD86 stimulation on a B cell increase IgG_1. Stimulation of the β_2AR is associated with an increase in the level of intracellular cAMP, as well as with an increase in the level of phosphorylated CREB (PO^4-CREB) and OCA-B expression, both of which appear to be dependent on PKA activation. Stimulation of CD86 is associated with an increase in NF-kB activation, resulting in an increase in p50 and phosphorylated p65 nuclear-localization, as well as Oct-2 expression. Activated Oct-2 binds to the octamer sequences within the IgH locus in conjunction with its coactivator OCA-B to allow for increased transactivating activity. Together, the CD86-induced signaling pathway cooperates with the β_2AR-induced signaling pathway to regulate the rate of mature IgG_1 transcription.

level of the transcription factor Oct-2, along with the increased level of its coactivator protein OCA-B, then bind cooperatively to the 3'-IgH enhancer region to increase the rate at which IgG_1 mRNA is produced. CD86 and the β_2AR appear to induce these effects via the activation of NF-κB and protein kinase A (PKA), respectively (Podojil et al., 2004). These findings were replicated in vivo; that is, in mice depleted of NE but exposed to different combinations of anti-CD40 antibody, IL-4, a β_2AR agonist, and/or an anti-CD86 antibody, a β_2AR-induced increase in OCA-B, a CD86-induced increase in Oct-2, increased binding of both factors to the 3'-IgH enhancer in the IgH locus, and an increase in IgG_1 mRNA and protein were measured (Podojil et al., 2004). As was also seen in vitro, each receptor stimulated individually in vivo induced an increase in IgG_1, but together they induced an additive effect (Podojil and Sanders, 2003; Podojil et al., 2004). These findings suggest that signaling pathways activated in a B cell by both an immunoreceptor (CD86) and a neuroreceptor (β_2AR) converge to regulate the IgG_1 response. For a more detailed description of how the study of the role played by the β_2AR in enhancing the IgG_1 response led to the discovery of the signaling pathway activated by CD86, please refer to the review by Podojil and Sanders (2005).

An understanding of the regulatory mechanism by which CD86 and/or the β_2AR increases the level of IgG_1 produced by a murine B cell may help to explain how acute versus chronic stress contributes to either enhancing or suppressing the level of immunity in humans. Also, an understanding of the mechanism by which IgG_1 is regulated by CD86 and the β_2AR will be useful in the design of targeted therapeutic approaches for individuals who might manifest changes in the level of these receptors and/or the ligands that stimulate these receptors during vaccination (e.g., in aged individuals).

6. Summary

An important function of the sympathetic nervous system is to maintain the internal environment of the body by making modest adjustments in cellular activities. In keeping with this function, many of the modest changes induced in immune cell function by sympathetic neurotransmitters dwarf in comparison with the large changes induced by regulatory molecules associated with the immune system itself. Although these modest changes in immune function render data interpretation difficult, they are relevant and necessary for maintaining immune homeostasis. By pursuing the study of the cellular, biochemical, and molecular mechanisms by which NE regulates immune homeostasis, we will better understand how these mechanisms may be influencing the etiology or progression of immune and nervous system–related disease states.

To date, convincing data are lacking to support a role for a neuroimmune interrelationship in the etiology or progression of a disease state. For example, if the sympathetic neurotransmitter NE plays a role in modulating immune function, then an age-related decline in lymphoid tissue innervation (Felten *et al.*, 1987b, 1989; Bellinger *et al.*, 1990; Ackerman *et al.*, 1991a) may contribute to the age-associated increase in the incidence of autoimmunity, cancer, and susceptibility to infection (Biondi and Zannino, 1997). On the other hand, if cytokines play a role in modulating nervous system function, then an age-related decline in immune function, coupled with changes mediated via the NE impact on immune function, may contribute to the age-associated increase in behavioral and cognitive dysfunctions (Forster and Lal, 1991). Although these possibilities are speculative, they do emphasize a need to understand the mechanisms by which one system influences the functioning of the other.

The knowledge gained from such studies will contribute to a better understanding of the apparently conflicting role played by the sympathetic nervous system in regulating immune homeostasis. For example, the level of immunocompetence may change as a result of either a change in the level of locally secreted NE in lymphoid organs, a change in the level of expression of the β_2AR on lymphocytes, a change in the ratio of CD^{4+} T-cell subsets participating in a particular immune response, or a change in the state of T-cell or B-cell activation when the β_2AR is stimulated. Although such changes in immunocompetence may not be immediately life-threatening to an individual, they may alter long-term health status and quality of life. The finding that NE stimulation of the β_2AR on a T cell or B cell appears to affect the level of protein produced per cell, as opposed to affecting the number of cells producing the effector protein, may indicate that a common mechanism is used by immune cells to translate the message sent by the brain to maintain immune homeostasis in the periphery.

References

Ackerman, K.D., Bellinger, D.L., Felten, S.Y., and Felten, D.L. (1991a). Ontogeny and senescence of noradrenergic innervation of the rodent thymus and spleen. In R. Ader, N. Cohen, and D.L. Felten (eds.), *Psychoneuroimmunology*. New York: Academic Press, pp. 71–125.

Ackerman, K.D., Madden, K.S., Livnat, S., Felten, S.Y., and Felten, D.L. (1991b). Neonatal sympathetic denervation alters the development of in vitro spleen cell proliferation and differentiation. *Brain Behav. Immun.* 5:235–261.

Ader, R. (2000). On the development of psychoneuroimmunology. *Eur. J. Pharmacol.* 405:167–176.

Ader, R., and Cohen, N. (1975). Behaviorally conditioned immunosuppression. *Psychosom. Med.* 37:333–340.

Alaniz, R.C., Thomas, S.A., Perez-Melgosa, M., Mueller, K., Farr, A.G., Palmiter, R.D., and Wilson, C.B. (1999). Dopamine beta-hydroxylase deficiency impairs cellular immunity. *Proc. Natl. Acad. Sci. U.S.A.* 96:2274–2278.

Bellinger, D.L., Lorton, D., Felten, S.Y., and Felten, D.L. (1990). Noradrenergic and peptidergic neural-immune interactions in aging. In Stress and the Aging Brain G. Nappi (ed.), New York: Raven Press, pp. 143–140.

Besedovsky, H.O., and Del Rey, A. (1996). Immune-neuro-endocrine interactions: Facts and hypotheses. *Endocr. Rev.* 17:64–102.

Besedovsky, H.O., Del Rey, A., Sorkin, E., Da Prada, M., and Keller, H.H. (1979). Immunoregulation mediated by the sympathetic nervous system. *Cell. Immunol.* 48:346–355.

Betz, M., and Fox, B.S. (1991). Prostaglandin E2 inhibits production of Th1 lymphokines but not of Th2 lymphokines. *J. Immunol.* 146:108–113.

Biondi, M., and Zannino, L.-G. (1997). Psychological stress, neuroimmunomodulation, and susceptibility to infectious diseases in animals and man: A review. *Psychother. Psychosom.* 66:3–26.

Blomhoff, H.K., Smeland, E.B., Beiske, K., Blomhoff, R., Ruud, E., Bjoro, T., Pfeifer-Ohlsson, S., Watt, R., Funderud, S., Godal, T., and Ohlsson, R. (1987). Cyclic AMP-mediated suppression of normal and neoplastic B cell proliferation is associated with regulation of *myc* and Ha-*ras* protooncogenes. *J. Cell. Physiol.* 131:426–433.

Borger, P., Kauffman, H.F., Postma, D.S., and Vellenga, E. (1996a). Interleukin-4 gene expression in activated human T lymphocytes is regulated by the cyclic adenosine monophosphate-dependent signaling pathway. *Blood* 87:691–698.

Borger, P., Kauffman, H.F., Postma, D.S., and Vellenga, E. (1996b). Interleukin-4 gene expression in activated human T lymphocytes is regulated by the cyclic adenosine monophosphate-dependent signaling pathway. *Blood* 87:691–698.

Cazaux, C.A., Sterin-Borda, L., Gorelik, G., and Cremaschi, G.A. (1995). Down-regulation of beta-adrenergic receptors induced by mitogen activation of intracellular signaling events in lymphocytes. *FEBS Lett.* 364:120–124.

Chelmicka-Schorr, E., Checinski, M., and Arnason, B.G.W. (1988). Chemical sympathectomy augments the severity of experimental allergic encephalomyelitis. *J. Neuroimmunol.* 17:347–350.

Chen, D., and Rothenberg, E.V. (1994). Interleukin 2 transcription factors as molecular targets of cAMP inhibition: Delayed inhibition kinetics and combinatorial transcription roles. *J. Exp. Med.* 179:931–942.

Clerc, R.G., Corcoran, L.M., LaBowitz, J.H., Baltimore, D., and Sharpe, P.A. (1988). The B-cell-specific Oct-2 protein contains POU box- and homeo box-type domains. *Genes Dev.* 2:1570–1581.

Cohen, D.P., and Rothstein, T.L. (1989). Adenosine 3′,5′-cyclic monphosphate modulates the mitogenic responses of murine B lymphocytes. *Cell. Immunol.* 121:113–124.

Corcoran, L.M., and Karvelas, M. (1994). Oct-2 is required in T cell-independent B cell activation for G1 progression and for proliferation. *Immunity* 1:635–645.

Crocker, I.C., Townley, R.G., and Khan, M.M. (1996). Phosphodiesterase inhibitors suppress proliferation of peripheral blood mononuclear cells and interleukin-4 and -5 secretion by human T-helper type 2 cells. *Immunopharmacolopy* 31:223–235.

Cross, R.J., Jackson, J.C., Brooks, W.H., Sparks, D.L., Markesbery, W.R., and Roszman, T.L. (1986). Neuroimmunomodulation: impairment of humoral immune responsiveness by 6-hydroxydopamine treatment. *Immunology* 57:145–152.

Diamantstein, T., and Ulmer, A. (1975). The antagonistic action of cyclic GMP and cyclic AMP on proliferation of B and T lymphocytes. *Immunology* 28:113–119.

Diaz, M., and Casali, P. (2002). Somatic immunoglobulin hypermutation. *Curr. Opin. Immunol.* 14:235–240.

96   V.M. Sanders

Feldman, R.D., Hunninghake, G.W., and McArdle, W. (1987). Beta-adrenergic receptor-mediated suppression of interleukin 2 receptors in human lymphocytes. *J. Immunol.* 139(10):3355–3359.

Felten, D.L., Felten, S.Y., Bellinger, D.L., Carlson, S.L., Ackerman, K.D., Madden, K.S., Olschowki, J.A., and Livnat, S. (1987a). Noradrenergic sympathetic neural interactions with the immune system: Structure and function. *Immunol. Rev.* 100: 225–260.

Felten, S.Y., Bellinger, D.L., Collier, T.J., Coleman, P.D., and Felten, D.L. (1987b). Decreased sympathetic innervation of spleen in aged Fischer 344 rats. *Neurobiol. Aging* 8:159–165.

Felten, D.L., Felten, S.Y., Madden, K.S., Ackerman, K.D., and Bellinger, D.L. (1989). *Development, Maturation and Senescence of Sympathetic Innervation of Secondary Immune Organs.* New York: Academic Press, pp. 381–397.

Felten, S.Y., Madden, K.S., Bellinger, D.L., Kruszewska, B., Moynihan, J.A., and Felten, D.L. (1998). The role of the sympathetic nervous system in the modulation of immune responses. *Adv. Pharmacol.* 42:583–587.

Forster, M. J., and Lal, H. (1991). Autoimmunity and cognitive decline in aging and Alzheimer's disease. In: Psychoneuroimmunology R. Ader, N. Cohen, and D.L. Felten (eds.), New York: Academic Press, pp. 709–748.

Fuchs, B.A., Campbell, K.S., and Munson, A.E. (1988). Norepinephrine and serotonin content of the murine spleen: Its relationship to lymphocyte beta-adrenergic receptor density and the humoral immune response in vivo and in vitro. *Cell. Immunol.* 117:339–351.

Gajewski, T.F., Schell, S.R., and Fitch, F.W. (1990). Evidence implicating utilization of different T cell receptor-associated signaling pathways by Th-1 and Th-2 clones. *J. Immunol.* 144:4110–4120.

Goin, J.C., Sterin-Borda, L., Borda, E.S., Finiasz, M., Fernandez, J., and de Bracco, M.M. (1991). Active alpha 2 and beta adrenoceptors in lymphocytes from patients with chronic lymphocytic leukemia. *Int. J. Cancer* 49:178–181.

Hall, N.R., McClure, J.E., Hu, S., Tare, S., Seals, C.M., and Goldstein, A.L. (1982). Effects of 6-hydroxydopamine upon primary and secondary thymus dependent immune responses. *Immunopharmacology.* 5:39–48.

Holte, H., Torjesen, P., Blomhoff, H.K., Ruud, E., Funderud, S., and Smeland, E.B. (1988). Cyclic AMP has the ability to influence multiple events during B cell stimulation. *Eur. J. Immunol.* 18:1359–1366.

Humbert, P.O., and Corcoran, L.M. (1997). Oct-2 gene disruption eliminates the peritoneal B-1 lymphocytes linage and attenuates B-2 cell maturation and function. *J. Immunol.* 159:5273–5285.

Jeannin, P., Delneste, Y., Lecoanet-Henchoz, S., Gauchat, J.-F., Ellis, J., and Bonnefoy, J.-Y. (1997). CD86 (B7-2) on human B cells: A functional role in proliferation and selective differentiation into IgE- and IgG4-producing cells. *J. Biol. Chem.* 272: 15613–15619.

Johnson, D.L., Ashmore, R.C., and Gordon, M.A. (1981). Effects of beta-adrenergic agents on the murine lymphocyte response to mitogen stimulation. *J. Immunopharmacol.* 3:205–219.

Kasahara, K., Tanaka, S., and Hamashima, Y. (1977a). Suppressed immune response to T-cell dependent antigen in chemically sympathectomized mice. *Res. Commun. Chem. Pathol. Pharmacol.* 18:533–542.

Kasahara, K., Tanaka, S., Ito, T., and Hamashima, Y. (1977b). Suppression of the primary immune response by chemical sympathectomy. *Res. Common. Chem. Pathol. Pharmacol.* 16:687–694.

Kasprowicz, D.J., Kohm, A.P., Berton, M.T., Chruscinski, A.J., Sharpe, A.H., and Sanders, V.M. (2000). Stimulation of the B cell receptor, CD86 (B7-2), and the beta-2-adrenergic receptor intrinsically modulates the level of IgG1 produced per B cell. *J. Immunol.* 165:680–690.

Katamura, K., Shintaku, N., Yamauchi, Y., Fukui, T., Ohshima, Y., Mayumi, M., and Furusho, K. (1995). Prostaglandin E2 at priming of naive CD4+ T cells inhibits acquisition of ability to produce IFN-gamma and IL-2, but not IL-4 and IL-5. *J. Immunol.* 155:4604–4612.

Khamlichi, A.A., Pinaud, E., Decourt, C., Chauveau, C., and Cogne, M. (2000). The 3′ IgH regulatory region: A complex structure in a search for a function. *Adv. Immunol.* 75:317–345.

Kim, U., Qin, X.F., Gong, S., Stevens, S., Luo, Y., Nussenzweig, M., and Roeder, R.G. (1996). The B-cell-specific transcription coactivator OCA-B/OBF-1/Bob-1 is essential for normal production of immunoglobulin isotypes. *Nature* 383:542–547.

Kohm, A.P., and Sanders, V.M. (1999). Suppression of antigen-specific Th2 cell-dependent IgM and IgG1 production following norepinephrine depletion in vivo. *J. Immunol.* 162:5299–5308.

Kohm, A.P., and Sanders, V.M. (2001). Norepinehrine and beta-2-adrenergic receptor stimulation regulate CD4+ T and B lymphocyte function in vitro and in vivo. *Pharmacol. Rev.* 53:487–525.

Kouassi, E., Boukhris, W., Descotes, J., Zukervar, P., Li, Y.S., and Revillard, J.P. (1987). Selective T cell defects induced by dopamine administration in mice. *Immunopharmacol. Immunotoxicol.* 9:477–488.

Krause, D.S., and Deutsch, C. (1991). Cyclic AMP directly inhibits IL-2 receptor expression in human T cells: Expression of both p55 and p75 subunits is affected. *J. Immunol.* 146:2285–2294.

Kruszewska, B., Felten, S.Y., and Moynihan, J.A. (1995). Alterations in cytokine and antibody production following chemical sympathectomy in two strains of mice. *J. Immunol.* 155:4613–4620.

Lacour, M., Arrighi, J., Muller, K.M., Carlberg, C., Saurat, J., and Hauser, C. (1994a). cAMP up-regulates IL-4 and IL-5 production from activated CD4+ T cells while decreasing IL-2 release and NF-AT induction. *Int. Immunol.* 6:1333–1343.

Lacour, M., Arrighi, J.F., Muller, K.M., Carlberg, C., Saurat, J.H., and Hauser, C. (1994b). cAMP up-regulates IL-4 and IL-5 production from activated CD4+ T cells while decreasing IL-2 release and NF-AT induction. *Int. Immunol.* 6:1333–1343.

Li, Y.S., Kouassi, E., and Revillard, J.-P. (1989). Cyclic AMP can enhance mouse B cell activation by regulating progression into late G_1/S phase. *Eur. J. Immunol.* 19:1721–1725.

Livnat, S., Felten, S.Y., Carlson, S.L., Bellinger, D.L., and Felten, D.L. (1985). Involvement of peripheral and central catecholamine systems in neural-immune interactions. *J. Neuroimmunol.* 10:5–30.

Madden, K.S., Felten, S.Y., Felten, D.L., Sundaresan, P.R., and Livnat, S. (1989). Sympathetic neural modulation of the immune system. I. Depression of T cell immunity *in vivo* and *in vitro* following chemical sympathectomy. *Brain Behav. Immun.* 3:72–89.

Mary, D., Peyron, J.-F., Auberger, P., Aussel, C., and Fehlmann, M. (1989). Modulation of T cell activation by differential regulation of the phosphorylation of two cytosolic proteins. *J. Biol. Chem.* 264:14498–14502.

McPherson, G.A., and Summers, R. J. (1982). Characteristics and localization of 3H-clonidine binding in membranes prepared from guinea pig spleen. *Clin. Exp. Pharmacol. Physiol.* 9:77–87.

Miles, K., Quintans, J., Chelmicka-Schorr, E., and Arnason, B.G.W. (1981). The sympathetic nervous system modulates antibody response to thymus-independent antigens. *J. Neuroimmunol.* 1:101–105.

Mosmann, T.R., Cherwinski, H., Bond, M.W., Giedlin, M.A., and Coffman, R.L. (1986). Two types of murine helper T cell clones. I. Definition according to profiles of lymphokine activities and secreted proteins. *J. Immunol.* 136:2348–2340.

Mosmann, T.R., and Coffman, R.L. (1989). TH1 and TH2 cells: Different patterns of lymphokine secretion lead to different functional properties. *Annu. Rev. Immunol.* 7:145–173.

Munoz, E., Zubiaga, A.M., Merrow, M., Sauter, N.P., and Huber, B.T. (1990). Cholera toxin discriminates between T helper 1 and 2 cells in T cell receptor-mediated activation: role of cAMP in T cell proliferation. *J. Exp. Med.* 172:95–103.

Muraguchi, A., Miyazaki, K., Kehrl, J.H., and Fauci, A.S. (1984). Inhibition of human B cell activation by diterpine forskolin: interference with B cell growth factor-induced G_1 to S transition of the B cell cycle. *J. Immunol.* 133:1283–1287.

Muthusamy, N., Baluyut, A.R., and Subbarao, B. (1991). Differential regulation of surface Ig- and Lyb2-mediated B cell activation by cyclic AMP. I. Evidence for alternative regulation of signaling through two different receptors linked to phosphatidylinositol hydrolysis in murine B cells. *J. Immunol.* 147:2483–2492.

Naito, Y., Endo, H., Arai, K., Coffman, R.L., and Arai, N. (1996). Signal transduction in Th clones: target of differential modulation by PGE2 may reside downstream of the PKC-dependent pathway. *Cytokine* 8:346–356.

Nielsen, P. J., Georgiev, O., Lorenz, B., and Schaffner, W. (1996). B lymphocytes are impaired in mice lacking the transcriptional co-activator Bob1/OCA-B/OBF1. *Eur. J. Immunol.* 26:3214–3218.

Novak, T. J., and Rothenberg, E.V. (1990). cAMP inhibits induction of interleukin 2 but not of interleukin 4 in T cells. *Proc. Natl. Acad. Sci. U.S.A.* 87:9353–9357.

Paliogianni, F., and Boumpas, D.T. (1996). Prostaglandin E2 inhibits the nuclear transcription of the human interleukin 2, but not the Il-4, gene in human T cells by targeting transcription factors AP-1 and NF-AT. *Cellular Immunology* 171: 95–101.

Parker, C.W., Huber, M.G., and Godt, S.M. (1995). Modulation of IL-4 production in murine spleen cells by prostaglandins. *Cell. Immunol.* 160:278–285.

Pinaud, E., Khamlichi, A.A., Le Morvan, C., Drouet, M., Nelsso, V., Le Bert, M., and Cogne, M. (2001). Localization of the 3' IgH locus elements that effect long-distance regulation of class switching recombination. *Immunity* 15:187–199.

Pochet, R., and Delespesse, G. (1983). Beta-adrenoceptors display different efficiency on lymphocyte subpopulations. *Biochem. Pharmacol.* 32:1651–1655.

Podojil, J., and Sanders, V. (2005). CD86 and Beta2-Adrenergic Receptor Stimulation Regulate B cell Activity Cooperatively. *Trends in Immunology* In Press.

Podojil, J.R., Kin, N.W., and Sanders, V.M. (2004). CD86 and beta2-adrenergic receptor signaling pathways, respectively, increase Oct-2 and OCA-B Expression and binding to the 3'-IgH enhancer in B cells. *J. Biol. Chem* 279:23394–23404.

Podojil, J.R., and Sanders, V.M. (2003). Selective regulation of mature IgG1 transcription by CD86 and beta2-adrenergic receptor stimulation. *J. Immunol.* 170:5143–5151.

Radojcic, T., Baird, S., Darko, D., Smith, D., and Bulloch, K. (1991). Changes in beta-adrenergic receptor distribution on immunocytes during differentiation: An analysis of T cells and macrophages. *J. Neurosci. Res.* 30:328–335.

Ramer-Quinn, D.S., Baker, R.A., and Sanders, V.M. (1997). Activated Th1 and Th2 cells differentially express the beta-2-adrenergic receptor: A mechanism for selective modulation of Th1 cell cytokine production. *J. Immunol.* 159:4857–4867.

Romagnani, S. (1991). Human TH1 and TH2 subsets: doubt no more. *Immunol. Today* 12:256–257.

Roper, R.L., Conrad, D.H., Brown, D.M., Warner, G.L., and Phipps, R.P. (1990). Prostaglandin E2 promotes IL-4-induced IgE and IgG1 synthesis. *J. Immunol.* 145:2644–2651.

Roper, R.L., Graf, B., and Phipps, R.P. (2002). Prostaglandin E2 and cAMP promote B lymphocyte class switching to IgG1. *Immunol. Lett.* 84:191–198.

Sanders, V.M., and Munson, A.E. (1984a). Beta-adrenoceptor mediation of the enhancing effect of norepinephrine on the murine primary antibody response in vitro. *J. Pharmacol. Exp. Ther.* 230(1):183–192.

Sanders, V.M., and Munson, A.E. (1984b). Kinetics of the enhancing effect produced by norepinephrine and terbutaline on the murine primary antibody response in vitro. *J. Pharmacol. Exp. Ther.* 231(3):527–531.

Sanders, V.M., and Munson, A.E. (1985a). Norepinephrine and the antibody response. *Pharmacol. Rev.* 37(3):229–248.

Sanders, V.M., and Munson, A.E. (1985b). Role of alpha adrenoceptor activation in modulating the murine primary antibody response in vitro. *J. Pharmacol. Exp. Ther.* 232(2):395–400.

Sanders, V.M., Baker, R.A., Ramer-Quinn, D.S., Kasprowicz, D.J., Fuchs, B.A., and Street, N.E. (1997). Differential expression of the beta-2-adrenergic receptor by Th1 and Th2 clones: Implications for cytokine production and B cell help. *J. Immunol.* 158:4200–4210.

Sanders, V.M., Kasprowicz, D.J., Kohm, A.P., and Swanson, M.A. (2001). Neurotransmitter receptors on lymphocytes and other lymphoid cells. In R. Ader, D. Felten, and N. Cohen (eds.), *Psychoneuroimmunology*, 3rd ed., Vol. 2. San Diego: Academic Press, pp. 161–196.

Schubart, D.B., Rolink, A., Kosco-Vilbois, M.H., Botteri, F., and Matthias, P. (1996). B-cell-specific coactivator OBF-1/OCA-B/Bob1 required forimmune response and germinal centre formation. *Nature* 383:538–542.

Stavnezer-Nordgren, J., and Sirlin, S. (1986). Specificity of immunoglobulin heavy chain switch correlates with activity of germline heavy chain genes prior to switching. *EMBO J.* 5:95–102.

Stein, S.H., and Phipps, R.P. (1991a). Antigen-specific IgG2a production in response to prostaglandin E$_2$, immune complexes, and IFN-gamma. *J. Immunol.* 147:2500–2506.

Stein, S.H., and Phipps, R.P. (1991b). Elevated levels of intracellular cAMP sensitize resting B lymphocytes to immune complex-induced unresponsiveness. *Eur. J. Immunol.* 21:313–318.

Stein, S.H., and Phipps, R.P. (1992). Anti-class II antibodies potentiate IgG2a production by lipopolysaccharide-stimulated B lymphocytes treated with prostaglandin E2 and IFN-gamma. *J. Immunol.* 148:3943–3949.

Stevens, S., Ong, J., Kim, U., Eckhardt, L.A., and Roeder, R.G. (2000). Role of OCA-B in 3′-IgH enhancer function. *J. Immunol.* 164:5306–5312.

Suvas, S., Singh, V., Sahdev, S., Vohra, H., and Agrewala, J.A. (2002). Distinct role of CD80 and CD86 in the regulation of the activation of B cell and B cell lymphoma. *J. Biol. Chem.* 277:7766–7775.

Swain, S.L., Croft, M., Dubey, M., Haynes, L., Rogers, P., Zhang, X., and Bradley, L.M. (1996). From naive to memory T cells. *Immunol. Rev.* 150:143–167.

Swanson, M.A., Lee, W.T., and Sanders, V.M. (2001). IFN-gamma production by Th1 cells generated from naive CD4(+) T cells exposed to norepinephrine. *J. Immunol.* 166:232–240.

Tamir, A., and Isakov, N. (1994). Cyclic AMP inhibits phosphatidylinositol-coupled and -uncoupled mitogenic signals in T lymphocytes. Evidence that cAMP alters PKC-induced transcription regulation of members of the jun and fos family of genes. *J. Immunol.* 152:3391–3399.

Tang, H., and Sharp, P.A. (1999). Transcriptional regulation of the murine 3′ IgH enhancer by OCT-2. *Immunity* 11:517–526.

Titinchi, S., and Clark, B. (1984). Alpha-2 -adrenoceptors in human lymphocytes: Direct characterization by [3H]Yohimbine binding. *Biochem. Biophys. Res. Commun.* 121(1):1–7.

Vischer, T.L. (1976). The differential effect of cAMP on lymphocyte stimulation by T- or B-cell mitogens. *Immunology* 30:735–739.

Wacholtz, M.C., Minakuchi, R., and Lipsky, P.E. (1991). Characterization of the 3′,5′-cyclic adenosine monophosphate-mediated regulation of IL2 production by T cells and Jurkat cells. *Cell. Immunol.* 135:285–298.

Watson, J. (1975). The influence of intracellular levels of cyclic nucleotides on cell proliferation and the induction of antibody synthesis. *J. Exp. Med.* 141:97–111.

Westly, H. J., and Kelley, K.W. (1987). Down-regulation of glucocorticoid and beta-adrenergic receptors on lectin-stimulated splenocytes. *Proc. Soc. Exp. Biol. Med.* 185:211–218.

Whisler, R.L., Beiqing, L., Grants, I.S., and Newhouse, Y.G. (1992). Cyclic AMP modulation of human B cell proliferative responses: Role of cAMP-dependent protein kinases in enhancing B cell responses to phorbol diesters and ionomycin. *Cell. Immunol.* 142:398–415.

Wirth, S., Lacour, M., Jaunin, F., and Hauser, C. (1996). Cyclic adenosine monophosphate (cAMP) differentially regulates IL-4 in thymocyte subsets. *Thymus* 24: 101–109.

Yoshimura, T., Nagao, T., Nakao, T., Watanabe, S., Usami, E., Kobayashi, J., Yamazaki, F., Tanaka, H., Inagaki, N., and Nagai, H. (1998). Modulation of Th1- and Th2-like cytokine production from mitogen- stimulated human peripheral blood mononuclear cells by phosphodiesterase inhibitors. *Gen. Pharmacol.* 30:175–180.

6
Gender Dimorphism and the Use of Sex Steroid/Receptor Antagonist After Trauma

MASHKOOR A. CHOUDHRY AND IRSHAD H. CHAUDRY

1. Introduction

Trauma remains the major cause of deaths in the United States and in other developing countries. Moreover, a significant number of trauma victims who survive initial injury succumb subsequently because of sepsis and multiple organ failure (Bone, 1992; Nathens and Marshall, 1996; Baue *et al.*, 1998; Marshall, 1999; Angele *et al.*, 2000; Baue, 2000; Choudhry *et al.*, 2003). Thus, sepsis and organ dysfunction continue to be the major cause of morbidity and mortality in trauma patients. Although intensive investigations during the past three decades have helped identify some of the mechanisms responsible for sepsis and organ dysfunction, despite all these efforts the prognosis of trauma patients remains elusive. Furthermore, these studies suggest that the postinjury pathogenesis is complex and is influenced by multiple factors. Among these, gender is suspected to be a major factor that plays a significant role in shaping the host response to injury (Schroder *et al.*, 1998; Angele *et al.*, 2000; Schroder *et al.*, 2000; Croce *et al.*, 2002; Yokoyama et al., 2002; Chaudry et al., 2003; George *et al.*, 2003b; Choudhry *et al.*, 2004). The primary aim of this article is to present a comprehensive summary of the studies dealing with the role of gender in response to trauma as well as to discuss potential targets that can be used to modulate endogenous levels of sex hormones to improve organ functions after experimental trauma.

2. Gender Dimorphism in Trauma Patients

In recent years, recognition of gender-based differences in patient response to injury/disease has been the subject of much interest (Bone, 1992; Diodato *et al.*, 2001; Verthelyi, 2001; Muller *et al.*, 2002; Orshal and Khalil, 2004; Morales-Montor *et al.*, 2004). Although the findings of some clinical studies do not support the role of gender in the overall outcome of trauma patients, others have provided evidence in support of the suggestion that gender plays a significant role in the outcome of trauma patients (Schroder *et al.*, 1998; Eachempati *et al.*, 1999; Offner *et al.*, 1999; Oberholzer *et al.*, 2000; McGwin *et al.*, 2002; Bowles *et al.*, 2003; George *et al.*, 2003a, 2003b; Gannon *et al.*, 2004). Recently, a retrospective analysis of more than 150,000 blunt or penetrating trauma patients (George *et al.*, 2003a, 2003b) suggested that after blunt trauma, male patients had a significantly higher risk of death as compared with female patients. In addition, these findings suggested that premenopausal women had a survival advantage in blunt trauma patients; however, the opposite pattern prevailed in patients with penetrating trauma (George *et al.*, 2003a, 2003b). Similarly, findings from a prospective analysis of septic patients suggested significantly more deaths in males compared with females (Schroder *et al.*, 1998). Consistent with these findings, another study (Wichmann *et al.*, 2000) concluded that while the overall mortality was not different between males and females after sepsis, only a few female patients required intensive care. In addition, the severity of sepsis/septic shock in females was much lower in intensive care patients. Similar conclusions that the male gender is associated with increased risk of major infection after trauma was drawn in yet another study (Offner *et al.*, 1999). Collectively, these findings suggest that gender plays a role in the outcome of trauma patients. In contrast, some other studies failed to establish the relationship between gender and the outcome of trauma patients (Eachempati *et al.*, 1999; Croce *et al.*, 2002; Bowles *et al.*, 2003; Gannon *et al.*, 2004). For instance, Eachempati *et al.* (1999) did not find significant differences in mortality between males and females among patients admitted to intensive care units with symptoms of systemic inflammatory response syndrome. Bowles *et al.* (2003) enrolled 15,170 trauma patients over a 5-year period (1993–1997) and compared outcomes based on gender, age, and severity of injury. They found that age, mechanism, and severity of injury but not gender influenced survival, thus the role of gender in the outcome of trauma patients remains controversial.

Although the cause for the observed differences in clinical studies remains to be established, a series of experimental studies of trauma suggest that the response to injury is different in males and females (Yao *et al.*, 1998; Angele *et al.*, 2000; Kahlke *et al.*, 2000a, 2000b; Samy *et al.*, 2001; Kovacs *et al.*, 2002; Yokoyama *et al.*, 2002; Chaudry *et al.*, 2003). These studies have shown that alterations in immune and cardiovascular functions after trauma-hemorrhage are more severe in mature males, ovariectomized and aged females, whereas both immune and cardiac functions are maintained in proestrus females under those conditions (Yao *et al.*, 1998; Angele *et al.*, 2000; Kahlke *et al.*, 2000a, 2000b; Samy *et al.*, 2001; Kovacs *et al.*, 2002; Yokoyama *et al.*, 2002; Chaudry *et al.*, 2003). Similarly, liver functions after trauma-hemorrhage were found to be depressed in males but were maintained in proestrus females (Remmers *et al.*, 1997, 1998a). Moreover, the survival rate of proestrus females subjected to sepsis after trauma-hemorrhage is significantly higher than age-matched males or ovariectomized females (Zellweger *et al.*, 1997). Thus, the results obtained in experimental model of trauma clearly suggest that the alterations in immune and other organ functions are gender specific. However, the findings from these experimental studies suggest that sex hormone levels and not the gender itself play roles in shaping the host response to an injury such as trauma. Because in patient studies the levels of sex hormones were not determined at the time of injury, it is difficult to ascertain the role of gender in the previously published studies. Therefore, in order to determine the role of gender in post-trauma morbidity and mortality, more patient studies should be planned. These studies should enroll a more homogenous patient population (e.g., age-matched), and the patient outcome should be correlated with the levels of sex hormones rather to the gender of the patients.

3. Sex Hormones and Response to Experimental Trauma

The action of sex hormones starts immediately after their synthesis when they migrate from the bloodstream to the cell across the cell membrane by a simple diffusion mechanism. Once inside the cells, the sex hormones like other steroids form complexes with their cytosolic and/or nuclear receptors. This complex then binds to chromatin and stimulates the transcription of a set of genes with a specific sex steroid–responsive regulatory element (Landers and Spelsberg, 1992; Olsen and Kovacs, 2001; Orshal and Khalil, 2004). Thus, both androgen and estrogen mediate their actions by activating the transcription factors and accordingly are expected to alter signaling at the nuclear level. However, there is mounting evidence indicating that sex steroids can also induce a nongenomic response within cells, a response that is not mediated through nuclear receptors but rather initiated at the plasma membrane, presumably through unconventional surface receptors (Landers and Spelsberg, 1992; Benten *et al.*, 1999; Zhang and Shapiro, 2000; Jarrar *et al.*, 2002; Wunderlich *et al.*, 2002; Zhang *et al.*, 2002; Mize *et al.*, 2003; Orshal and Khalil, 2004). Previous studies have utilized multiple ways to modulate the actions of sex hormones and then determined their role in post-trauma organ dysfunction. Some of these approaches are discussed in the following section.

3.1. Sex Hormones and Immune Response to Trauma

Previous studies have shown that the suppression of immune response is apparent immediately after trauma-hemorrhage in males and persists for a prolonged period of time, despite fluid resuscitation (Chaudry *et al.*, 1990; Zellweger *et al.*, 1995; Xu *et al.*, 1998; Samy *et al.*, 2000). The suppression of immune functions after trauma-hemorrhage is characterized by a decrease in macrophage antigen presentation and T-cell proliferation (Fig. 6.1). In addition, a decrease in Th1 cytokines (IL-2 and IFN-γ) and increase in Th2 cytokines (IL-4 and IL-10) was commonly reported under those injury conditions (Chaudry *et al.*, 1990; Ayala *et al.*, 1992; Wichmann *et al.*, 1996a, 1997; Xu *et al.*, 1998; Angele *et al.*, 2000; Kahlke *et al.*, 2000a; Knoferl *et al.*, 2001, 2002; Jarrar *et al.*, 2002). Females in the proestrus stage of estrus cycle, however, have normal/maintained immune responses after trauma-hemorrhage (Fig. 6.1). Surgical removal of ovaries 2 weeks prior to trauma-hemorrhage, however, resulted in suppressed immune responses similar to those observed in male mice after trauma-hemorrhage (Kahlke *et al.*, 2000a, 2000b; Jarrar *et al.*, 2000c; Knoferl *et al.*, 2001). Furthermore, castration 2 weeks prior to trauma-hemorrhage prevents the suppression of immune functions in male mice (Wichmann *et al.*, 1996b). Administration of male sex hormones 5α-dihydrotestosterone (5α-DHT) in castrated males results in suppression of splenic and peritoneal macrophage cytokine production (Angele *et al.*, 2001). Additionally, pretreatment of female mice with a physiological amount of 5α-DHT for 2 weeks prior to trauma-hemorrhage also caused suppressed splenic and peritoneal macrophage cytokine production (Angele *et al.*, 2001). On the other hand, female sex steroids have been shown to prevent the suppression in immune cell functions after trauma-hemorrhage (Kahlke *et al.*, 2000b; Angele *et al.*, 2001; Knoferl *et al.*, 2001, 2002). Treatment of male mice with 17β-estradiol prevented the decrease

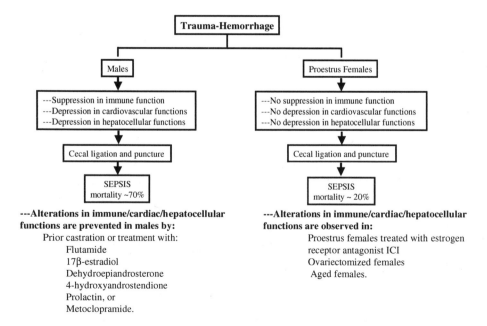

FIGURE 6.1. Gender-specific response to trauma-hemorrhage.

in splenic and peritoneal macrophage production of cytokines (Knoferl *et al.*, 2000). The decrease in Th1 cytokine after trauma-hemorrhage was also prevented in 17β-estradiol–treated male mice. In addition, administration of 17β-estradiol in ovariectomized females after trauma-hemorrhage restores immune functions similar to those observed in sham-injured animals. These findings suggest that changes in immune functions after trauma-hemorrhage are gender specific. In particular, male sex hormones 5α-DHT contribute to the suppression of immune cell functions, whereas the female sex hormones 17β-estradiol maintain immune function after trauma-hemorrhage.

3.2. Sex Hormones and Alterations in Cardiac Function

Similar to immune response, sexual dimorphism has also been reported with respect to cardiac functions (Fig. 6.1). Findings from previous studies have shown that cardiac output, stroke volume, +dP/dt, −dP/dt, and total peripheral resistance were markedly altered after trauma-hemorrhage in males (Wang *et al.*, 1993; Remmers *et al.*, 1997; Angele *et al.*, 1998c; Remmers *et al.*, 1998a; Yao *et al.*, 1998; Mizushima *et al.*, 2000; Ba *et al.*, 2001; Ancey *et al.*, 2002; Kuebler *et al.*, 2002; Ba *et al.*, 2003; Yang *et al.*, 2004). However, in proestrus females, which have high circulating levels of estrogen, cardiac functions are maintained after trauma-hemorrhage and fluid resuscitation (Fig. 6.1) (Ba *et al.*, 2003). In contrast, cardiac functions are suppressed in ovariectomized females (Yang *et al.*, 2004). Administration of a single dose of 17β-estradiol in ovariectomized females prevented the depression in cardiac function after trauma-hemorrhage (Mizushima *et al.*, 2000; Yang *et al.*, 2004). Similar administration of 17β-estradiol in males also prevented cardiac dysfunction after trauma-hemorrhage (Chaudry *et al.*, 2003; Yang *et al.*, 2004). Additional findings suggest that decreasing androgen levels by performing castration 2 weeks prior to trauma-hemorrhage significantly improved/restored cardiac function under those conditions (Chaudry *et al.*, 2003; Yang *et al.*, 2004). On the other hand, administration of male sex hormones 5α-DHT in castrated males and in females resulted in suppressed cardiac function after trauma-hemorrhage (Chaudry *et al.*, 2003; Yang *et al.*, 2004). Thus, similar to immune function, male sex hormones are found to suppress cardiac function; female hormones on the other hand prevent the suppression of cardiac function after trauma-hemorrhage. However, the mechanism of the beneficial effects of female or harmful effect of male steroids after trauma-hemorrhage remains to be established. Previous studies have shown that 17β-estradiol may mediate its salutary effect on cardiac function via modulation of proinflammatory cytokines such as IL-6 and TNF-α (Levine *et al.*, 1990; O'Neill *et al.*, 1994; Yamauchi-Takihara *et al.*, 1995; Kerger *et al.*, 1999; Mizushima *et al.*, 2000; Yang *et al.*, 2004). In addition, administration of 17β-cstradiol also prevented the decrease in nitric oxide synthase and increased leukocyte infiltration (Angele *et al.*,

1998c). Both diminished nitric oxide synthase activity and increased leukocyte infiltration potentially contribute to altered cardiac functions after trauma hemorrhage (Miyao et al., 1993; Hierholzer et al., 1998; Kerger et al., 1999). Thus, estradiol-mediated prevention of both the decrease in nitric oxide synthase and increase of leukocyte infiltration is likely another potential mechanism by which estrogen mediates its salutary action on cardiac function after trauma-hemorrhage. Altogether, one or more than one of the above mechanisms could be responsible for the protective effects of 17β-estradiol observed in the males after trauma-hemorrhage and resuscitation.

3.3. Sex Hormones and Alterations in Hepatocellular Function

In addition to impaired cardiac and immune functions, alterations in hepatic functions have also been found to be gender-specific after trauma-hemorrhage. These findings suggest that liver functions are markedly depressed in male rats after trauma-hemorrhage and resuscitation (Remmers et al., 1998a, 1998b; Jarrar et al., 2000c; Kuebler et al., 2002; Chaudry et al., 2003; Jarrar et al., 2004). On the other hand, female animals during the proestrus stage of their estrus cycle showed normal liver functions after trauma-hemorrhage and resuscitation (Jarrar et al., 2000b). The findings also showed that ovariectomized females had depressed hepatocellular function after trauma-hemorrhage (Jarrar et al., 2000c; Kuebler et al., 2001). However, administration of 17β-estradiol in ovariectomized females after trauma-hemorrhage restored liver functions under those conditions (Knoferl et al., 2001). These results collectively suggest that, as with immune and cardiac functions, male sex steroids have a deleterious effect on liver function, whereas estrogen maintains liver function after trauma-hemorrhage.

3.4. Sex Hormones and Mortality in Animals After Trauma-Hemorrhage from Subsequent Sepsis

In addition to organ function, studies were also performed to determine whether or not modulation of sex hormones after trauma-hemorrhage influences the mortality in animals from subsequent sepsis. Findings from these studies suggested that more than 70% of male animals die after sepsis induced by cecal ligation and puncture (CLP) after trauma-hemorrhage (Zellweger et al., 1997). In contrast, morality in proestrus females was found to be ~20%. Furthermore, studies have shown that prior castration or treatment of male animals with 17β-estradiol resulted in increased survival of trauma-hemorrhage after sepsis (Zellweger et al., 1997). In contrast, surgical removal of ovaries in female mice prior to trauma-hemorrhage increased the lethality from subsequent sepsis (Zellweger et al., 1997; Chaudry et al., 2003).

4. Sex Hormone Receptors

Androgen and estrogen receptors are expressed in almost all the cells, including cells from the immune system, heart, and liver (Chaudry et al., 2003; Orshal and Khalil, 2004). Two major subtypes of estrogen receptors (ERs) (i.e., ER-α and ER-β) have been identified. Furthermore, several subtypes of ER-α, such as ER-αA, ER-αC, ER-αE, and ER-αF, and of ER-β, such as ER-β1, ER-β2, ER-β4, and ER-β5, have been described (Kos et al., 2002; Orshal and Khalil, 2004). Although some studies suggest that ER-α promotes the protective effects of estrogen, others have shown that ER-β is more critical for the beneficial effect of estrogen (Case and Davison, 1999; Kos et al., 2002; Pare et al., 2002; Scobie et al., 2002; Orshal and Khalil, 2004). A recent study showed that ER-α is essential for thymic and splenic development in males, whereas expression of ER-β is required for estrogen-mediated thymic cortex atrophy and thymocyte phenotypic shift in females (Erlandsson et al., 2001). Previous studies have demonstrated that B-cell lymphopoiesis is normal in female ER-α–disrupted mice (Smithson et al., 1998), suggesting that ER-β might be responsible for regulating B-cell formation in bone marrow (Kincade et al., 2000). The overall distribution of AR and ER expression was not different in T cells of male and female animals with and without trauma. Recent findings suggested that the T-cell expression of ER-α and ER-β in response to trauma-hemorrhage is different (Samy et al., 2000; Samy et al., 2001; Samy et al., 2003). These findings have shown that while ER-α expression did not change after trauma-hemorrhage, the expression of ER-β was significantly decreased in the proestrus females after trauma-hemorrhage. Such alterations in ER-β expression may contribute to the observed changes in T-cell functions.

4.1. Use of Male Sex Hormone Receptor Antagonist

The most widely used male sex hormone receptor antagonist in the treatment of experimental trauma is flutamide. Flutamide blocks receptors of testosterone, dihydrotestosterone, and 3α-androstenediol (Chaudry et al., 2003). Flutamide is a nonsteroidal agent and is known to inhibit androgen uptake or nuclear binding of the activated androgen receptor to nuclear response elements in the nucleus (Kolvenbag and Nash, 1999; Chaudry et al., 2003). It is used clinically for the treatment of androgen-sensitive prostatic carcinoma. Animal studies have shown that treatment of animals with flutamide is effective in preventing the deleterious effects of hemorrhagic shock on cell-mediated immunity, as well as in preventing the cardiovascular and hepatocellular depression (Angele et al., 1997; Angele et al., 1998a, 2000, 2001; Yokoyama et al., 2002; Chaudry et al., 2003). Another common antiandrogenic compound is bicalutamide (AstraZeneca Chicago, IL). Bicalutamide, similar to flutamide, is a nonsteroidal and shares flutamide in its mechanism of action (Furr, 1996; Kolvenbag and Nash, 1999; Angele *et al.*, 2000; Chaudry *et al.*, 2003). It competitively inhibits the binding of androgen to its cytosolic receptor in target tissues and is used clinically in the treatment of prostate cancer.

4.2. Use of Female Sex Hormone Receptor Antagonist

Similarly, there are a number of pharmacological agents that have been used to block estrogen receptors (Martel *et al.*, 1998; Tremblay *et al.*, 1998; Angele *et al.*, 2000; Katzenellenbogen and Katzenellenbogen, 2000; Stauffer *et al.*, 2000; Meyers *et al.*, 2001; Yokoyama *et al.*, 2002; Chaudry *et al.*, 2003). Among them, ICI 182,780 and EM-800 were used in many studies dealing with the role of estrogen in trauma (Angele *et al.*, 2000; Yokoyama *et al.*, 2002; Chaudry *et al.*, 2003). EM-800 is an estrogen receptor antagonist that blocks the transcriptional functions of estrogen receptor α and β (Luo *et al.*, 1997; Martel *et al.*, 1998; Tremblay *et al.*, 1998). ICI 182,780 on the other hand inhibits estrogen binding to the receptor complex. ICI is 10 times more potent than EM-800 and can be administered orally or subcutaneously. In addition, there are compounds that work as an agonist and are recognized for their specificity toward ER-α and ER-β. Propyl pyrazole triol (PPT) is a potent ER-α agonist that does not activate ER-β (Martel *et al.*, 1998; Tremblay *et al.*, 1998; Angele *et al.*, 2000; Katzenellenbogen and Katzenellenbogen, 2000; Stauffer *et al.*, 2000; Meyers *et al.*, 2001; Yokoyama *et al.*, 2002; Chaudry *et al.*, 2003). In contrast, the compound diarylpropionitrile (DPN) is a potency-selective agonist for ER-β with a more than 70-fold higher binding affinity for ER-β than ER-α. Tamoxifen is another agent that has been used in breast-cancer patients with estrogen receptor–positive tumors. These agents can be used to block specific receptors and thus will allow for evaluating their role both in cancer as well as in experimental models of trauma.

5. Synthesis of Sex Steroids

Sex hormones are primarily synthesized in the gonads; but recent studies have provided evidence that these hormones can be synthesized in nongonadal tissues (Samy *et al.*, 2000, 2001, 2003; Chaudry *et al.*, 2003). Cholesterol, which is the starting material for sex steroid biosynthesis, is converted to testosterone (Fig. 6.2) in the presence of various enzymes. Testosterone is common to both male and female sex hormones, but its conversion into 5α-DHT or 17β-estradiol is highly dependent on the availability of two critical enzymes, 5α-reductase and aromatase, respectively (Brodie and Njar, 2000; Chaudry *et al.*, 2003; Flores *et al.*, 2003; Arora and Potter, 2004; Occhiato *et al.*, 2004). 5α-Reductase enables the conversion of testosterone into male 5α-DHT, whereas aromatase metabolizes the same testosterone

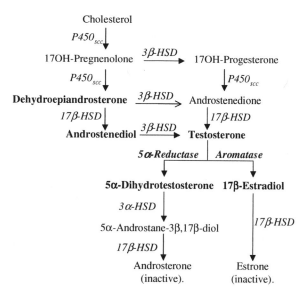

FigURE 6.2. Enzymatic pathway for the sex steroid synthesis in healthy conditions. Abbreviations: 3α-HSD, 3α-hydroxysteroid dehydrogenase; 17β-HSD, 17β-hydroxysteroid dehydrogenase.

into female sex steroid 17β-estradiol. Previous studies have shown that the enzymes [5α-reductase, aromatase, 3α-hydroxysteroid dehydrogenase (3α-HSD), 3β-hydroxysteroid dehydrogenase (3β-HSD), and 17β-hydroxysteroid dehydrogenase (17β-HSD)] responsible for the conversion of cholesterol to male or female sex hormones are present in peripheral tissues including spleen and T cells (Samy *et al.*, 2000, 2001, 2003; Chaudry *et al.*, 2003) (Fig. 6.2). In recent studies, we evaluated whether trauma-hemorrhage affects the expression of these enzymes in immune cells. Findings from these studies have shown that trauma-hemorrhage differentially regulates the expression and activity of 5α-reductase, aromatase, and 17β-HSD in male and female T cells (Samy *et al.*, 2000, 2001, 2003; Chaudry *et al.*, 2003). In males, there was an increase in T-cell 5α-reductase activity after trauma-hemorrhage; aromatase activity on the other hand is relatively low and is not altered after trauma-hemorrhage in T cells derived from male animals (Fig. 6.3A). Thus, a trauma-hemorrhage–mediated increase in T-cell 5α-reductase activity is likely to contribute to increased synthesis of 5α-DHT by these cells. 5α-DHT can be metabolized by 3α-HSD and 17β-HSD, sequentially, into androsterone, a metabolite that is inactive because of its inability to bind to the androgen receptor. However, studies show no alteration in the activity of 3α-HSD and a decrease in the expression of 17β-HSD in male T cells after trauma-hemorrhage (Samy *et al.*, 2000, 2001, 2003; Chaudry *et al.*, 2003). This suggests a lack of 5α-DHT catabolism in T cells of trauma-hemorrhaged males and consequently results in elevated levels

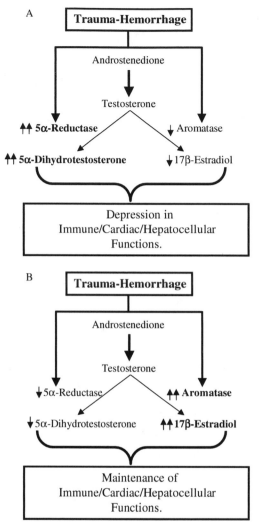

FIGURE 6.3. Regulation of 5α-reductase and aromatase activity in T cell after trauma-hemorrhage in males (A) and proestrus females (B).

of 5α-DHT within the cells. 5α-DHT binding affinity to the androgen receptor is severalfold higher than that of testosterone and is considered to be the most potent androgen (Samy *et al.*, 2000, 2001, 2003; Chaudry *et al.*, 2003). Interestingly, findings from these studies further suggested that in castrated animals, trauma-hemorrhage decreases 5α-reductase activity and increases expression of 17β-HSD, suggesting inactivation of 5α-DHT in castrated males. Altogether, these findings suggest that increased synthesis and decreased catabolism of 5α-DHT is likely the principal cause for the loss of T-cell functions in intact males after trauma-hemorrhage (Samy *et al.*,

2000, 2001, 2003; Zheng *et al.*, 2002; Chaudry *et al.*, 2003; Schneider *et al.*, 2003).

T cells in the proestrus females, on the other hand, have low 5α-reductase activity and thus low 5α-DHT levels (Fig. 6.3B). In contrast, aromatase as well as 17β-HSD activities significantly increase in proestrus females after trauma-hemorrhage (Fig. 6.3B), suggesting increased synthesis of 17β-estradiol (Samy *et al.*, 2000, 2001, 2003; Zheng *et al.*, 2002; Chaudry *et al.*, 2003). The lack of change in 17β-HSD oxidative activities leads to less conversion of 17β-estradiol to estrone, which is an inactive metabolite and does not bind to estrogen receptors. In contrast, activities of aromatase and 17β-HSD are low in the T cells of the ovariectomized animals as compared with the proestrus females after trauma-hemorrhage (Samy *et al.*, 2000, 2001, 2003; Zheng *et al.*, 2002; Chaudry *et al.*, 2003). This results in lower 17β-estradiol production and consequent immunosuppression. Whether similar alteration in steroidogenesis occurs in other organs such as heart and liver remains to be established. Findings obtained from immune cells suggest that steroidogenic enzymes differ between males and proestrus females and that the trauma-hemorrhage further influences the activity of these enzymes. Because the synthesis of active steroids markedly influences cytokine production, these enzymes may serve as potential targets for therapeutic modulation after trauma-hemorrhage.

5.1. Use of Inhibitors of Sex Steroid Synthesis

5α-Reductase is an attractive target for the modulation of sex hormone synthesis. In this regard, studies have shown that treatment of animals with 4-hydroxyandrostenedione (4-OHA), a potent inhibitor of 5α-reductase activity, attenuated alterations in cytokine production after trauma-hemorrhage (Schneider *et al.*, 2003). Other 5α-reductase inhibitors that can be used are 4-azasteroids, 6-azasteroids, and 10-azasteroids (Flores *et al.*, 2003; Occhiato *et al.*, 2004). Finasteride is among the first 5α-reductase inhibitors approved in the United States for the treatment of benign prostatic hyperplasia. In humans, finasteride decreases prostatic DHT levels by 70–90% and reduces prostate size (Flores *et al.*, 2003; Occhiato *et al.*, 2004) while testosterone tissue levels remain constant. The use of finasteride demonstrated a sustained improvement in the treatment of benign prostatic hyperplasia (Flores *et al.*, 2003; Occhiato *et al.*, 2004). Other potent and selective inhibitors of 5α-reductase, such as androstandiene-3-carboxylic, Episteride, and estratriene carboxylic acid, have also demonstrated a high inhibitory activity (Flores *et al.*, 2003; Occhiato *et al.*, 2004). Episteride has been shown to lower serum DHT levels by 50% in clinical trials (Flores *et al.*, 2003; Occhiato *et al.*, 2004).

Aromatase inhibitors are another class of antiestrogen medications that suppress estrogen levels in the blood by inhibiting one of the enzymes needed to produce the hormone. These drugs, which include letrozole, anastrozole, and exemestane, work best in postmenopausal women (Brodie and Njar, 2000; Arora and Potter, 2004). Furthermore, formestane and examestane have been shown to be effective in breast cancer patients with advanced disease (Brodie and Njar, 2000; Arora and Potter, 2004). A number of other inhibitors such as 7α-substituted androstenediones evaluated clinically are exemestane, atemestane, and 10-propagylandrostenedione (Brodie and Njar, 2000; Arora and Potter, 2004). Although both atemestane and 10-propagylandrostenedione are potent aromatase inhibitors and highly effective in lowering estrogen levels in breast cancer patients, only exemestane currently remains an available treatment option.

5.2. Use of Sex Steroid Metabolites as Therapeutic Regimens After Trauma-Hemorrhage

In addition to the use of inhibitors of sex hormone synthesis pathways, studies have also utilized approaches in which animals were treated with sex steroid metabolites such as dehydroepiandrosterone (DHEA) and androstenediol (adiol). Both DHEA and adiol are intermediate metabolites of sex hormone synthesis (Fig. 6.2). DHEA, synthesized in the adrenal glands, is present both as a biologically active free form and as an inactive sulfated form (Nestler et al., 1991; Brown et al., 1999; Baulieu et al., 2000; Jarrar et al., 2000a, 2001). DHEA is the most abundant sex steroid in the circulation and is an intermediate in the pathway for the synthesis of testosterone and estrogen. However, depending on the hormonal milieu, androgenic and estrogenic effects of DHEA have been reported (Nestler et al., 1991; Jarrar et al., 2001). DHEA has been shown to have immunoenhancing effects on splenocytes harvested from normal animals (Brown et al., 1999; Jarrar et al., 2001). Studies have shown that administration of DHEA in male animals prevented the depression in immune, cardiac, and hepatocellular function after trauma-hemorrhage (Angele et al., 1998b; Jarrar et al., 2001). Moreover, the survival rate also markedly improved in trauma-hemorrhaged animals after polymicrobial sepsis. To determine whether the salutary effects of DHEA on immune, cardiovascular, and hepatocellular function after trauma-hemorrhage were owing to a direct effect of DHEA or its conversion into 17β-estradiol or testosterone, DHEA and the specific estrogen receptor antagonist ICI 182,780 or androgen receptor antagonist flutamide were administered simultaneously in separate experiments. The results indicated that the salutary effects of DHEA on the depressed organ functions after trauma-hemorrhage were abolished in the presence of the estrogen receptor antagonist ICI 182,780. However, the beneficial effects of DHEA were not affected in animals treated with androgen receptor blocker, flutamide (Angele et al., 1998b; Catania et al., 1999; Jarrar et al., 2001). These findings suggest that the salutary effects of DHEA on the depressed organ and immune functions after trauma-hemorrhage are mediated via the estrogen receptor.

Similar to DHEA, recent studies have evaluated the role of another metabolite, adiol, in altered cardiac and hepatocellular function after trauma-hemorrhage (Shimizu et al., 2004, 2005). Adiol, one of the metabolites of DHEA, has been reported to have greater protective effects than DHEA against lethal bacterial infections and endotoxin shock. Furthermore, adiol has also been reported to produce protective effects after ionizing radiation in mice (Whitnall et al., 2002). Recently, studies from our laboratory have shown that adiol administration after trauma-hemorrhage improves cardiac and hepatic function in male animals. Additional recent findings from our laboratory suggested that adiol ameliorates hepatic functions via activation of PPAR-γ (unpublished observations). However, whether adiol affects the cardiac/hepatic function either directly or via modulation of sex hormones remains to be established.

6. Additional Target for the Modulation of Sex Hormones and Their Influence on Post-Trauma Organ Functions

Previous studies have indicated an interaction between the hypothalamic-pituitary-adrenocortical axis and the hypothalamic-pituitary-gonadal axis are likely the pathway that the brain uses to modulate immune responses in many diseases conditions (Neill, 1970, 1972; Neill and Smith, 1974; Eskandari and Sternberg, 2002; Chaudry et al., 2003; Eskandari et al., 2003). Furthermore, studies have also demonstrated that plasma levels of catecholamines such as norepinephrine, epinephrine, and dopamine are elevated early after the onset of hemorrhage and that sustained levels correlate with irreversibility of shock (Tarnoky and Nagy, 1983; Chaudry et al., 2003). Additionally, studies have shown that administration of the anterior pituitary hormone prolactin enhances immune responses after severe hemorrhage (Zhu et al., 1996; Zellweger et al., 1996a) and decreases mortality from subsequent sepsis (Zhu et al., 1996; Zellweger et al., 1996a, 1996b). Moreover, treatment of animals with the dopamine antagonist metoclopramide (MCP), which is known to increase prolactin secretion, had similar beneficial effects on the depressed immune responses after severe hemorrhage (Zellweger et al., 1998; Jarrar et al., 2000d). MCP administration after trauma-hemorrhage also was found to recover the depressed cardiac output and hepatocellular function (Lanza et al., 1987). Furthermore, MCP administration significantly increased circulating levels of prolactin and decreased the plasma levels of the proinflammatory cytokine IL-6. Thus, administration of MCP, which increased prolactin secretion, appears to be a useful adjunct for restoring the depressed cardiac and hepatocellular functions and downregulating inflammatory cytokine release after trauma and hemorrhagic shock.

7. Conclusion

In summary, while the role of gender in the outcome of trauma patients remains to be established, the results obtained in experimental models of trauma-hemorrhage support the suggestion that gender does play a significant role in shaping the host response after injury. Furthermore, from the experimental studies, it appears that steroids that could be useful adjuncts in the treatment of trauma include 17β-estradiol and flutamide. In addition, 5α-reductase inhibitors that prevent the conversion of testosterone into 5α-dihydrotestosterone could be used in the treatment of shock. The hormones/agonists, which appear particularly useful, include dehydroepiandrosterone, androstenediol, prolactin, and metoclopramide. The concept for the use of such agents comes from experimental studies demonstrating that administration of estrogen, flutamide, prolactin, metoclopramide, or DHEA restored the depressed immune, cardiac, and liver functions and decreased the lethality from subsequent sepsis.

Acknowledgments. This work was supported by National Institutes of Health grants R21 AA12901 (M.A.C.), R37 GM39519, and R01 GM37127 (I.H.C.).

References

Ancey, C., Corbi, P., Froger, J., Delwail, A., Wijdenes, J., Gascan, H., Potreau, D., and Lecron, J.C. (2002). Secretion of IL-6, IL-11 and LIF by human cardiomyocytes in primary culture. *Cytokine* 18:199–205.

Angele, M.K., Wichmann, M.W., Ayala, A., Cioffi, W.G., and Chaudry, I.H. (1997). Testosterone receptor blockade after hemorrhage in males. Restoration of the depressed immune functions and improved survival following subsequent sepsis. *Arch. Surg.* 132:1207–1214.

Angele, M.K., Ayala, A., Cioffi, W.G., Bland, K.I., and Chaudry, I.H. (1998a). Testosterone: The culprit for producing splenocyte immune depression after trauma hemorrhage. *Am. J. Physiol.* 274:C1530–C1536.

Angele, M.K., Catania, R.A., Ayala, A., Cioffi, W.G., Bland, K.I., and Chaudry, I.H. (1998b). Dehydroepiandrosterone: An inexpensive steroid hormone that decreases the mortality due to sepsis following trauma-induced hemorrhage. *Arch. Surg.* 133:1281–1288.

Angele, M.K., Smail, N., Wang, P., Cioffi, W.G., Bland, K.I., and Chaudry, I.H. (1998c). L-arginine restores the depressed cardiac output and regional perfusion after trauma-hemorrhage. *Surgery* 124:394–401.

Angele, M.K., Schwacha, M.G., Ayala, A., and Chaudry, I.H. (2000). Effect of gender and sex hormones on immune responses following shock. *Shock* 14:81–90.

Angele, M.K., Knoferl, M.W., Ayala, A., Bland, K.I., and Chaudry, I.H. (2001). Testosterone and estrogen differently effect Th1 and Th2 cytokine release following trauma-haemorrhage. *Cytokine* 16:22–30.

Arora, A., and Potter, J.F. (2004). Aromatase inhibitors: current indications and future prospects for treatment of postmenopausal breast cancer. *J. Am. Geriatr. Soc.* 52:611–616.

Ayala, A., Perrin, M.M., Ertel, W., and Chaudry, I.H. (1992). Differential effects of hemorrhage on Kupffer cells: Decreased antigen presentation despite increased inflammatory cytokine (IL-1, IL-6 and TNF) release. *Cytokine* 4:66–75.

Ba, Z.F., Wang, P., Koo, D.J., Ornan, D.A., Bland, K.I., and Chaudry, I.H. (2001). Attenuation of vascular endothelial dysfunction by testosterone receptor blockade after trauma and hemorrhagic shock. *Arch. Surg.* 136:1158–1163.

Ba, Z.F., Kuebler, J.F., Rue, L.W., III, Bland, K.I., Wang, P., and Chaudry, I.H. (2003). Gender dimorphic tissue perfusion response after acute hemorrhage and resuscitation: Role of vascular endothelial cell function. *Am. J. Physiol. Heart Circ. Physiol.* 284:H2162–H2169.

Baue, A.E. (2000). A debate on the subject "Are SIRS and MODS important entities in the clinical evaluation of patients?" The con position. *Shock* 14:590–593.

Baue, A.E., Durham, R., and Faist, E. (1998). Systemic inflammatory response syndrome (SIRS), multiple organ dysfunction syndrome (MODS), multiple organ failure (MOF): Are we winning the battle? *Shock* 10:79–89.

Baulieu, E.E., Thomas, G., Legrain, S., Lahlou, N., Roger, M., Debuire, B., Faucounau, V., Girard, L., Hervy, M.P., Latour, F., Leaud, M.C., Mokrane, A., Pitti-Ferrandi, H., Trivalle, C., de Lacharriere, O., Nouveau, S., Rakoto-Arison, B., Souberbielle, J.C., Raison, J., Le Bouc, Y., Raynaud, A., Girerd, X., and Forette, F. (2000). Dehydroepiandrosterone (DHEA), DHEA sulfate, and aging: Contribution of the DHEAge Study to a sociobiomedical issue. *Proc. Natl. Acad. Sci. U.S.A.* 97:4279–4284.

Benten, W.P., Lieberherr, M., Stamm, O., Wrehlke, C., Guo, Z., and Wunderlich, F. (1999). Testosterone signaling through internalizable surface receptors in androgen receptor-free macrophages. *Mol. Biol. Cell* 10:3113–3123.

Bone, R.C. (1992). Toward an epidemiology and natural history of SIRS (systemic inflammatory response syndrome). *JAMA* 268:3452–3455.

Bowles, B.J., Roth, B., and Demetriades, D. (2003). Sexual dimorphism in trauma? A retrospective evaluation of outcome. *Injury* 34:27–31.

Brodie, A.M., and Njar, V.C. (2000). Aromatase inhibitors and their application in breast cancer treatment. *Steroids* 65:171–179.

Brown, G.A., Vukovich, M.D., Sharp, R.L., Reifenrath, T.A., Parsons, K.A., and King, D.S. (1999). Effect of oral DHEA on serum testosterone and adaptations to resistance training in young men. *J. Appl. Physiol.* 87:2274–2283.

Case, J., and Davison, C.A. (1999). Estrogen alters relative contributions of nitric oxide and cyclooxygenase products to endothelium-dependent vasodilation. *J. Pharmacol. Exp. Ther.* 291:524–530.

Catania, R.A., Angele, M.K., Ayala, A., Cioffi, W.G., Bland, K.I., and Chaudry, I.H. (1999). Dehydroepiandrosterone restores immune function following trauma-haemorrhage by a direct effect on T lymphocytes. *Cytokine* 11:443–450.

Chaudry, I.H., Ayala, A., Ertel, W., and Stephan, R.N. (1990). Hemorrhage and resuscitation: Immunological aspects. *Am. J. Physiol.* 259:R663–R678.

Chaudry, I.H., Samy, T.S., Schwacha, M.G., Wang, P., Rue, L.W., III, and Bland, K.I. (2003). Endocrine targets in experimental shock. *J. Trauma* 54:S118–S125.

Choudhry, M.A., Schwacha, M.G., Matsutani, T., Bland, K.I., and Chaudry, I.H. (2003). Cellular, molecular and sexual dimorphic response to trauma-hemorrhage. In M. Ogawa, H. Yamamoto, and M. Hirota (eds.), *The Biological Response to Planned and Unplanned Injuries: Cellular, Molecular and Genetic Aspects.* New York: Elsevier, pp. 25–38.

Choudhry, M.A., Rana, S.N., Kavanaugh, M.J., Kovacs, E.J., Gamelli, R.L., and Sayeed, M.M. (2004). Impaired intestinal immunity and barrier function: A cause for enhanced bacterial translocation in alcohol intoxication and burn injury. *Alcohol* 33:199–208.

Croce, M.A., Fabian, T.C., Malhotra, A.K., Bee, T.K., and Miller, P.R. (2002). Does gender difference influence outcome? *J. Trauma* 53:889–894.

Diodato, M.D., Knoferl, M.W., Schwacha, M.G., Bland, K.I., and Chaudry, I.H. (2001). Gender differences in the inflammatory response and survival following haemorrhage and subsequent sepsis. *Cytokine* 14:162–169.

Eachempati, S.R., Hydo, L., and Barie, P.S. (1999). Gender-based differences in outcome in patients with sepsis. *Arch. Surg.* 134:1342–1347.

Erlandsson, M.C., Ohlsson, C., Gustafsson, J.A., and Carlsten, H. (2001). Role of oestrogen receptors alpha and beta in immune organ development and in oestrogen-mediated effects on thymus. *Immunology* 103:17–25.

Eskandari, F., and Sternberg, E.M. (2002). Neural-immune interactions in health and disease. *Ann. N.Y. Acad. Sci.* 966:20–27.

Eskandari, F., Webster, J.I., and Sternberg, E.M. (2003). Neural immune pathways and their connection to inflammatory diseases. *Arthritis Res. Ther.* 5:251–265.

Flores, E., Bratoeff, E., Cabeza, M., Ramirez, E., Quiroz, A., and Heuze, I. (2003). Steroid 5alpha-reductase inhibitors. *Mini. Rev. Med. Chem.* 3:225–237.

Furr, B.J. (1996). The development of Casodex (bicalutamide): Preclinical studies. *Eur. Urol.* 29(Suppl 2):83–95.

Gannon, C.J., Pasquale, M., Tracy, J.K., McCarter, R.J., and Napolitano, L.M. (2004). Male gender is associated with increased risk for postinjury pneumonia. *Shock* 21:410–414.

George, R.L., McGwin, G., Jr., Metzger, J., Chaudry, I.H., and Rue, L.W., III (2003a). The association between gender and mortality among trauma patients as modified by age. *J. Trauma* 54:464–471.

George, R.L., McGwin, G., Jr., Windham, S.T., Melton, S.M., Metzger, J., Chaudry, I.H., and Rue, L.W., III (2003b). Age-related gender differential in outcome after blunt or penetrating trauma. *Shock* 19:28–32.

Hierholzer, C., Kalff, J.C., Omert, L., Tsukada, K., Loeffert, J.E., Watkins, S.C., Billiar, T.R., and Tweardy, D.J. (1998). Interleukin-6 production in hemorrhagic shock is accompanied by neutrophil recruitment and lung injury. *Am. J. Physiol.* 275:L611–L621.

Jarrar, D., Wang, P., Cioffi, W.G., Bland, K.I., and Chaudry, I.H. (2000a). Mechanisms of the salutary effects of dehydroepiandrosterone after trauma-hemorrhage: Direct or indirect effects on cardiac and hepatocellular functions? *Arch. Surg.* 135:416–422.

Jarrar, D., Wang, P., Cioffi, W.G., Bland, K.I., and Chaudry, I.H. (2000b). The female reproductive cycle is an important variable in the response to trauma-hemorrhage. *Am. J. Physiol. Heart Circ. Physiol.* 279:H1015–H1021.

Jarrar, D., Wang, P., Knoferl, M.W., Kuebler, J.F., Cioffi, W.G., Bland, K.I., and Chaudry, I.H. (2000c). Insight into the mechanism by which estradiol improves organ functions after trauma-hemorrhage. *Surgery* 128:246–252.

Jarrar, D., Wang, P., Song, G. Y., Knoferl, M. W., Cioffi, W. G., Bland, K. I., and Chaudry, I. H. (2000d). Metoclopramide: A novel adjunct for improving cardiac and hepatocellular functions after trauma-hemorrhage. *Am. J. Physiol. Endocrinol. Metab.* 278:E90–E95.

Jarrar, D., Kuebler, J. F., Wang, P., Bland, K. I., and Chaudry, I. H. (2001). DHEA: A novel adjunct for the treatment of male trauma patients. *Trends Mol. Med.* 7: 81–85.

Jarrar, D., Kuebler, J. F., Rue, L. W., III, Matalon, S., Wang, P., Bland, K. I., and Chaudry, I. H. (2002). Alveolar macrophage activation after trauma-hemorrhage and sepsis is dependent on NF-kappaB and MAPK/ERK mechanisms. *Am. J. Physiol. Lung Cell Mol. Physiol.* 283:L799–L805.

Jarrar, D., Song, G. Y., Kuebler, J. F., Rue, L. W., Bland, K. I., and Chaudry, I. H. (2004). The effect of inhibition of a major cell signaling pathway following trauma hemorrhage on hepatic injury and interleukin 6 levels. *Arch. Surg.* 139:896–901.

Kahlke, V., Angele, M. K., Ayala, A., Schwacha, M. G., Cioffi, W. G., Bland, K. I., and Chaudry, I. H. (2000a). Immune dysfunction following trauma-haemorrhage: Influence of gender and age. *Cytokine* 12:69–77.

Kahlke, V., Angele, M. K., Schwacha, M. G., Ayala, A., Cioffi, W. G., Bland, K. I., and Chaudry, I. H. (2000b). Reversal of sexual dimorphism in splenic T lymphocyte responses after trauma-hemorrhage with aging. *Am. J. Physiol. Cell Physiol.* 278: C509–C516.

Katzenellenbogen, B. S., and Katzenellenbogen, J. A. (2000). Estrogen receptor transcription and transactivation: Estrogen receptor alpha and estrogen receptor beta: Regulation by selective estrogen receptor modulators and importance in breast cancer. *Breast Cancer Res.* 2:335–344.

Kerger, H., Waschke, K. F., Ackern, K. V., Tsai, A. G., and Intaglietta, M. (1999). Systemic and microcirculatory effects of autologous whole blood resuscitation in severe hemorrhagic shock. *Am. J. Physiol.* 276:H2035–H2043.

Kincade, P. W., Medina, K. L., Payne, K. J., Rossi, M. I., Tudor, K. S., Yamashita, Y., and Kouro, T. (2000). Early B-lymphocyte precursors and their regulation by sex steroids. *Immunol. Rev.* 175:128–137.

Knoferl, M. W., Diodato, M. D., Angele, M. K., Ayala, A., Cioffi, W. G., Bland, K. I., and Chaudry, I. H. (2000). Do female sex steroids adversely or beneficially affect the depressed immune responses in males after trauma-hemorrhage? *Arch. Surg.* 135:425–433.

Knoferl, M. W., Jarrar, D., Angele, M. K., Ayala, A., Schwacha, M. G., Bland, K. I., and Chaudry, I. H. (2001). 17 beta-Estradiol normalizes immune responses in ovariectomized females after trauma-hemorrhage. *Am. J. Physiol. Cell Physiol.* 281: C1131–C1138.

Knoferl, M. W., Angele, M. K., Schwacha, M. G., Bland, K. I., and Chaudry, I. H. (2002). Preservation of splenic immune functions by female sex hormones after trauma-hemorrhage. *Crit. Care Med.* 30:888–893.

Kolvenbag, G. J., and Nash, A. (1999). Bicalutamide dosages used in the treatment of prostate cancer. *Prostate* 39:47–53.

Kos, M., Denger, S., Reid, G., and Gannon, F. (2002). Upstream open reading frames regulate the translation of the multiple mRNA variants of the estrogen receptor alpha. *J. Biol. Chem.* 277:37131–37138.

Kovacs, E. J., Messingham, K. A., and Gregory, M. S. (2002). Estrogen regulation of immune responses after injury. *Mol. Cell Endocrinol.* 193:129–135.

Kuebler, J.F., Jarrar, D., Wang, P., Bland, K.I., and Chaudry, I.H. (2001). Dehydroepiandrosterone restores hepatocellular function and prevents liver damage in estrogen-deficient females following trauma and hemorrhage. *J. Surg. Res.* 97: 196–201.

Kuebler, J.F., Jarrar, D., Toth, B., Bland, K.I., Rue, L., III, Wang, P., and Chaudry, I.H. (2002). Estradiol administration improves splanchnic perfusion following trauma-hemorrhage and sepsis. *Arch. Surg.* 137:74–79.

Landers, J.P., and Spelsberg, T.C. (1992). New concepts in steroid hormone action: Transcription factors, proto-oncogenes, and the cascade model for steroid regulation of gene expression. *Crit. Rev. Eukaryot. Gene Expr.* 2:19–63.

Lanza, V., Palazzadriano, M., Scardulla, C., Mercadante, S., Valdes, L., and Bellanca, G. (1987). Hemodynamics, prolactin and catecholamine levels during hemorrhagic shock in dogs pretreated with a prolactin inhibitor (bromocriptine). *Pharmacol. Res. Commun.* 19:307–318.

Levine, B., Kalman, J., Mayer, L., Fillit, H.M., and Packer, M. (1990). Elevated circulating levels of tumor necrosis factor in severe chronic heart failure. *N. Engl. J. Med.* 323:236–241.

Luo, S., Martel, C., Chen, C., Labrie, C., Candas, B., Singh, S.M., and Labrie, F. (1997). Daily dosing with flutamide or Casodex exerts maximal antiandrogenic activity. *Urology* 50:913–919.

Marshall, J.C. (1999). Organ dysfunction as an outcome measure in clinical trials. *Eur. J. Surg. Suppl.* 62–67.

Martel, C., Labrie, C., Belanger, A., Gauthier, S., Merand, Y., Li, X., Provencher, L., Candas, B., and Labrie, F. (1998). Comparison of the effects of the new orally active antiestrogen EM-800 with ICI 182 780 and toremifene on estrogen-sensitive parameters in the ovariectomized mouse. *Endocrinology* 139:2486–2492.

McGwin, G., Jr., George, R.L., Cross, J.M., Reiff, D.A., Chaudry, I.H., and Rue, L.W., III (2002). Gender differences in mortality following burn injury. *Shock* 18: 311–315.

Meyers, M.J., Sun, J., Carlson, K.E., Marriner, G.A., Katzenellenbogen, B.S., and Katzenellenbogen, J.A. (2001). Estrogen receptor-beta potency-selective ligands: Structure-activity relationship studies of diarylpropionitriles and their acetylene and polar analogues. *J. Med. Chem.* 44:4230–4251.

Miyao, Y., Yasue, H., Ogawa, H., Misumi, I., Masuda, T., Sakamoto, T., and Morita, E. (1993). Elevated plasma interleukin-6 levels in patients with acute myocardial infarction. *Am. Heart J.* 126:1299–1304.

Mize, A.L., Shapiro, R.A., and Dorsa, D.M. (2003). Estrogen receptor-mediated neuroprotection from oxidative stress requires activation of the mitogen-activated protein kinase pathway. *Endocrinology* 144:306–312.

Mizushima, Y., Wang, P., Jarrar, D., Cioffi, W.G., Bland, K.I., and Chaudry, I.H. (2000). Estradiol administration after trauma-hemorrhage improves cardiovascular and hepatocellular functions in male animals. *Ann. Surg.* 232:673–679.

Morales-Montor, J., Chavarria, A., De Leon, M.A., Del Castillo, L.I., Escobedo, E.G., Sanchez, E.N., Vargas, J.A., Hernandez-Flores, M., Romo-Gonzalez, T., and Larralde, C. (2004). Host gender in parasitic infections of mammals: An evaluation of the female host supremacy paradigm. *J. Parasitol.* 90:531–546.

Muller, V., Losonczy, G., Heemann, U., Vannay, A., Fekete, A., Reusz, G., Tulassay, T., and Szabo, A.J. (2002). Sexual dimorphism in renal ischemia-reperfusion injury in rats: Possible role of endothelin. *Kidney Int.* 62:1364–1371.

Nathens, A.B., and Marshall, J.C. (1996). Sepsis, SIRS, and MODS: What's in a name? *World J. Surg.* 20:386–391.

Neill, J.D. (1970). Effect of "stress" on serum prolactin and luteinizing hormone levels during the estrous cycle of the rat. *Endocrinology* 87:1192–1197.

Neill, J.D. (1972). Sexual differences in the hypothalamic regulation of prolactin secretion. *Endocrinology* 90:1154–1159.

Neill, J.D., and Smith, M.S. (1974). Pituitary-ovarian interrelationships in the rat. *Curr. Top. Exp. Endocrinol.* 2:73–106.

Nestler, J.E., Clore, J.N., and Blackard, W.G. (1991). Metabolism and actions of dehydroepiandrosterone in humans. *J. Steroid Biochem. Mol. Biol.* 40:599–605.

O'Neill, P.J., Ayala, A., Wang, P., Ba, Z.F., Morrison, M.H., Schultze, A.E., Reich, S.S., and Chaudry, I.H. (1994). Role of Kupffer cells in interleukin-6 release following trauma-hemorrhage and resuscitation. *Shock* 1:43–47.

Oberholzer, A., Keel, M., Zellweger, R., Steckholzer, U., Trentz, O., and Ertel, W. (2000). Incidence of septic complications and multiple organ failure in severely injured patients is sex specific. *J. Trauma* 48:932–937.

Occhiato, E.G., Guarna, A., Danza, G., and Serio, M. (2004). Selective non-steroidal inhibitors of 5alpha-reductase type 1. *J. Steroid Biochem. Mol. Biol.* 88:1–16.

Offner, P.J., Moore, E.E., and Biffl, W.L. (1999). Male gender is a risk factor for major infections after surgery. *Arch. Surg.* 134:935–938.

Olsen, N.J., and Kovacs, W.J. (2001). Effects of androgens on T and B lymphocyte development. *Immunol. Res.* 23:281–288.

Orshal, J.M., and Khalil, R.A. (2004). Gender, sex hormones, and vascular tone. *Am. J. Physiol. Regul. Integr. Comp. Physiol.* 286:R233–R249.

Pare, G., Krust, A., Karas, R.H., Dupont, S., Aronovitz, M., Chambon, P., and Mendelsohn, M.E. (2002). Estrogen receptor-alpha mediates the protective effects of estrogen against vascular injury. *Circ. Res.* 90:1087–1092.

Remmers, D.E., Wang, P., Cioffi, W.G., Bland, K.I., and Chaudry, I.H. (1997). Testosterone receptor blockade after trauma-hemorrhage improves cardiac and hepatic functions in males. *Am. J. Physiol.* 273:H2919–H2925.

Remmers, D.E., Cioffi, W.G., Bland, K.I., Wang, P., Angele, M.K., and Chaudry, I.H. (1998a). Testosterone: The crucial hormone responsible for depressing myocardial function in males after trauma-hemorrhage. *Ann. Surg.* 227:790–799.

Remmers, D.E., Wang, P., Cioffi, W.G., Bland, K.I., and Chaudry, I.H. (1998b). Chronic resuscitation after trauma-hemorrhage and acute fluid replacement improves hepatocellular function and cardiac output. *Ann. Surg.* 227:112–119.

Samy, T.S., Schwacha, M.G., Cioffi, W.G., Bland, K.I., and Chaudry, I.H. (2000). Androgen and estrogen receptors in splenic T lymphocytes: Effects of flutamide and trauma-hemorrhage. *Shock* 14:465–470.

Samy, T.S., Knoferl, M.W., Zheng, R., Schwacha, M.G., Bland, K.I., and Chaudry, I.H. (2001). Divergent immune responses in male and female mice after trauma-hemorrhage: Dimorphic alterations in T lymphocyte steroidogenic enzyme activities. *Endocrinology* 142:3519–3529.

Samy, T.S., Zheng, R., Matsutani, T., Rue, L.W., III, Bland, K.I., and Chaudry, I.H. (2003). Mechanism for normal splenic T lymphocyte functions in proestrus females after trauma: Enhanced local synthesis of 17beta-estradiol. *Am. J. Physiol. Cell Physiol.* 285:C139–C149.

Schneider, C.P., Schwacha, M.G., Samy, T.S., Bland, K.I., and Chaudry, I.H. (2003). Androgen-mediated modulation of macrophage function after trauma-hemorrhage: Central role of 5alpha-dihydrotestosterone. *J. Appl. Physiol.* 95:104–112.

Schroder, J., Kahlke, V., Staubach, K.H., Zabel, P., and Stuber, F. (1998). Gender differences in human sepsis. *Arch. Surg.* 133:1200–1205.

Schroder, J., Kahlke, V., Book, M., and Stuber, F. (2000). Gender differences in sepsis: Genetically determined? *Shock* 14:307–310.

Scobie, G.A., Macpherson, S., Millar, M.R., Groome, N.P., Romana, P.G., and Saunders, P.T. (2002). Human oestrogen receptors: Differential expression of ER alpha and beta and the identification of ER beta variants. *Steroids* 67:985–992.

Shimizu, T., Choudhry, M.A., Szalay, L., Rue, L.W., III, Bland, K.I., and Chaudry, I.H. (2004). Salutary effects of androstenediol on cardiac function and splanchnic perfusion after trauma-hemorrhage. *Am. J. Physiol. Regul. Integr. Comp. Physiol.* 287:R386–R390.

Shimizu, T., Szalay, L., Choudhry, M.A., Schwacha, M.G., Rue, L.W., III, Bland, K.I., and Chaudry, I.H. (2005). Mechanism of salutary effects of androstenediol on hepatic function after trauma-hemorrhage: Role of endothelial and inducible nitric oxide synthase. *Am. J. Physiol. Gastrointest. Liver Physiol.* 288:G244–G250.

Smithson, G., Couse, J.F., Lubahn, D.B., Korach, K.S., and Kincade, P.W. (1998). The role of estrogen receptors and androgen receptors in sex steroid regulation of B lymphopoiesis. *J. Immunol.* 161:27–34.

Stauffer, S.R., Coletta, C.J., Tedesco, R., Nishiguchi, G., Carlson, K., Sun, J., Katzenellenbogen, B.S., and Katzenellenbogen, J.A. (2000). Pyrazole ligands: Structure-affinity/activity relationships and estrogen receptor-alpha-selective agonists. *J. Med. Chem.* 43:4934–4947.

Tarnoky, K., and Nagy, S. (1983). Relationship to survival of catecholamine levels and dopamine-beta-hydroxylase activity in experimental haemorrhagic shock. *Acta Physiol. Hung.* 61:59–68.

Tremblay, A., Tremblay, G.B., Labrie, C., Labrie, F., and Giguere, V. (1998). EM-800, a novel antiestrogen, acts as a pure antagonist of the transcriptional functions of estrogen receptors alpha and beta. *Endocrinology* 139:111–118.

Verthelyi, D. (2001). Sex hormones as immunomodulators in health and disease. *Int. Immunopharmacol.* 1:983–993.

Wang, P., Ba, Z.F., Burkhardt, J., and Chaudry, I.H. (1993). Trauma-hemorrhage and resuscitation in the mouse: Effects on cardiac output and organ blood flow. *Am. J. Physiol.* 264:H1166–H1173.

Whitnall, M.H., Wilhelmsen, C.L., McKinney, L., Miner, V., Seed, T.M., and Jackson, W.E., III (2002). Radioprotective efficacy and acute toxicity of 5-androstenediol after subcutaneous or oral administration in mice. *Immunopharmacol. Immunotoxicol.* 24:595–626.

Wichmann, M.W., Zellweger, R., DeMaso, C.M., Ayala, A., and Chaudry, I.H. (1996a). Enhanced immune responses in females, as opposed to decreased responses in males following haemorrhagic shock and resuscitation. *Cytokine* 8:853–863.

Wichmann, M.W., Zellweger, R., DeMaso, C.M., Ayala, A., and Chaudry, I.H. (1996b). Mechanism of immunosuppression in males following trauma-hemorrhage. Critical role of testosterone. *Arch. Surg.* 131:1186–1191.

Wichmann, M.W., Ayala, A., and Chaudry, I.H. (1997). Male sex steroids are responsible for depressing macrophage immune function after trauma-hemorrhage. *Am. J. Physiol.* 273:C1335–C1340.

Wichmann, M.W., Inthorn, D., Andress, H.J., and Schildberg, F.W. (2000). Incidence and mortality of severe sepsis in surgical intensive care patients: The influence of patient gender on disease process and outcome. *Intensive Care Med.* 26:167–172.

Wunderlich, F., Benten, W.P., Lieberherr, M., Guo, Z., Stamm, O., Wrehlke, C., Sekeris, C.E., and Mossmann, H. (2002). Testosterone signaling in T cells and macrophages. *Steroids* 67:535–538.

Xu, Y.X., Ayala, A., and Chaudry, I.H. (1998). Prolonged immunodepression after trauma and hemorrhagic shock. *J. Trauma* 44:335–341.

Yamauchi-Takihara, K., Ihara, Y., Ogata, A., Yoshizaki, K., Azuma, J., and Kishimoto, T. (1995). Hypoxic stress induces cardiac myocyte-derived interleukin-6. *Circulation* 91:1520–1524.

Yang, S., Zheng, R., Hu, S., Ma, Y., Choudhry, M.A., Messina, J.L., Rue, L.W., III, Bland, K.I., and Chaudry, I.H. (2004). Mechanism of cardiac depression after trauma-hemorrhage: Increased cardiomyocyte IL-6 and effect of sex steroids on IL-6 regulation and cardiac function. *Am. J. Physiol. Heart Circ. Physiol.* 287: H2183–H2191.

Yao, Y.M., Redl, H., Bahrami, S., and Schlag, G. (1998). The inflammatory basis of trauma/shock-associated multiple organ failure. *Inflamm. Res.* 47:201–210.

Yokoyama, Y., Schwacha, M.G., Samy, T.S., Bland, K.I., and Chaudry, I.H. (2002). Gender dimorphism in immune responses following trauma and hemorrhage. *Immunol. Res.* 26:63–76.

Zellweger, R., Ayala, A., DeMaso, C.M., and Chaudry, I.H. (1995). Trauma-hemorrhage causes prolonged depression in cellular immunity. *Shock* 4:149–153.

Zellweger, R., Wichmann, M.W., Ayala, A., DeMaso, C.M., and Chaudry, I.H. (1996a). Prolactin: A novel and safe immunomodulating hormone for the treatment of immunodepression following severe hemorrhage. *J. Surg. Res.* 63:53–58.

Zellweger, R., Zhu, X.H., Wichmann, M.W., Ayala, A., DeMaso, C.M., and Chaudry, I.H. (1996b). Prolactin administration following hemorrhagic shock improves macrophage cytokine release capacity and decreases mortality from subsequent sepsis. *J. Immunol.* 157:5748–5754.

Zellweger, R., Wichmann, M.W., Ayala, A., and Chaudry, I.H. (1998). Metoclopramide: A novel and safe immunomodulating agent for restoring the depressed macrophage immune function after hemorrhage. *J. Trauma* 44:70–77.

Zellweger, R., Wichmann, M.W., Ayala, A., Stein, S., DeMaso, C.M., and Chaudry, I.H. (1997). Females in proestrus state maintain splenic immune functions and tolerate sepsis better than males. *Crit. Care Med.* 25:106–110.

Zhang, C.C., and Shapiro, D.J. (2000). Activation of the p38 mitogen-activated protein kinase pathway by estrogen or by 4-hydroxytamoxifen is coupled to estrogen receptor-induced apoptosis. *J. Biol. Chem.* 275:479–486.

Zhang, Z., Maier, B., Santen, R.J., and Song, R.X. (2002). Membrane association of estrogen receptor alpha mediates estrogen effect on MAPK activation. *Biochem. Biophys. Res. Commun.* 294:926–933.

Zheng, R., Samy, T.S., Schneider, C.P., Rue, L.W., III, Bland, K.I., and Chaudry, I.H. (2002). Decreased 5alpha-dihydrotestosterone catabolism suppresses T lymphocyte functions in males after trauma-hemorrhage. *Am. J. Physiol. Cell Physiol.* 282:C1332–C1338.

Zhu, X.H., Zellweger, R., Ayala, A., and Chaudry, I.H. (1996). Prolactin inhibits the increased cytokine gene expression in Kupffer cells following haemorrhage. *Cytokine* 8:134–140.

Part II
Neural and Neuroendocrine Mechanisms in Host Defense: Physiological Effects of Stress on Host Susceptibility to Infection and Autoimmunity

7

The Impact of Psychological Stress on the Immune Response to and Pathogenesis of Herpes Simplex Virus Infection

ROBERT H. BONNEAU AND JOHN HUNZEKER

1. Introduction

There is extensive anecdotal evidence supporting an association between psychological stress and one's susceptibility to a variety of infectious pathogens. These pathogens include a number of viruses with significant short- and long-term health consequences. Infections with one such virus, herpes simplex virus (HSV), have long been recognized to be linked to a variety of life stressors. These HSV infections, both primary and recurrent, have been thought to be, in part, a function of a decreased immune surveillance and a suppression of antiviral immune defense mechanisms. Such stress-induced alterations in immune capacity may be mediated by one or more products of the nervous and endocrine systems. A number of human and animal studies have provided data to support this hypothesis and represent a subset of a larger group of studies that have established a solid link among psychological stress, immune function, and diseases caused by pathogenic microorganisms (reviewed in Moynihan and Ader, 1996; Sheridan et al., 1998; Bonneau et al., 2001; Bailey et al., 2003; Moynihan and Stevens, 2004). Recent advances in experimental immunology have broadened our knowledge of immunological processes that, in turn, have facilitated the design of studies to better examine the interactions among the nervous, endocrine, and immune systems at both the cellular and molecular levels. The information provided in this chapter will review some of the studies that have established a link between stress and HSV, and the impact of this link on human health will be discussed.

1.1. Herpes Simplex Virus

HSV is an icosahedral double-stranded DNA virus belonging to the Herpesviridae family of viruses. Herpes simplex virus type 1 (HSV-1) and type 2 (HSV-2) represent the two antigenically distinct serotypes of HSV.

125

Although HSV-1 is generally considered to be associated with orofacial infections and HSV-2 with genital infections, both serotypes can affect any region of the body. Both HSV-1 and HSV-2 cause a wide variety of clinical syndromes in humans (e.g., herpes labialis, gingivostomatitis, keratoconjunctivitis, herpes genitalis) with the basic lesion being an intraepithelial vesicle from which progeny virus is shed. It is during the primary infection that virus particles first enter sensory nerve endings that innervate the lesion and are then transported to the local sensory ganglia (e.g., trigeminal ganglia for orofacial infections; dorsal root ganglia for genital infections). Once the virus reaches these sensory ganglia, the virus may either produce an acute infection resulting in subsequent neuronal cell death or enter a dormant or latent phase (reviewed in Jones, 2003). During this latent state, there is restricted viral gene expression that is limited to the production of what are referred to as latency-associated transcripts or LATs (Block and Hill, 1997) and the minimal or absent expression of any viral proteins. Although the primary lesions generally resolve as antibody- and cell-mediated immune responses develop, it is in the ganglia where the virus remains for the lifetime of the host. In some of the neurons harboring latent HSV, the virus is periodically reactivated. The resulting infectious HSV virion is then transported intra-axonally to the cells at or near the site of the initial infection (Holland *et al.*, 1999). This viral reactivation may be asymptomatic or, in some cases, result in debilitating lesions. Although the precise mechanisms underlying latent HSV reactivation and the development of recurrent herpetic disease are poorly understood, there is evidence that psychological stress plays a role, perhaps through an alteration in the host's immune competency.

1.2. Immunity to HSV Infection

The role of the immune response in controlling HSV infections is multifaceted and complex. This complexity is a function, in part, of HSV being distributed in both neuronal and extraneuronal sites during the primary and latent stages of infection. In addition, HSV persists within the host even in the presence of concomitant immunity by establishing itself as a latent infection within sensory neurons that innervate the sites of peripheral infection. Despite these challenges, considerable progress has been made in defining the immune components that play a key role in the resolution of an HSV infection and, in the case of adaptive immune responses, in identifying the specific virus-encoded epitopes that are recognized by the B and T lymphocytes. Because a comprehensive discussion of the immune response to HSV infections is beyond the scope to this chapter, the reader is directed to recent reviews on this topic for more detailed information (Koelle and Corey, 2003; Khanna *et al.*, 2004).

1.2.1. Innate and Adaptive Immunity

Both the innate and adaptive arms of the immune response have been shown to play a role in immunity to HSV in both immune and naïve individuals. Furthermore, both arms have been shown to function not only at the site of the primary infection in the periphery but also in the initial control of HSV-1 replication in the sensory ganglia and the establishment of a latent infection. Although the immune response at the primary site of HSV infection has been well characterized, less has been known about the immunological events that occur in the sensory ganglia during latency. However, as described below, there is recent evidence of a role for T lymphocytes in controlling HSV-1 latency (Liu *et al.*, 2000; Khanna *et al.*, 2003, 2004). Such findings have challenged the long-held view that HSV is able to "hide" from the immune system during latency and suggest that the immune system controls recurrent herpetic disease both during active infection in the periphery and during latent infection in the nervous system. Latency represents the stage of infection during which stress, in part, may be associated with HSV reactivation and recurrent disease.

The innate immune system possesses two primary functions in mediating a defense against pathogenic microorganisms: the direct destruction of the pathogens and the initiation of specific types of adaptive immunity that will follow. Studies in both humans and animals have been valuable in deciphering the components of the innate immune response that play important roles in defense against HSV. Innate immune components that are elicited in response to HSV-1 infection include: the activation of the complement cascade; the activation of macrophages; and the recruitment, activation, and maturation of natural killer (NK) cells, dendritic cell precursors, and γ/δ extrathymic-derived T cells. Activation of the innate immune response also results in the generation and secretion of a variety of cytokines and chemokines that orchestrate the above activities. These responses are designed to limit the initial infection and to abrogate further virus propagation. As equally important is the role of these responses in the generation of the adaptive immune response, which is comprised of the activation of both HSV-specific helper and cytotoxic T cells as well as the maturation of B cells for HSV-specific antibody production.

The adaptive immune response to HSV infection has been shown to play a key role in primary and, more recently, in latent HSV infections. The production of HSV-specific antibodies has been known for quite some time. These antibodies function in both a virus neutralizing capacity as well as in conjunction with other cell types (e.g., polymorphonuclear leukocytes, monocytes, and NK-like cells) in mediating antibody-dependent cell-mediated cytotoxicity (ADCC). There is also compelling evidence that antibody alone is able to provide protection in neonates exposed to HSV at or near the time of birth (Brown *et al.*, 1991). However, despite the presence of high levels of circulating HSV-specific antibodies in latently infected adults, HSV reactivation and recrudescent herpetic disease appear to go unchecked. More recently, a number of studies have focused on the specific roles of CD4$^+$ and CD8$^+$ T lymphocytes in anti-HSV immunity and the specificity of these T-lymphocyte responses. There are several lines of evidence indicating that both of these subsets of T lymphocytes are functionally important in controlling HSV infections. CD8$^+$ T cells mediate their protective ability via their cytotoxic effects on virus-infected target cells and their synthesis of cytokines—cytokines that may be able to inhibit viral replication. The recent finding of a role for CD8$^+$ T cells for controlling HSV reactivation in latently infected ganglia (Khanna et al., 2003, 2004) is intriguing since it was once thought that latently infected cells were essentially hidden from immunosurveillance mechanisms. CD4$^+$ T cells are important in that they release numerous cytokines that can both orchestrate the inflammatory response that is associated with HSV infection as well as play a key role in determining the magnitude of the CD8$^+$ T-cell response. Overall, components of the adaptive immune response have the potential to control HSV infections and thus may form the basis for the development of an effective anti-HSV vaccine.

1.2.2. Memory Cytotoxic T Lymphocytes

An individual who has previously recovered from a primary HSV infection exhibits an adaptive immune response upon HSV reinfection that differs primarily in its rapidity and magnitude as compared with an individual who has never been infected. Given the relative insignificance of high titers of HSV-specific antibody in the development of recurrent herpetic diseases in adults, most attention has focused on memory T lymphocyte responses, particularly the cytotoxic T lymphocyte (CTL) responses.

Memory cytotoxic T lymphocytes (CTLm) play an important role in mediating long-term protective immunity against a variety of intracellular pathogens (reviewed in Doherty et al., 1992; Sher and Coffman, 1992; Gray, 1993; Mackay, 1993; Pamer, 1993; Ahmed, 1994; Sprent, 1994). Although these CTLm do not prevent recurrence or reinfection per se, they do limit the severity of infection/disease by becoming activated in an accelerated fashion following reexposure to the particular pathogen (Gray, 1993; Mackay, 1993; Ahmed, 1994; Sprent, 1994). This activation occurs as memory T cells move continually from blood to tissues, scanning cell surfaces for the appropriate antigen to which they are programmed to recognize. Upon an encounter with this antigen, CTLm are readily triggered to become effector CTL. In the case of viral infections, the encounter of the CTLm with a virus-specific antigen triggers these memory CTL to proliferate rapidly, differentiate into effector cells, and localize to the site of infection. It is here where these CTL destroy virus-infected cells, thus eliminating the virus "factories" that produce infectious viral progeny. These CTL also synthesize cytokines such as IFN-γ and TNF-α that have both a direct antiviral activity as well as the ability to enhance components of antigen processing and presentation.

A role of CTLm in mediating protection against recurrent HSV infections has been documented in both humans and in animals (Schmid and Rouse, 1992; Rinaldo and Torpey, 1993; Ashley *et al.*, 1994; Koelle *et al.*, 1998). Although HSV reactivation occurs quite frequently, it usually does not lead to an apparent clinical infection (Wald *et al.*, 1995, 1997) especially when the immune system is intact. Our knowledge of CTL-mediated defense against HSV has grown significantly. This knowledge, driven by advances in experimental immunology, has provided a solid foundation on which to decipher the mechanisms underlying the impact of stress on the immunity to HSV and to other viral infections.

1.3. Psychological Stress and Immunity

Stress can be defined as a state of altered homeostasis resulting from either an external or an internal stimulus. The host's response to stress is designed to restore homeostasis through a variety of adaptive neuroendocrine-mediated mechanisms (Ramsey, 1982). This restoration of homeostasis is accomplished, in part, through the increased synthesis of a variety of neuroendocrine-derived peptides and hormones. For example, the perception of a psychological stressor activates the hypothalamic-pituitary-adrenal (HPA) axis and the sympathetic branch of the autonomic nervous system. This activation culminates in the synthesis of both cortisol (corticosterone in rodents) by the adrenal glands and epinephrine and norepinephrine by the sympathetic nervous system. Interestingly, in addition to their roles in maintaining homeostasis, these and other products of the nervous and endocrine system pathways can also modulate immune function through their binding to specific receptors that are expressed by the many cell types that comprise the immune system. As a result, most aspects of immune function, including immunity to HSV, have the potential to be modulated under conditions of psychological stress.

Although studies in humans have established a clear relationship between stress and susceptibility to HSV infection, these studies have not been able to address each of the components of the immune response that underlie this relationship. However, the use of animal models has provided the opportunity to determine these immune components given the ability to establish an HSV infection and assess the responses at various sites throughout the body. Using a murine model, the effects of stress on susceptibility to HSV infections have been known for nearly 50 years (Rasmussen *et al.*, 1957). Although these studies were important in establishing an association between stress and viral infection, they were unable to identify specific antiviral immune defense mechanisms that were modulated by stress and that contributed to viral pathogenesis, because the components and mechanisms of antiviral immune responses were unknown at that time. Since then, however, significant progress has been made in understanding the complex immune interactions that are necessary for the development of protective immunity to viral infections, including HSV. Thus, it is now possible to study the effects of stress on defined components of the antiviral immune response.

A number of animal-based studies have been conducted in an attempt to decipher the impact of psychological stress on immune function. Although many of the early studies had simply focused on aspects of non-specific immune function such as lymphocyte proliferation and cytokine production, more recent studies have examined the effects of stress on the specific immune response to viral pathogens including HSV (Bonneau *et al.*, 1991a, 1991b; Bonneau *et al.*, 1993; Dobbs *et al.*, 1993; Bonneau, 1996; Bonneau *et al.*, 1997; Brenner and Moynihan, 1997; Glaser and Kiecolt-Glaser, 1997; Pariante *et al.*, 1997; Bonneau *et al.*, 1998; Carr *et al.*, 1998; Delano and Mallery, 1998; Leo *et al.*, 1998; Noisakran *et al.*, 1998; Padgett *et al.*, 1998a; Cruess *et al.*, 2000; Leo and Bonneau, 2000a, 2000b; Buske-Kirschbaum *et al.*, 2001; Wonnacott and Bonneau, 2002; Anglen *et al.*, 2003; Ortiz *et al.*, 2003), influenza virus (Sheridan *et al.*, 1991; Feng *et al.*, 1991; Hermann *et al.*, 1993, 1994a, 1994b, 1995; Dobbs *et al.*, 1996; Padgett *et al.*, 1998b; Cohen *et al.*, 1999; Sheridan *et al.*, 2000; Konstantinos and Sheridan, 2001; Hunzeker *et al.*, 2004), adenovirus (Cohen *et al.*, 1997), pseudorabies virus (deGroot *et al.*, 1999a, 1999b), and Theiler's virus (Johnson *et al.*, 2004; Mi *et al.*, 2004; Sieve *et al.*, 2004; Welsh *et al.*, 2004). Some of these latter studies have also focused on the contribution of glucocorticoids and sympathetic nervous system products and have used a variety of stressors including physical restraint, social reorganization, social disruption, hypothermia, and acoustic stress to induce increased neuroendocrine levels. Despite the known widespread effects of stress on antiviral immunity, the studies described in this chapter focus specifically on the impact of stress on immunity to HSV.

2. Stress and the Innate Immune Response to HSV Infection

2.1. Inflammatory Response

An inflammatory response to many pathogens is generally characterized by redness, swelling, heat, and pain at the site of infection. These symptoms depend on a constellation of events including, but not limited to, proin-flammatory cytokines (e.g., IL-1, IL-6, tumor necrosis factor-α), and leukocyte trafficking. There have been numerous studies that have examined the influence of psychological stress on the early, inflammatory events that are initiated upon HSV infection. For example, using a murine model, restraint stress was shown to decrease both type I and II interferon production in response to a primary, HSV-1 dermal infection (Ortiz *et al.*, 2003). Furthermore, this decrease in interferon production correlated with an increase in virus titer at the site of primary infection. In another study, hyperthermic stress was shown to increase the levels of both IL-6 mRNA and the protein itself in the trigeminal ganglia of HSV-1 latently infected mice (Noisakran *et al.*, 1998). The observed stress-induced increase in the expression of this proinflammatory cytokine was shown to be mediated by glucocorticoids, because the administration of cyanoketone, a corticosterone synthesis inhibitor, blocked this stress-induced increase in IL-6 (Noisakran *et al.*, 1998). More importantly, additional studies demonstrated a potential role for stress-induced increases in IL-6 in HSV pathogenesis. In these studies, latently infected mice, as a result of a previous corneal infection, demonstrated reactivated virus after exposure to hyperthermic stress (Kriesel *et al.*, 1997), and this reactivation was due, in part, to IL-6. In contrast, other studies (Bonneau *et al.*, 1998) have shown that stress, administered in the form of restraint during the primary footpad HSV infection, actually suppresses splenic IL-6 production in an adrenal-dependent manner.

The above studies suggest that psychological stress can modulate the proinflammatory response to an HSV infection. The precise effects of stress on this response may depend on a number of variables including the location at which the response is being measured (peripheral, lymphoid, central nervous system) and the stage of infection (primary vs. latent) of the response. However, it is important to note that, in some circumstances, an inflammatory response may be desirable, whereas in other circumstances it may result in immunopathology and thus be detrimental to the host.

2.2. Leukocyte Trafficking and Recruitment

Leukocyte recruitment to sites of inflammation is crucial for initiating immune-mediated control of a viral infection. An inhibition of leukocyte trafficking may impair the ability of the immune system to control viral replication, thus leading to an increase in the pathology associated with viral replication. Restraint stress has been shown to alter the pattern of leukocyte recruitment and retention. For example, during a primary intranasal HSV infection, restraint stress delays both CD4+ and CD8+ T-cell recruitment into the brain of mice with symptoms of HSV encephalitis (Anglen *et al.*, 2003). In a corneal infection model, hyperthermic stress was found to reduce the numbers of NK cells in the trigeminal ganglia of mice that were latently infected with HSV-1 (Noisakran *et al.*, 1998). Restraint stress was found to suppress lymphadenopathy in HSV-1 infected mice. This suppression was glucocorticoid dependent as receptor blockade with the glucocorticoid receptor antagonist RU486 restored lymph node cellularity (Sheridan *et al.*, 1998). In contrast, studies by Dhabhar and colleagues (Dhabhar and

McEwen, 1997, 1999) have demonstrated that an acute restraint stress session (2 h) can actually enhance a lymphocyte trafficking–dependent delayed-type hypersensitivity (DTH) response in an adrenal-dependent manner. However, such a stress-induced enhancement of cell trafficking has yet to be demonstrated when using an HSV model. Alterations in leukocyte trafficking induced by psychological stress may have a profound impact on controlling viral replication and disease pathogenesis by altering the available pool of responding cells (e.g. CD8[+] T cells), leading to prolonged viral replication and increased morbidity and mortality. Additional studies are needed to support this hypothesis.

2.3. Cells of the Innate Immune System

The nonspecific innate immune response to a viral infection consists of a variety of different cell types that contribute to the early defense against infection. To date, only a few studies have examined the influence of stress on these cell types in the context of an HSV infection.

Macrophages produce cytokines that influence the adaptive immune response, may present antigen to B and T lymphocytes, and eliminate virus infected cells. The role of macrophages during a primary HSV infection is not completely understood. However, both adoptive transfer and depletion studies have shown that macrophages can indeed contribute to host resistance to an HSV infection (Morahan and Morse, 1979; Pinto *et al.*, 1997). In an *ex vivo* activation assay, Koff and Dunegan (1986) demonstrated that the sympathetic nervous system products epinephrine and norepinephrine can reduce the ability of macrophages to lyse HSV-1-infected target cells. In support of these *in vitro* findings, Davis and colleagues have recently shown that exercise stress can reduce the anti-HSV activity of macrophages (Davis *et al.*, 2004). However, the effects of stress-induced alterations in macrophage activity on HSV pathogenesis has not yet been determined.

Natural killer (NK) cells serve as a bridge between the innate and adaptive arms of the immune response by limiting viral replication as well as producing cytokines that aid in the adaptive response. As with macrophages, the role of the NK cell during a primary HSV infection is not completely understood. For example, conflicting studies have suggested that NK cells are not necessary (Bukowski and Welsh, 1986), whereas others have shown that NK cells are required for the resolution of an HSV infection (Biron *et al.*, 1989). The differences between these studies may primarily be due to species differences (human vs. mouse). Consistent with these conflicting reports, stress has been shown to have differing effects on NK cell activity during a primary HSV infection. For example, restraint stress has been shown to reduce splenic NK cell lytic activity after an intradermal HSV-1 infection (Bonneau *et al.*, 1991a). However, in a model of exercise stress, running to fatigue did not have any effect on splenic NK cell activity (Davis *et al.*, 2004). Additional studies need to be conducted to better elucidate the role of NK cells during a primary infection and how stress may affect their antiviral activities. In interpreting the findings of these and other studies, it is important not to generalize among studies but instead to consider differences such as the species being examined, the site of HSV infection, and the stage of infection (e.g., primary, latent, recurrent) being examined.

3. Stress and the Adaptive Immune Response to HSV Infection

3.1. Humoral Immunity

As noted above, antibodies play a role in clearing HSV infections in a direct manner by viral neutralization and by acting in concert with other cells (e.g., NK cells) to reduce the viral load. As a result of these activities, antibodies have the potential to significantly reduce the extent of viral spread and pathogenesis. To date, most of the studies that have examined the influence of psychological stress on the B cells during an HSV infection have focused on measuring antibody titers. For example, as compared with individually housed mice, group-housed mice were shown to produce lower levels of some, but not all, cytokines in response to HSV-1 infection; the kinetics of this response was not altered. However, these group housing conditions were shown not to alter levels of circulating IgM or IgG antibody. Therefore, these stress–induced changes in cytokine production did not translate into altered circulating IgM or IgG antibody titers (Karp et al., 1997). In contrast, another study showed that footshock stress altered cytokine production and increased circulating anti-HSV IgM titers in response to a subcutaneous HSV challenge (Brenner and Moynihan, 1997). These footshock stress–induced increases in antibody titers were found in both BALB/c and C57BL/6 mice. In contrast, another study showed that footshock stress administered at the time of HSV-1 challenge via footpad scarification reduced IgM titers (Kusnecov et al., 1992). The differences in the results obtained between these studies may be primarily due to differences in shock intensity, suggesting that the more stressful the event, the more likely that immunosuppression is to occur. The observed relationship between stress and anti-HSV antibodies in murine models has also been extended to studies in humans. For example, Glaser and Kiecolt-Glaser (1997) demonstrated that caregivers of individuals suffering from dementia had increased antibody titers to whole viral antigen but exhibited no difference in neutralizing antibody titers to a latent HSV-1 infection.

The ability of an individual to generate HSV-specific antibodies after a primary infection may be important, to a limited extent, in protecting the host against reinfection either as a consequence of peripheral reexposure to HSV or from reactivation of the latent virus. Although a less common route of transmission, the acquisition of HSV by a neonate, most often from their mother during delivery, is still a significant health concern and results in a number of Cesarean sections annually. In the case of neonatal HSV infections, the transfer of HSV-specific antibody from mother to child is extremely important in protecting the health of the newborn infant. Thus, any change in this transfer via either the placental (prenatal transfer) or transmammary (postnatal transfer) routes could compromise the resistance of the newborn to herpetic infection. Using a murine model, recent studies from our laboratory provided evidence that a previous HSV-2 infection in female mice protects their offspring against HSV-2 infection (Yorty and Bonneau, 2003, 2004a) and that this protection is associated with the transfer of high titers of HSV-specific antibodies. Furthermore, these studies demonstrated that an acute stressor does not affect either transmammary (Yorty and Bonneau 2004b) or placental transfer of anti-HSV antibodies to newborn mice despite high circulating levels of corticosterone (Yorty and Bonneau, 2004a). In contrast, sustained elevated levels of corticosterone in the mother immediately after parturition diminishes the serum and milk levels of HSV-2 antibodies and thus reduces the amount of antibody available to the neonates. This corticosterone-mediated reduction in antibody transfer is accompanied by increased HSV-2 susceptibility in neonatal mice who were nursed by these mothers (Yorty *et al.*, 2004).

The above findings demonstrate that how stress modulates the immune response depends on a variety of factors, including the type of stressor as well as the stage of infection (e.g., primary or latent). Furthermore, these studies demonstrate that stress can affect both the active generation as well as the passive transfer of antibodies.

3.2. Cellular Immunity

In contrast with the humoral immune response to HSV infection, there has been significant focus on the study of the effects of stress on the T-lymphocyte response. One reason for this focus is the essential role that T cells play in controlling viral replication during the infection, as noted above. Initial studies of the effect of stress on the primary immune response to HSV infection demonstrated that stress, applied in the form of physical restraint, suppresses the lymphoproliferative response in the popliteal lymph nodes after footpad infection with HSV (Bonneau *et al.*, 1991a; Dobbs *et al.*, 1993). Using this model, it was also shown that stress suppresses the extent of HSV-specific CTL and natural killer (NK) cell activity generated in response to local HSV infection (Bonneau *et al.*, 1991a). This suppression in the generation of CTL was shown to occur early in the sequence of events that are involved in the differentiation and maturation of HSV-specific precursor CTL (CTLp) to a phenotypically lytic state (Bonneau *et al.*, 1991a). Stress-induced suppression of lymphproliferation and HSV-specific CTL activity was also seen in studies in which a footshock model of stress was employed (Kusnecov *et al.*, 1992). By using a combination of surgical (adrenalectomy) and pharmaceutical approaches (β-adrenergic and glucocorticoid receptor antagonists,corticosterone-releasing pellets), a role for both adrenal-dependent and catecholamine-mediated mechanisms in stress-induced suppression of the primary cellular immune response to HSV infection was defined (Bonneau *et al.*, 1993; Dobbs *et al.*, 1993). Lymphocyte trafficking was found to be corticosterone dependent as the administration of RU486 restored their accumulation. CD8[+] T-cell lytic activity was restored by addition of nadalol, a nonspecific adrenergic receptor antagonist. These results demonstrate that psychological stress can affect all aspects of T-cell function, including proliferation, trafficking, and lytic activity. Given the role that T cells play in controlling HSV infection, stress-induced modulation of their function should result in prolonged viral replication. Paradoxically, there may also be decreased disease pathogenesis associated with stress, because the immune response itself can contribute significantly to the virus-associated damage. The development of this immunopathology may be a function of the stage of infection during which the T-cell response is activated.

Recent studies in our laboratory have begun to examine the impact of corticosterone, a major stress-associated hormone, on dendritic cell function. Dendritic cells are specialized for the uptake, transport, processing, and MHC class I–restricted presentation of antigen to naïve CD8[+] T cells—critical events that are an absolute requirement for the generation of both effector and memory CD8[+] cytotoxic T lymphocyte (CTL) responses. Dendritic cells play a pivotal role in generating these CD8[+] T-cell immune responses by presenting MHC class I–peptide complexes on their surface via the classical MHC class I pathway. We have shown (Truckenmiller *et al.*, 2005) that physiologically relevant stress levels of corticosterone, acting via the glucocorticoid receptor, suppresses the formation of specific peptide–MHC class I complexes on the surface of virus-infected dendritic cells. We also determined that the mechanism of this suppression is via the action of corticosterone on components of the class I pathway involved in the processing of protein to form antigenic peptide. This corticosterone-mediated suppression of complexes on dendritic cells also results in a marked reduction in their ability to activate a specific T-cell hybridoma, thus suggesting that a suppression of complex formation may have a functional effect *in vivo* by suppressing the induction of HSV-specific CTL.

As illustrated above, the impact of stress on the functioning of dendritic cells and other antigen-presenting cells in T-cell activation is an important component in our understanding of stress-neuroendocrine-immune interactions. Recent studies from our laboratory have expanded this aspect of T-cell activation by examining the effects of stress and corticosterone on the proliferation and function of microglia in the brains of both noninfected and HSV-infected mice (Nairand Bonneau 2006; Nair and Bonneau, unpublished observations). Microglia are thought to function as a type of antigen-presenting cell within the brain. Briefly, these studies have demonstrated that stress-associated corticosterone increases the numbers and activation status of microglia in the brain. These findings suggest the stress may be able to induce a proinflammatory response within the central nervous system that, in turn, may be a contributing factor in the development of various stress-induced inflammatory conditions in the central nervous system, including those caused by HSV infection. A better understanding of the interrelationship between antigen presentation and T-cell activation in not only secondary lymphoid tissues (e.g., spleen, lymph nodes) and the brain but also in other sites within the body will provide the knowledge to better define stress-induced, neuroendocrine-mediated modulation of the cellular immune response.

4. The Impact of Stress on the Pathogenesis of HSV Infection

4.1. Introduction

HSV is a natural pathogen of humans and, as noted above, is characterized by the ability to cause an acute infection at a peripheral site and to establish a latent infection in the local sensory ganglia that innervate the site of the initial infection (Blyth and Hill, 1984). The hallmark of HSV is its ability to spontaneously reactivate from this quiescent, noninfectious latent state and cause a recurrent infection in the periphery, at, or near, the site of the original infection (Hill, 1985). In humans, HSV reactivation and recurrent infections typically occur spontaneously. However, correlations have been made between reactivation and physical or emotional stress, fever, exposure to ultraviolet light, tissue damage, and immune suppression (Stevens, 1975; Hill, 1985). More recent efforts by us and others have focused on the use of mouse models to decipher the roles of psychological stress and its associated neuroendocrine components in the development of both primary and recurrent HSV infection. These models have provided a foundation for understanding the effects of stress on HSV infection.

4.2. The Impact of Stress on Primary HSV Infection

The ability to limit the amount of HSV replication during a primary infection may, in turn, reduce the amount of latent virus that colonizes the local sensory ganglia (Bonneau and Jennings, 1989). Therefore, by destroying HSV-infected cells before progeny virions can be produced may actually limit the extent of neuronal involvement and reduce the amount of latent virus within these ganglia. Therefore, the impact of a stressor on the immune response during the primary HSV infection may have consequences on not only one's initial encounter with HSV but also on the development of recurrent HSV infections throughout the lifetime of the host.

Studies in humans have indicated that stress can affect the frequency, severity, and duration of HSV infection. A number of murine studies during the past 15 years have substantiated these findings by providing compelling experimental evidence that stress and stress-induced, neuroendocrine-derived products increase the development and severity of HSV infection in both the periphery (Bonneau et al., 1991a; Kusnecov et al., 1992; Bonneau et al., 1997; Brenner and Moynihan, 1997; Wonnacott and Bonneau, 2002) and the central nervous system (Anglen et al., 2003). For example, mouse models in which either a restraint (Bonneau et al., 1991a; Wonnacott and Bonneau, 2002; Ortiz et al., 2003) or footshock (Kusnecov et al., 1992) model of stress were used were shown to result in increased HSV titers at a site of local, HSV infection. Stress was also shown to suppress the ability of adoptively transferred lymphocytes with HSV-specific lytic activity to confer protection against lethal HSV infection in an immunocompromised host as well as inhibiting the restoration of immune responsiveness to HSV infection after sublethal gamma irradiation (Bonneau et al., 1997). These latter findings are important because adoptive immunotherapy represents a potentially effective approach by which to control the extent of viral infections in an immunocompromised host. Other studies using the glucocorticoid antagonist androstenediol (AED) have shown that AED can significantly improve the survival of mice with HSV-1–induced encephalitis. Moreover, it was shown that this increased level of protection was mediated by an augmentation in type I interferon production (Daigle and Carr, 1998). As noted above, recent studies from our laboratory have focused on the impact of maternal stress on the transfer of antibody from mother to fetus/neonate. In these studies, we have demonstrated that increased levels of postpartum maternal corticosterone increases neonatal susceptibility to HSV-2–associated infection by reducing the amount of HSV-specific antibody transferred via the transmammary route (Yorty et al., 2004b).

As was described earlier, these studies have shown that stress suppresses components of the primary CTL responses to HSV infection (Bonneau et al., 1991a; Kusnecov et al., 1992; Bonneau et al., 1993; Dobbs et al., 1993; Bonneau, 1997; Brenner and Moynihan, 1997; Leo et al., 1998; Leo and Bonneau, 2000a,b) and memory (Bonneau et al., 1991b; Bonneau, 1996, 1998; Leo et al., 1998; Leo and Bonneau, 2000a; Wonnacott and Bonneau, 2002). By using a combination of surgical and pharmacological approaches, a role for both the HPA axis (Bonneau et al., 1993, 1998) and the sympathetic nervous system (Leo and Bonneau, 2000a,b) has clearly been demonstrated as one of the underlying mechanisms mediating stress-induced modulation of immunity and HSV-associated pathology.

4.3. The Impact of Stress on HSV Reactivation and Recurrent HSV Infection

4.3.1. Human Studies

In all but a few cases, a primary infection with HSV resolves within a few weeks and without any long-term consequences. However, the ability of HSV to establish itself in a latent state and to reactivate periodically during the lifetime of the host and cause recurrent lesions is a significant challenge to the development of therapeutic strategies against HSV. There has long been evidence that acute and chronic psychosocial stress can trigger recurrences of both oral and genital infections. Support for this hypothesis has been provided by studies that have demonstrated that emotional stress is associated with the development of recurrent HSV infections (Katcher et al., 1973; Young et al., 1976; Friedmann et al., 1977; Bierman, 1983; Schmidt et al., 1985; Glaser et al., 1987) and that personality may play a role in the frequency of these recurrences (Stout and Bloom, 1986). The severity of HSV infection has also been correlated with psychosocial factors (Silver et al., 1986; Longo and Clum, 1989), which suggests that the control and resolution of recurrent HSV lesions by the memory components of the immune response may be compromised. There is also evidence that social support can moderate the relationship between stress and HSV infection (Glaser et al., 1985; VanderPlate et al., 1988) and that high levels of depressive mood may result in a higher rate of HSV recurrence (Kemeny et al., 1989). More recent studies have shown that stress may be a significant predictor of recurrent genital herpes in HIV-infected women (Pereira et al., 2003). Together, these studies strongly support the existence of a relationship between psychological stress and the development of herpesvirus infections.

4.3.2. Animal Studies

It has long been known that psychological stress is associated with bouts of recurrent HSV infection. However, only recently has such a relationship been demonstrated in mice and has the immunological and neuroendocrine mechanisms underlying virus reactivation and recrudescent infection been investigated. For many years, it has been known that local trauma (e.g., hair plucking, tape stripping, chemical-induced trauma, ultraviolet light) administered to a previously infected area can result in HSV reactivation and recurrent disease. However, the ability of each of these methods to consistently induce recurrent disease is quite low (Hurd and Robinson, 1977; Hill et al., 1978; Harbour et al., 1983; Hill et al., 1983).

The immune system of the mouse is very well characterized and is similar, in many respects, to that of humans. This similarity, coupled with the wide and continually expanding variety of reagents for studying an immune response in mice has continued to make the mouse the focal point of immunological studies. However, the lack of a mouse model of *spontaneous* reactivation has limited the usefulness of mice in the study of HSV reactivation and the associated immune response. Despite this limitation, two models of psychological stress have been shown to be effective in inducing HSV reactivation and recurrent disease. These models have been useful in investigating the neuroendocrine mechanisms that underlie behaviorally mediated reactivation of latent HSV infection. For example, the use of a hyperthermic stress model (Sawtell and Thompson, 1992; Thompson and Sawtell, 1997; Sawtell, 1998) has been used to show that HPA axis activation plays an important role in HSV-1 reactivation in the trigeminal ganglion and that this reactivation may be associated with an increase in IL-6 expression in the ganglia itself (Noisakran *et al.*, 1998). In addition, the use of a social stress model in mice has been shown to be effective in causing recurrent HSV infection in mice (Padgett *et al.*, 1998a). This correlation between neuroendocrine activity, immune function, and latent HSV reactivation (Halford *et al.*, 1996; Carr *et al.*, 1998; Noisakran *et al.*, 1998; Padgett *et al.*, 1998a) is particularly interesting in light of recent findings that HSV-specific memory CD8[+] T cells are selectively activated and retained in latently infected sensory ganglia (Khanna *et al.*, 2003). Expanding these studies to determine the neuroendocrine mechanisms mediating these effects are in progress.

5. The Impact of Stress on the Memory Immune Response to HSV Infection

Immunological memory is a key component of the overall immune response and is important in reducing the severity of recurrent HSV infection. Because recovery from recurrent HSV infections is largely controlled by T lymphocytes, the ability to generate a population of HSV-specific CTLm during the initial infection with HSV is thought to be critical for the long-term defense against recurrent HSV infection. The psychological stress that is often associated with the primary episode of infection, as well as the development of recurrent disease, emphasizes the potential significance of the effect of stress on the development of HSV-specific T cell-mediated immunity.

In studies to date, restraint stress has been shown not to suppress the generation of CTLm in response to systemic HSV infection. However, such a stressor has been shown to be effective at inhibiting the activation of CTLm (Bonneau *et al.*, 1991b, 1996; Leo *et al.*, 1998) and their migration to the site of recurrent HSV infection (Bonneau *et al.*, 1991b). Further studies into the mechanisms underlying the suppression of CTLm activation revealed that stress does not the suppress the expression of the T-cell receptor, IL-2 receptor, and other accessory molecules required for T-lymphocyte activation. However, it does suppress the production of cytokines involved in HSV-specific CTLm activation, which is likely to be the mechanism underlying stress-induced suppression of CTLm activation (Bonneau *et al.*, 1996). These studies were later extended to show that products of adrenal function associated with stress are responsible for suppression of cytokine production and the development of lytic activity, but they do not affect splenic cellularity (Bonneau *et al.*, 1997). As in the studies of the primary immune response, adrenal-dependent mechanisms do not appear to be solely responsible for stress-induced suppression of memory responses to HSV infection.

Although much has been learned about the effects of stress on the generation and activation of mCTL, there is nothing known about the effects of stress, either acute or chronic, on the long-term maintenance of memory T cells *in vivo* and their ability to survey the body for sites of HSV infections.

6. The Role of Stress in Preventing HSV Infections

Given the role that the immune system plays in the resistance to many viral infections, including HSV, vaccination remains an often desired strategy by which to confer protection. However, identifying the salient immune components that are involved in this protection may represent an obstacle to the design of an effective vaccine. The complex nature of HSV infections may also provide a formidable challenge to the development of a vaccine that would be effective at all stages of infection and at multiple sites within the body. Even if a vaccine were to be developed, the fact that stress has been shown to suppress the induction of vaccine-induced immunity (Glaser *et al.*, 1992; Kiecolt-Glaser *et al.*, 1996; Vedhara *et al.*, 1999; Glaser *et al.*, 2000; Burns *et al.*, 2002; Miller *et al.*, 2004) suggests that not all individuals who are vaccinated may necessarily be protected. Moreover, the impact of stress on the reactivation of immunological memory may suppress vaccine-mediated protection against infection.

Despite long-term efforts, there is currently not a vaccine to prevent either a primary HSV infection or an infection arising from reactivation of HSV from the latent state. However, there are a variety of therapeutic agents that have been designed to interfere directly with viral replication (acyclovir, penciclovir), are converted *in vivo* to acyclovir or penciclovir (valacyclovir HCl and famciclovir, respectively), or prevent the virus from fusing to cell membranes, thus barring virus entry into the cell (docosanol). These drugs may be beneficial by reducing the severity and shortening the course of recurrent infections and may aid in lowering the incidence of future outbreaks by reducing the amount of latent virus in the neural ganglia. However, they do not prevent recurrences entirely and therefore do not eliminate the possibility of spreading an HSV infection from one individual to another. However, a reduced viral load and shorter duration of infection may reduce this possibility to some extent. The effectiveness of these drugs may be enhanced by an effectively functioning immune system. Thus, the effects of stress on immune function may be important even when drug therapy is the chosen treatment strategy.

Any effective strategy to eliminate HSV infections from the general population will need to not only prevent a primary infection but also the establishment of a latent infection. For those individuals already harboring latent HSV, the ability of this strategy to either eliminate the latent virus, inhibit its reactivation, and/or block the development of recrudescent disease after reactivation is desirable. However, it is unlikely that any one treatment strategy will be effective in accomplishing each of the above objectives in preventing HSV infection.

What role could controlling stress play in the prevention of recurrent HSV? First, despite the many facets of an HSV infection, it is clear that the immune system plays a role, to some degree, in one's ability to prevent and/or recover from infection. Second, psychological stress has been shown to be intimately linked to the development of recurrent HSV infections— most likely through the suppression of both HSV-specific and nonspecific immune defense mechanisms. As is outlined above, a number of life events, some of which are inherently stressful, have been shown to be associated with outbreaks of recurrent HSV infection. Such events may, in part, dictate why some individuals have frequent reactivation events and other individuals have few. Therefore, limiting the degree of psychological stress may, in some cases, be adequate for maintaining a sufficient immune defense against HSV infection. This limiting of stress may be accomplished though a variety of behavioral modification and social support mechanisms.

7. Concluding Remarks

Much has recently been learned about how the innate and adaptive immune components serve to either eliminate or control many viral infections. Although a role for most components of the immune response have been shown to contribute to the control of HSV, how each of these components function in a coordinated fashion to prevent and/or eliminate HSV infections remains to be fully elucidated. This is particularly true as the components of the immune system become better defined (e.g., subsets of T helper cells), enjoy renewed attention (e.g., regulatory T cells), or are dissected at a molecular level (e.g., T cell activation). New methodologies in detecting and quantifying antigen-specific immune responses will also make it possible to better define the components of the immune system that are activated in response to HSV infection and their roles in either preventing or, in some cases, eliciting the pathology that is associated with HSV infections. It is this knowledge that continues to provide the foundation that is necessary for the development of an effective vaccine against HSV. Together, an increased knowledge of both innate and adaptive immune components will also allow us to further define the relationship between psychological stress and the interactions among the nervous, endocrine, and immune systems.

The information presented in this chapter has focused exclusively on the effects of stress on HSV infection. However, it is important to remember that stress and its associated neuroendocrine products represent only one of many factors that can modulate immune function and, thus, one's susceptibility to HSV infection. Therefore, other products of the nervous and endocrine systems, such as those associated with gender and metabolic disease conditions (e.g., diabetes), should be evaluated for their effects on immune function. The impact of genetics on neuroendocrine-immune interactions also remains a largely unexplored area that could have a significant impact on our overall understanding of the effects of stress on antiviral immunity. Technological advances in the areas of genomics and proteomics provide an impetus to better understand the relationship among psychological stress, neuroendocrine activation, and immune function.

The impact of stress on the immune response to and pathogenesis of HSV infection represents only one small component of the widespread interest and need to decipher the cellular and molecular mechanisms underlying stress-neuroendocrine-immune interactions. A better understanding of these mechanisms will allow us to understand this relationship as it pertains to not only HSV infections but also to other diseases that may be immunologically resisted, such as those caused by a variety of pathogenic microorganisms or even cancer.

References

Ahmed, R. (1994). Viral persistence and immune memory. *Semin. Virol.* 5:319–324.

Anglen, C.S., Truckenmiller, M.E., Schell, T.D., and Bonneau, R.H. (2003). The dual role of CD8[+] T lymphocytes in the development of stress-induced herpes simplex virus encephalitis. *J. Neuroimmunol.* 140:13–27.

Ashley, R., Wald, A., and Corey, L. (1994). Cervical antibodies in patients with oral herpes simplex virus type-1 (HSV-1) infection: Local anamnestic responses after genital HSV-2 infection. *J. Virol.* 68:5284–5286.

Bailey, M., Engler, H., Hunzeker, J., and Sheridan, J.F. (2003). The hypothalamic-pituitary-adrenal axis and viral infection. *Viral Immunol.* 16:141–157.

Bierman, S.M. (1983). A proposed biologic cure for recurrent genital herpes simplex through injection of neurolytic agents into cutaneous sensory nerves. *Med. Hypotheses.* 10:97–103.

Biron, C.A., Byron, K.S., and Sullivan, J.L. (1989). Severe herpesvirus infections in an adolescent without natural killer cells. *N. Engl. J. Med.* 320:1731–1735.

Block, T.M., and Hill, J.M. (1997). The latency associated transcripts (LAT) of herpes simplex virus: still no end in sight. *J. Neurovirol.* 3:313–321.

Blyth, W.A., and Hill, T.J. (1984). Establishment, maintenance, and control of herpes simplex virus (HSV). Latency. In B.T. Rouse, and C. Lopez (eds.), *Immunobiology of Herpes Simplex Virus Infection.* Boca Raton, FL: CRC Press, pp. 9–32.

Bonneau, R.H. (1996). Stress-induced effects on integral immune components involved in herpes simplex virus (HSV)-specific memory cytotoxic T lymphocyte activation. *Brain Behav. Immun.* 10:139–163.

Bonneau, R.H., and Jennings, S.R. (1989). Modulation of acute and latent herpes simplex virus infection in C57BL/6 mice by adoptive transfer of immune lymphocytes with cytolytic activity. *J. Virol.* 63:1480–1484.

Bonneau, R.H., Sheridan, J.F., Feng, N., and Glaser, R. (1991a). Stress-induced suppression of herpes simplex virus (HSV)-specific cytotoxic T lymphocyte and natural killer cell activity and enhancement of acute pathogenesis following local HSV infection. *Brain Behav. Immun.* 5:170–192.

Bonneau, R.H., Sheridan, J.F., Feng, N., and Glaser, R. (1991b). Stress-induced effects on cell-mediated innate and adaptive memory components of the murine immune response to local and systemic herpes simplex virus (HSV) infection. *Brain Behav. Immun.* 5:274–295.

Bonneau, R.H., Sheridan, J.F., Feng, N., and Glaser, R. (1993). Stress-induced modulation of the primary cellular immune response to herpes simplex virus infection is mediated by both adrenal-dependent and adrenal-independent mechanisms. *J. Neuroimmunol.* 42:167–176.

Bonneau, R.H., Brehm, M.A., and Kern, A.M. (1997). The impact of psychological stress on the efficacy of anti-viral adoptive immunotherapy in an immunocompromised host *J. Neuroimmunol.* 78:19–33.

Bonneau, R.H., Zimmerman, K.M., Ikeda, S.C., and Jones, B.C. (1998), Differential effects of stress-induced adrenal function on components of the herpes simplex virus-specific memory cytotoxic T lymphocyte response. *J. Neuroimmunol.* 82: 199–207.

Bonneau, R.H., Padgett, D.A., and Sheridan, J.F. (2001). Psychoneuroimmune interactions in infectious disease: Studies in animals. In R. Ader, D.L. Felten, and N. Cohen (eds.), *Psychoneuroimmunology*, 3rd ed. San Diego: Academic Press, pp. 483–497.

Brenner, G.J., and Moynihan, J.A. (1997). Stressor-induced alterations in immune response and viral clearance following infection with herpes simplex virus-type 1 in BALB/c and C57BL/6 mice. *Brain Behav. Immun.* 11:9–23.

Brown, Z.A., Benedetti, J., Ashley, R., Burchett, S., Selke, S., Berry, S., Vontver, L.A., and Corey, L. (1991). Neonatal herpes simplex virus infection in relation to asymptomatic maternal infection at the time of labor. *N. Engl. J. Med.* 324: 1247–1252.

Bukowski, J.F., and Welsh, R.M. (1986). The role of natural killer cells and interferon in resistance to acute infection of mice with herpes simplex virus type 1. *J. Immunol.* 136:3481–3485.

Burns, V.E., Drayson, M., Ring, C., and Carroll, D. (2002). Perceived stress and psychological well-being are associated with antibody status after meningitis C conjugate vaccination. *Psychosom. Med.* 64:963–970.

Buske-Kirschbaum, A., Geiben, A., Wermke, C., Pirke, K.M., and Hellhammer, D. (2001). Preliminary evidence for Herpes labialis recurrence following experimentally induced disgust. *Psychother. Psychosom.* 70:86–91.

Carr, D.J., Noisakran, S., Halford, W.P., Lukacs, N., Asensio.V., and Campbell, I.L. (1998). Cytokine and chemokine production in HSV-1 latently infected trigeminal ganglion cell cultures: effects of hyperthermic stress *J. Neuroimmunol.* 85: 111–121.

Cohen, S., Line, S., Manuck, S.B., Rabin, B.S., Heise, E.R., and Kaplan, J.R. (1997). Chronic social stress, social status, and susceptibility to upper respiratory infections in nonhuman primates. *Psychosom. Med.* 59:213–221.

Cohen, S., Doyle, W.J., and Skoner, D.P. (1999). Psychological stress, cytokine production, and severity of upper respiratory illness. *Psychosom. Med.* 61:175–180.

Cruess, S., Antoni, M., Cruess, M., Fletcher, D., Ironson, M.A., Kumar, G., Lutgendorf, M., Hayes, S., Klimas, A., and Schneidermann, N. (2000). Reductions in herpes simplex virus type 2 antibody titers after cognitive behavioral stress management and relationships with neuroendocrine function, relaxation skills, and social support in HIV-positive men. *Psychosom. Med.* 62:828–837.

Daigle, J., and Carr, D.J. (1998). Androstenediol antagonizes herpes simplex virus type 1-induced encephalitis through the augmentation of type I IFN production. *J. Immunol.* 160:3060–3066.

Davis, J.M., Murphy, E.A., Brown, A.S., Carmichael, M.D., Ghaffar, A., and Mayer, E.P. (2004). Effects of oat beta-glucan on innate immunity and infection after exercise stress. *Med. Sci. Sports Exerc.* 36:1321–1327.

de Groot, J., Moonen-Leusen, H.W., Thomas, G., Bianchi, A.T., Koolhaas, J.M., and van Milligen, F.J. (1999a). Effects of mild stress on the immune response against pseudorabies virus in mice. *Vet. Immunol. Immunopathol.* 67:153–160.

de Groot, J., van Milligen, F.J., Moonen-Leusen, B.W., Thomas, G., and Koolhaas, J.M. (1999b). A single social defeat transiently suppresses the anti-viral immune response in mice. *J. Neuroimmunol.* 95:143–151.

DeLano, R.M., and Mallery, S.R. (1998). Stress-related modulation of central nervous system immunity in a murine model of herpes simplex encephalitis. *J. Neuroimmunol.* 89:51–58.

Dhabhar, F.S., and McEwen, B.S. (1997). Acute stress enhances while chronic stress suppresses cell-mediated immunity in vivo: A potential role for leukocyte trafficking. *Brain Behav. Immun.* 4:286–306.

Dhabhar, F.S., and McEwen, B.S. (1999). Enhancing versus suppressive effects of stress hormones on skin immune function *Proc. Natl. Acad. Sci. U.S.A.* 96:1059–1064.

Dobbs, C.M., Vasquez, M., Glaser, R., and Sheridan, J.F. (1993). Mechanisms of stress-induced modulation of viral pathogenesis and immunity. *J. Neuroimmunol.* 48:151–160.

Dobbs, C.M., Feng, N., Beck, F.M., and Sheridan, J.F. (1996). Neuroendocrine regulation of cytokine production during experimental influenza viral infection: Effects of restraint stress-induced elevation in endogenous corticosterone. *J. Immunol.* 157:1870–1877.

Doherty, P.C., Allan, W., Eichelberger, M., and Carding, S.R. (1992). Roles of α, β, and γ/δ T cell subsets in viral immunity. *Ann. Rev. Immunol.* 10:123–151.

Feng, N., Pagniano, R., Tovar, C.A., Bonneau, R.H., Glaser, R., and Sheridan, J.F. (1991). The effect of restraint stress on the kinetics, magnitude, and isotype of the humoral immune response to influenza virus infection. *Brain Behav. Immun.* 5:370–382.

Friedman, E., Katcher, A.H., and Brightman, V.J. (1977). Incidence of recurrent herpes labialis and upper respiratory infection: A prospective study of the influence of biological, social and psychological predictors. *Oral Surg. Oral Med. Oral Pathol.* 43:873–878.

Glaser, R., and Kiecolt-Glaser, J.K. (1997). Chronic stress modulates the virus-specific immune response to latent herpes simplex virus type 1. *Annal. Behav. Med.* 19:78–82.

Glaser, R., Kiecolt-Glaser, J.K., Speicher, C.E., and Holliday, J.E. (1985). Stress, loneliness, and changes in herpesvirus latency. *J. Behav. Med.* 8:249–260.

Glaser, R., Rice, J., Sheridan, J., Fertel, R., Stout, J., Speicher, C., Pinsky, D., Kotur, M., Post, A., Beck, M., and Kiecolt-Glaser, J.K. (1987). Stress-related immune suppression: health implications. *Brain Behav. Immun.* 1:7–20.

Glaser, R., Kiecolt-Glaser, J.K., Bonneau, R.H., Malarkey, W., and Hughes, J. (1992). Stress-induced modulation of the immune response to recombinant hepatitis B vaccine. *Psychosom. Med.* 54:22–29.

Glaser, R., Sheridan, J.F., Malarkey, W.B., MacCallum, R.C., and Kiecolt-Glaser, J.K. (2000). Chronic stress modulates the immune response to a pneumococcal pneumonia vaccine. *Psychosom. Med.* 62:804–807.

Gray, D. (1993). Immunological memory. *Ann. Rev. Immunol.* 11:49–77.

Harbour, D.A., Hill, T.J., and Blyth, W.A. (1983). Recurrent herpes simplex in the mouse: Inflammation in the skin and activation of virus in the ganglia following peripheral stimulation. *J. Gen. Virol.* 64:1491–1498.

Hermann, G., Tovar, C.A., Beck, F.M., Allen, C., and Sheridan, J.F. (1993). Restraint stress differentially affects the pathogenesis of an experimental influenza viral infection in three inbred strains of mice. *J. Neuroimmunol.* 47:83–94.

Hermann, G., Tovar, C.A., Beck, F.M., and Sheridan, J.F. (1994a). Kinetics of glucocorticoid response to restraint stress and/or experimental influenza viral infection in two inbred strains of mice. *J. Neuroimmunol.* 49:25–33.

Hermann, G., Beck, F.M., Tovar, C.A., Malarkey, W.B., Allen, C., and Sheridan, J.F. (1994b). Stress-induced changes attributable to the sympathetic nervous system during experimental influenza viral infection in DBA/2 inbred mouse strain. *J. Neuroimmunol.* 53:173–180.

Hermann, G., Beck, F.M., and Sheridan, J.F. (1995). Stress-induced glucocorticoid response modulates mononuclear cell trafficking during an experimental influenza viral infection. *J. Neuroimmunol.* 56:179–186.

Hill, T.J. (1985). Herpes simplex virus latentcy. In B. Roizman (ed.), *The Herpesviruses*. New York: Plenum, pp. 175–240.

Hill, T.J., Blyth, W.A., and Harbour, D.A. (1978). Trauma to the skin causes recurrence of herpes simplex in the mouse. *J. Gen. Virol.* 39:21–28.

Holland, D.J., Miranda-Saksena, M., Boadle, R.A., Armati, P., and Cunningham, A.L. (1999). Anterograde transport of herpes simplex virus proteins in axons of peripheral human fetal neurons: an immunoelectron microscopy study. *J. Virol.* 73:8503–8511.

Hunzeker, J., Padgett, D.A., Sheridan, P.A., Dhabhar, F.S., and Sheridan, J.F. (2004). Modulation of natural killer cell activity by restraint stress during an influenza A/PR8 infection in mice. *Brain Behav. Immun.* 18:526–535.

Hurd, J., and Robinson, T.W. (1977). Herpes virus reactivation in a mouse model. *J. Antimicrob. Chemother.* 3:99–106.

Johnson, R.R., Storts, R., Welsh, T.H., Jr., Welsh, C.J., and Meagher, M.W. (2004). Social stress alters the severity of acute Theiler's virus infection. *J. Neuroimmunol.* 148:74–85.

Jones, C. (2003). Herpes simplex virus type 1 and bovine herpesvirus 1 latency. *Clin. Microbiol. Rev.* 6:79–95.

Karp, J.D., Moynihan, J.A., and Ader, R. (1997). Psychosocial influences on immune responses to HSV-1 infection in BALB/c mice. *Brain Behav. Immun.* 11:47–62.

Katcher, A.H., Brightman, V., Luborsky, L., and Ship, I. (1973). Prediction of the incidence of recurrent herpes labialis and systemic illness from psychological measurements. *J. Dent. Res.* 52:49–58.

Khanna, K.M., Bonneau, R.H., Kinchington, P.R., and Hendricks, R.L. (2003). Herpes simplex virus-specific memory CD8+ T cells are selectively activated and retained in latently infected sensory ganglia. *Immunity* 18:593–603.

Khanna, K.M., Lepisto, A.J., Decman, V., and Hendricks, R.L. (2004). Immune control of herpes simplex virus during latency. *Curr. Opin. Immunol.* 16:463–469.

Kiecolt-Glaser, J.K., Glaser, R., Gravenstein, S., Malarkey, W.B., and Sheridan, J. (1996). Chronic stress alters the immune response to influenza virus vaccine in older adults. *Proc. Nat. Acad. Sci. U.S.A.* 93:3043–3047.

Koelle, D.M., and Corey, L. (2003). Recent progress in herpes simplex virus immunobiology and vaccine research. *Clin. Microbiol. Rev.* 16:96–113.

Koelle, D.M., Posavad, C.M., Barnum, G.R., Johnson, M.L., Frank, J.M., and Corey, L. (1998). Clearance of HSV-2 from recurrent genital lesions correlates with infiltration of HSV-specific cytotoxic T lymphocytes. *J. Clin. Invest.* 101:1500–1508.

Koff, W.C., and Dunegan, M.A. (1986). Neuroendocrine hormones suppress macrophage-mediated lysis of herpes simplex virus-infected cells. *J. Immun.* 136: 705–709.

Konstantinos, A.P., and Sheridan, J.F. (2001). Stress and influenza viral infection: Modulation of proinflammatory cytokine responses in the lung. *Respir. Physiol.* 128:71–77.

Kriesel, J.D., Gebhardt, B.M., Hill, J.M., Maulden, S.A., Hwang, I.P., Clinch, T.E., Cao, X., Spruance, S.L., and Araneo, B.A. (1997). Anti-interleukin-6 antibodies inhibit herpes simplex virus reactivation. *J. Infect. Dis.* 175:821–827.

Kusnecov, A.V., Grota, L.J., Schmidt, S.G., Bonneau, R.H., Sheridan, J.F., Glaser, R., and Moynihan, J.A. (1992) Decreased herpes simplex virus immunity and enhanced pathogenesis following stressor administration in mice. *J. Neuroimmunol.* 38:129–138.

Leo, N.A., and Bonneau, R.H. (2000a). Chemical sympathectomy alters cytotoxic T lymphocyte responses to herpes simplex virus infection. *Ann. N.Y. Acad. Sci.* 840: 803–808.

Leo, N.A., and Bonneau, R.H. (2000b). Mechanisms underlying chemical sympathectomy-induced suppression of herpes simplex virus-specific cytotoxic T lymphocyte activation and function. *J. Neuroimmunol.* 110:45–56.

Leo, N.A., Callahan, T.A., and Bonneau, R.H. (1998). Peripheral sympathetic denervation alters both the primary and memory cellular immune response to herpes simplex virus infection. *Neuroimmunomodulation* 5:22–35.

Liu, T., Khanna, K.M., Chen, X., Fink, D.J., and Hendricks, R.L. (2000). CD8(+) T cells can block herpes simplex virus type 1 (HSV-1) reactivation from latency in sensory neurons. *J. Exp. Med.* 191:1459–1466.

Longo, D.J., and Clum, G.A. (1989). Psychosocial factors affecting genital herpes recurrences: Linear vs mediating models. *J. Psychosom. Res.* 33:161–166.

Mackay, C.R. (1993). Immunological memory. *Adv. Immunol.* 53:217–265.

Mi, W., Belyavskyi, M., Johnson, R.R., Sieve, A.N., Storts, R., Meagher, M.W., and Welsh, C.J. (2004). Alterations in chemokine expression following Theiler's virus infection and restraint stress. *J. Neuroimmunol.* 151:103–115.

Miller, G.E., Cohen, S., Pressman, S., Barkin, A., Rabin, B.S., and Treanor J.J. (2004). Psychological stress and antibody response to influenza vaccination: When is the critical period for stress, and how does it get inside the body? *Psychosom. Med.* 66:215–223.

Morahan, P.S., and Morse, S. (1979). Macrophage-virus interactions. In M. Proffitt (ed.), *Virus-Lymphocyte Interactions: Implications for Disease.* New York: Elsevier, pp. 17–35.

Moynihan, J.A., and Ader, R. (1996). Psychoneuroimmunology: Animal models of disease. *Psychosom. Med.* 58:546–558.

Moynihan, J.A., and Stevens, S.Y. (2001). Mechanisms of stress-induced modulation in animals In R. Ader, D.L. Felten, and N. Cohen (eds.), *Psychoneuroimmunology*, 3rd ed., San Diego: Academic Press, pp. 227–250.

Nair, A., and Bonneau, R.H. (2006). Stress-induced elevation of glucocorticoids increases microglia proliferation through NMDA receptor activation. *J. Neuroimmunol,* 171:72–85.

Noisakran, S., Halford, W.P., and Carr, D.J. (1998). Role of the hypothalamic pituitary adrenal axis and IL-6 in stress-induced reactivation of latent herpes simplex virus type 1. *J. Immunol.* 160:5441–5447.

Ortiz, G.C., Sheridan, J.F., and Marucha, P.T. (2003). Stress-induced changes in pathophysiology and interferon gene expression during primary HSV-1 infection. *Brain Behav. Immun.* 17:329–338.

Padgett, D.A., Sheridan, J.F., Dorne, J., Berntson, G.G., Candelora, J., and Glaser, R. (1998a). Social stress and the reactivation of latent herpes simplex virus type 1. *Proc. Natl. Acad. Sci. U. S. A.* 95:7231–7235.

Padgett, D.A., MacCallum, R.C., and Sheridan, J.F. (1998b). Stress exacerbates age-related decrements in the immune response to an experimental influenza viral infection. *J. Gerontol. A. Biol. Sci. Med. Sci.* 53:B347–353.

Pamer, E.G. (1993). Cellular immunity to intracellular bacteria. *Curr. Opin. Immunol.* 5:492–496.

Pariante, C.M., Carpiniello, B., Orru, M.G., Sitzia, R., Piras, A., Farci, A.M., DelGiacco, G.S., Piludu, G., and Miller, A. H. (1997). Chronic caregiving stress alters peripheral blood immune parameters: The role of age and severity of stress. *Psychother. Psychosom.* 66:199–207.

Pereira, D.B., Antoni, M.H., Danielson, A., Simon, T., Efantis-Potter, J., Carver, C.S., Duran, R.E., Ironson, G., Klimas, N., Fletcher, M.A., and O'Sullivan, M.J. (2003). Stress as a predictor of symptomatic genital herpes virus recurrence in women with human immunodeficiency virus. *J. Psychosom. Res.* 54:237–244.

Pinto, A.J., Stewart, D., van Rooijen, N., and Morahan, P.S. (1991). Selective depletion of liver and splenic macrophages using liposomes encapsulating the drug dichloromethylene diphosphonate: Effects on antimicrobial resistance. *J. Leukoc. Biol.* 49:579–586.

Ramsey, J.M. (1982). *Basic Pathophysiology: Modern Stress and the Disease Process.* Menlo Park, CA; Addison-Wesley, pp. 30–73.

Rasmussen, A.F., Marsh, J.T., and Brill, N.O. (1957). Increased susceptibility to herpes simplex in mice subjected to avoidance-learning stress or restraint. *Proc. Soc. Exp. Biol. Med.* 96:183–189.

Rinaldo, C.R., and Torpey, D.J. (1993). Cell-mediated immunity and immunosuppression in herpes simplex virus infection. *Immunodeficiency.* 5:33–90.

Sawtell, N.M., and Thompson, R.L. (1992). Rapid in vivo reactivation of herpes simplex virus type 1 in latently infected murine ganglionic neurons after transient hyperthermia. *J. Virol.* 66:2150–2156.

Sawtell, N.M. (1998). The probability of in vivo reaction herpes simplex virus type 1 increases with the number of latently infected neurons in the ganglia. *J. Virol.* 72:6888–6892.

Schmid, D.S., and Rouse, B.T. (1992). The role of T cell immunity in control of herpes simplex virus. *Curr. Top. Microbiol. Immunol.* 179:57–74.

Schmidt, D.D., Zyzanski, S., Ellner, J., Kumar, M.L., and Arno, J. (1985). Stress as a precipitating factor in subjects with recurrent herpes labialis. *J. Fam. Pract.* 20: 359–366.

Sher, A., and Coffman, R.L. (1992). Regulation of immunity to parasites by T cells and T cell-derived cytokines. *Ann. Rev. Immunol.* 10:385–409.

Sheridan, J.F., Feng, N.G., Bonneau, R.H., Allen, C.M., Huneycutt, B.S., and Glaser, R. (1991). Restraint stress differentially affects anti-viral cellular and humoral immune responses in mice. *J. Neuroimmunol.* 31:245–255.

Sheridan, J.F., Dobbs, C., Jung, J., Chu, X., Konstantinos, A., Padgett D., and Glaser R. (1998). Stress-induced neuroendocrine modulation of viral pathogenesis and immunity. *Ann. N.Y. Acad. Sci.* 840:803–808.

Sheridan, J.F., Stark, J.L., Avitsur, R., and Padgett, D.A. (2000). Social disruption, immunity, and susceptibility to viral infection. Role of glucocorticoid insensitivity and NGF. *Ann. N.Y. Acad. Sci.* 917:894–905.

Sieve, A.N., Steelman, A.J., Young, C.R., Storts, R., Welsh, T.H., Welsh, C.J., and Meagher, M.W. (2004). Chronic restraint stress during early Theiler's virus infection exacerbates the subsequent demyelinating disease in SJL mice. *J. Neuroimmunol.* 155:103–118.

Silver, P.S., Auerbach, S.M., Vishniavsky, N., Kaplowitz, L.G. (1986). Psychological factors in recurrent genital herpes infection: stress, coping style, social support, emotional dysfunction, and symptom recurrence. *J. Psychosom. Res.* 30:163–171.

Sprent, J. (1994). T and B memory cells. *Cell* 76:315–322.

Stout, C.W., and Bloom L.J. (1986). Genital herpes and personality. *J. Human Stress* 12:119–124.

Stevens, J.G. (1975). Latent herpes simplex virus and the nervous system. *Curr. Top. Microbiol. Immunol.* 70:31–50.

Thompson, R.L., and Sawtell, N.M. (1997). The herpes simplex virus and type 1 latency-associated transcript gene regulates the establishment of latency. *J. Virol.* 71:5432–5440.

Truckenmiller, M.E., Princiotta, M.F., Norbury, C.C., and Bonneau, R.H. (2005). Corticosterone impairs MHC class I antigen presentation by dendritic cells via reduction of peptide generation. *J. Neuroimmunol.* 160:48–60.

Vedhara, K. Cox, N.K., Wilcock, G.K., Perks, P., Hunt, M., Anderson, S., Lightman, S.L., Shanks, N.M. (1999). Chronic stress in elderly caregivers of dementia patients and antibody response to influenza vaccination. *Lancet* 353:627–631.

Wald, A., Zeh, J., Selke, S., Ashley, R.L., and Corey, L. (1995). Virologic characteristics of subclinical and symptomatic genital herpes infections. *N. Engl. J. Med.* 333: 770–775.

Wald, A., Corey, L., Cone, R., Hobson, A., Davis, G., and Zeh, J. (1997). Frequent genital herpes simplex virus 2 shedding in immunocompetent women: Effect of acyclovir treatment. *J. Clin. Invest.* 99:1092–1097.

Welsh, C.J., Bustamante, L., Nayak, M., Welsh, T.H., Dean, D.D., and Meagher, M.W. (2004). The effects of restraint stress on the neuropathogenesis of Theiler's virus infection II: NK cell function and cytokine levels in acute disease. *Brain Behav. Immun.* 18:166–174.

Wonnacott, K.M., and Bonneau, R.H. (2002). The effects of stress on memory cytotoxic T lymphocyte (CTLm)-mediated protection against herpes simplex virus (HSV) infection at mucosal sites. *Brain Behav. Immun.* 16:104–117.

Yorty, J.L., and Bonneau, R.H. (2003). Transplacental transfer and subsequent neonate utilization of herpes simplex virus-specific immunity is resilient to acute maternal stress. *J. Virol.* 77:6613–6619.

Yorty, J.L., and Bonneau, R.H. (2004a). Prenatal transfer of low amounts of herpes simplex virus (HSV)-specific antibody protects newborn mice against HSV infection during acute maternal stress. *Brain Behav. Immun.* 18:15–23.

Yorty, J.L., and Bonneau, R.H. (2004b). The impact of maternal stress on the transmammary transfer and protective capacity of herpes simplex virus-specific immunity. *Am. J. Physiol. Regul. Integr. Comp. Physiol.* 287:R1316–1324.

Yorty, J.L. Schultz, S.A., and Bonneau, R.H. (2004). Postpartum maternal corticosterone decreases maternal and neonatal antibody levels and increases susceptibility of newborn mice to herpes simplex virus-associated mortality. *J. Neuroimmunol.* 150:48–58.

Young, E.J., Killam, A.P., and Greene, J.F., Jr. (1976). Disseminated herpesvirus infection. Association with primary genital herpes in pregnancy. *JAMA* 235: 2731–2733.

8
Influenza Viral Infection: Stress-induced Modulation of Innate Resistance and Adaptive Immunity

MICHAEL T. BAILEY, DAVID A. PADGETT, AND JOHN F. SHERIDAN

1. Introduction

If you believe what you read in the newspapers, the world is poised for a pandemic. The scourge is likely to be infection with the influenza A virus. Although there may be some doubt concerning these cataclysmic predictions, they are based on solid epidemiological and historical data. It is well documented that during the past several centuries, an influenza virus pandemic has raced through the human population every 20–40 years or so. In 1918–1919, a pandemic due to influenza virus infected one out of every five humans. This "Spanish Flu," which was also known as "La Grippe," is estimated to have killed more than 30 million people in less than 2 years (Mills *et al.*, 2004). To put this in perspective, this influenza pandemic killed one out of every four soldiers that died during World War I (Oxford *et al.*, 2005). Luckily, there has not been a repeat of the 1918–1919 pandemic. However, recent events such as the emergence of the avian influenza that is currently causing mortality in Asia may signal the evolution of a new, highly virulent influenza virus that might cause a serious worldwide influenza epidemic.

Quite a bit has changed since the 1918 influenza pandemic. First, the science of virology has advanced to the point that we have a very good understanding of the biology, molecular genetics, and virulence factors of the influenza virus. Second, immunology has revealed how the body combats infectious virus, limits disease, and stimulates protective immunologic memory. The advances in both virology and immunology have led to the development of effective vaccines that protect the human population against the yearly outbreaks of influenza viral infections. However, even with protective vaccines, influenza virus still kills 20,000 to 30,000 Americans each year (Anderson and Smith, 2005). Furthermore, the fear of a pandemic is very real. Our understanding of the virology of influenza suggests that the virus that will cause the next pandemic is already spreading in the bird population throughout Asia. The vaccines that have been used for decades will offer no protection against this "new" influenza virus. Instead, survival will depend on one's own innate and adaptive immune responses. Therefore, the purpose of this chapter is to detail the immune response to the influenza A virus and survey how the slings and arrows of the stressors in our lives can influence those immune responses that we rely upon for survival.

2. Immune Response to Influenza Viral Infection

The influenza A virus enters the body through the respiratory tract. Although an infected individual can sneeze in your face to transfer the virus, inoculation is typically a bit more surreptitious. The virus can exist in nasal droplets on the doorknob that you will use to open the door to your office, on the handset to the phone you use to tell your kids you are coming home early from work, on the handle of the shopping cart at your local grocery store, or on the hand that you will shake to say hello or goodbye. In fact, most people infect themselves hand-to-nose. Once the virus obtains entry to the body, it has to gain entry into specific cell populations for its own survival as it is an obligate intracellular parasite requiring the host cell's machinery for replication. The typical influenza A virus productively infects respiratory epithelial cells (Matrosovich et al., 2004). Once the virus binds to the surface of the first respiratory epithelial cell, the race is on pitting replication of the virus versus the host's immune system.

After attachment to the host cell, the influenza virus is internalized by endocytosis, with several important events occurring in the endosome. For the virus's sake, it uncoats its genetic material (eight single-stranded RNA molecules) and directs them to the nucleus of the cell where viral replication will ensue. This uncoating, however, exposes the viral genetic material of the virus to several important molecules. These include a subset of Toll-like receptors (TLR-3, TLR-7, and TLR-9) that serve as alarm signals that something untoward has been encountered by the host cell (Hornung et al., 2004; Lund et al., 2004; Barchet et al., 2005). The Toll-like receptors distinguish the RNA molecules of influenza virus from that of normal host DNA and RNA because of conserved genetic patterns common to viral pathogens (for review, see Finberg and Kurt-Jones, 2004; Kaisho and Akira, 2004; Pasare and Medzhitov, 2004). Once the Toll-like receptors encounter such a pathogenic insult, they trigger gene expression in the infected cell for induction of a rapid innate immune response (Diebold et al., 2004). From this initial ligation of the Toll-like receptors that occurs within minutes of viral entry, the timeline of the immune response to influenza A virus has been well characterized. The following sections outline the innate, adaptive, and memory immune responses to influenza virus.

2.1. Innate Immune Response to Influenza Virus (Type I Interferons)

Ligation of the Toll-like receptors in the endosome of the infected cell triggers an immune response in several ways. The first response is an attempt to protect surrounding cells from subsequent infection with new influenza viruses that will quickly begin to emerge from the infected cell. The infected cell has developed a method of telling its neighbors that it is infected with a "virus" and they need to protect themselves at all costs. Through this signal, the infected cell says "this virus will kill you if you allow it in and allow it to overtake your molecular machinery—raise your defenses, destroy your RNA molecules, and shut-down protein synthesis." This signal is provided in the form of the type I interferons, which include interferon-alpha and Interferon-beta (IFN-α and IFN-β) (Katze *et al.*, 2002; Proietti *et al.*, 2002).

IFN-β is encoded by a single gene, whereas there are more than 20 different genes that code for IFN-α in both humans and mice. IFN-α and IFN-β share a ubiquitously expressed heterodimeric receptor and, thus, share many, if not all, functions. IFN-α/β binding leads to receptor dimerization, activation of the Janus Activated Kinase (JAK):Signal Transducer and Activators of Transcription (STAT) pathway, and induction of genes containing an IFN-stimulated response element (ISRE) in their regulatory region (Darnell *et al.*, 1994; Barnes *et al.*, 2002). Thus, influenza virus induced IFN-α/β production drives new gene transcription in the healthy cells immediately surrounding the infected ones.

These interferons induce about 20–30 ISRE-containing genes, several of which have antiviral activities that can protect against virus replication. The first is 2′, 5′ oligoadenylate synthetase, which activates a ribonuclease that digests double-stranded RNA in the target cell (Hovanessian, 1991). A second gene that is activated is a protein kinase that inhibits the initiation step of protein synthesis (Hovanessian, 1991; Samuel, 1991). Presumably, by destroying all of the double-stranded RNA in a cell and by temporarily shutting down all new gene expression and protein synthesis, replication and subsequent spread of the infecting virus would be blocked. This early virus-induced, IFN activation of cellular transcription constitutes the initial antiviral immune response of the host.

2.2. Innate Immune Responses to Influenza Virus (Inflammation)

Whereas TLRs can mediate pathogen recognition and initiate an antiviral response in the tissue parenchyma, recruitment of cells (i.e., inflammation) is necessary for the development of effective innate host defenses and the subsequent development of adaptive immunity. In addition to driving the transcription of the type I interferons, ligation of the Toll-like receptors also stimulates an inflammatory response (Diebold *et al.*, 2004; Pasare and

Medzhitov, 2004). The inflammatory response is initiated to recruit a variety of leukocytes (e.g., macrophages, NK cells, and B and T lymphocytes) from the blood to further control virus replication and to begin initiation of the adaptive immune response that will be considered later. Again, almost coincident with infection of the first cell, Toll-like receptor activation results in the transcription of inflammatory cytokine genes. These include IL-1 and TNF-α (Julkunen *et al.*, 2001; Conn *et al.*, 1995). In turn, these proinflammatory cytokines amplify and engage additional cells in the response to invading microorganisms.

How do the virus-induced proinflammatory cytokines draw additional cells into the lung and into the regional lymph nodes? Analogous to erecting directional signs on highway exit ramps, the proinflammatory cytokines create *reactive endothelium* in the vicinity of the infected tissue. In response to IL-1 and TNF, the endothelial cells that line the blood vessels adjacent to the site of viral replication augment their expression of intracellular adhesion molecule-1 (ICAM-1) and vascular cell adhesion molecule-1 (VCAM-1) (McHale *et al.*, 1999; Streiter *et al.*, 2003). As these molecules are displayed on the circulatory side of the blood vessels, cells that flow by and express the appropriate cognate ligand become tethered to the vessel wall. Once loosely bound to the vessels in the vicinity of viral replication, the inflammatory cells themselves become activated. The same IL-1 and TNF that created the reactive endothelium now drive the expression of chemokine receptors on the surface of the tethered cells (Weber *et al.*, 1999; Rosseau *et al.*, 2000).

The display of chemokine receptors is a critical step in inflammation as the cells that are tethered to the vessel endothelium become competent to leave the circulation. Now, recognition of specific chemokine signals will enable the macrophages and natural killer cells to migrate in a directional manner (i.e., chemotax) toward the site where influenza virus is replicating and causing tissue destruction. Like the proinflammatory cytokines and type I interferons, multiple chemokines are also produced after Toll-like receptor activation by influenza virus. In fact, a chemotactic gradient emanates from the point of infection where virus titers are highest and serves as a beacon to attract macrophages and NK cells (Randolph and Furie, 1995; Nieto *et al.*, 1999). Binding to the newly expressed chemokine receptors on the tethered leukocytes is followed by a rapid rearrangement of the cell's cytoskeleton, thus changing the shape of the cell and enabling it to transmigrate from the bloodstream through the vascular endothelium into the site of inflammation (Hordijk, 2003; Walzer *et al.*, 2005).

Influenza A virus infection results in the secretion of multiple chemokines including Macrophage Inflammatory Protein (MIP)-1α, Macrophage Inflammatory Protein (MIP)-1β, Regulated on Activation, Normal T Expressed or Secreted (RANTES), Monocyte Chemotactic Protein (MCP)-1, Monocyte Chemotactic Protein (MCP)-3, Macrophage Inflammatory Protein (MIP)-3α, and Interferon-γ-inducible protein (IP) 10 (Kaufmann *et al.*, 2001; Julkunen *et al.*, 2000). During the first few days after infection, together with the proinflammatory cytokines, these chemokines recruit macrophages and NK cells into the tissue parenchyma where viral replication is taking place. Subsequently, after the adaptive immune response has been activated, many of these same chemokines will function to draw virus-specific T cells into the tissue to resolve the infection (Kaufmann *et al.*, 2001; Debes *et al.*, 2004).

In sum, the initial signals driven by Toll-like receptor activation drive two very important aspects of the immune response to influenza infection. First, the type I interferons attempt to provide local protection through the induction of an antiviral state in the parenchymal cells of the respiratory tract. And second, the proinflammatory cytokines and chemokines send out the alarm signal to recruit cells of the immune system to the site of infection where they will kill virus-infected cells and activate the adaptive immune response.

2.3. *Cell-mediated Immune Responses to Influenza Virus*

Under normal healthy conditions, there are relatively few lymphocytes in the lungs. However, upon infection, lymphocytes are recruited in large numbers. Natural killer (NK) cells constitute one of the body's first cells recruited into any tissue as a defense against viral infection. They play a key role in killing virus-infected cells that lack or have altered expression of cell surface MHC-I ligands, which is typical during the early phase of a viral infection (Chambers *et al.*, 1998; Lanier, 1998). The number and cytolytic activity of NK cells is highly correlated with natural resistance to viral challenge. Chronically low NK activity in humans is associated with increased incidence and duration of infection and decreased survival during influenza viral infections (Stein-Streilein *et al.*, 1989).

In addition to the NK cell, both T and B lymphocytes are involved in recovery from influenza infection and with protection against reinfection with the same strain of virus. However, in contrast with the NK cell, which does not recognize influenza viral antigens in an antigen-specific manner, the T- and B-cell responses are restricted to virus-specific subclones. Whereas the NK cell traffics to the lung by the initial alarm signal sounded by Toll-like receptor activation, the T and B cells have to be selected for activation based on their antigen specificity. This occurs in secondary lymph nodes that drain the site of infection. With regard to influenza infection of the lung, this is the mediastinal, the superficial cervical, and the deep cervical lymph nodes. These lymph nodes serve as the site where the antigen-specific lymphocytes are driven to proliferate (e.g., clonal expansion); it is also the site where they mature so that once antigen is encountered, they are competent to mediate their respective effector functions. Because of their killing power and potential ability to cause immunopathology if not adequately controlled, antigen-specific responses are tightly coordinated. It is generally acknowledged that the conductor of the adaptive immune response is the CD4[+] helper T cell which is the first antigen-specific cell activated in response to influenza virus.

In turn, it drives expansion and activation of the influenza-specific CD8[+] T cells and the antibody-producing B cells.

Clonal activation of naïve CD4[+] cells requires antigen-specific signaling through the T-cell receptor (TCR) as well as signaling through multiple costimulatory molecules (Chambers and Allison, 1997; Lanzavecchia *et al.*, 1999). This is provided by cells whose job is to acquire and process viral antigen. More specifically, dendritic cells at the site of infection in the lung engulf viral particles, mature in response to inflammatory signals, migrate to draining lymph nodes, and present antigen in the context of MHC class II molecules to activate naïve virus-specific CD4[+] cells (Fonteneau *et al.*, 2003; Lund *et al.*, 2004).

Depending on the type of antigen encountered and the microenvironment where the antigen is presented to the antigen-specific CD4[+] T cell, the immune response may differ dramatically. If antigen is encountered by the CD4[+] cell in the presence of IL-12, and IFN-γ, naïve CD4[+] cells differentiate into T helper 1 (Th1) cells that produce quantities of IFN-γ, IL-2, and TNF-α (Deng *et al.*, 2004). Alternatively, activation of the naïve CD4[+] T cell in the presence of IL-4 induces the development of Th2 effectors that secrete IL-4, IL-5, and IL-13 (Oran and Robinson, 2004). Although most infections will drive the development of both types of T cells, viral infections such as that with influenza A virus are known to predominantly induce Th1 or type 1 immunity (Cella *et al.*, 2000; Lopez *et al.*, 2002).

In turn, clonal expansion of influenza virus-specific CD4[+] T cells of the Th1 subtype promotes the activation of cytotoxic CD8[+] T cells (Maillard *et al.*, 2002; Fonteneau *et al.*, 2003). Such T cells play a major role in the clearance of influenza virus from the lungs via an MHC class I–dependent cytotoxic mechanism involving perforin and Fas (Topham *et al.*, 1997). These molecules serve simply to kill the virus infected cell. The CD8[+] T cell is important for recovery from infection, as there is elevated viral replication, morbidity, and mortality in mice that lack CD8[+] T cells (Liu *et al.*, 2003).

The Th1 CD4[+] T cell also drives B-cell differentiation and expansion into plasma cells that produce subclasses of antibody (i.e., IgA and IgG2a) for antibody-dependent cellular cytotoxicity mediated by NK cells (Hashimoto *et al.*, 1983; Jegerlehner *et al.*, 2004). Impairment of B-cell development slows clearance of influenza viral antigen but has no substantial influence on survival or recovery from infection (Gerhard *et al.*, 1997). Although not necessary for recovery from infection, the strength of antigen-specific B-cell expansion is revealed upon reinfection with antigenically similar strains of influenza infection. It showcases the power of immunological memory.

2.4. Immunological Memory to Influenza Virus

The two compartments of adaptive immunity (i.e., the T- and B-cell compartments) are also responsible for long-term memory to influenza. However, these cells evolved for different purposes, with B cells evolving to neutralize extracellular virions and cytotoxic T cells evolving to eradicate virally infected cells. For example, circulating antibody is able to neutralize influenza before it infects and spreads between host cells. If the infection progresses, memory cytotoxic T cells are needed to successfully eradicate the virus, with helper T-cells facilitating both antibody secretion and T-cell cytotoxicity.

2.5. Memory B-Cell Response to Influenza Virus

Two main B-cell types contribute to immunological memory: antibody secreting plasma cells and memory B cells. Costimulation during B-cell activation and affinity maturation of the B-cell receptor dictates whether these cells will differentiate into plasma cells or memory B cells. For example, interactions involving CD40L and the B cell–specific activator favor memory B-cell development, whereas interaction with OX-40, CD23, and Blimp-1 leads to plasma B-cell development. Activated B cells migrate to the germinal centers of lymphoid follicles in response to the CXC chemokine B lymphocyte chemoattractant (BLC) (Forester *et al.*, 1996; Pevzner *et al.*, 1999; Okada *et al.*, 2002). Follicular dendritic cells play a substantial role in recruiting B cells to lymphoid follicles by producing BLC and providing antigenic stimulation in the form of a bound immune complex (Qin *et al.*, 2000; Tew *et al.*, 2001). Once in the germinal center of a lymph node, B cells undergo rapid proliferation, which is accompanied by hypermutation of the immunoglobulin variable region genes. As replication progresses, only B cells expressing surface antibody with high affinity for antigen are allowed to survive; all others die through apoptosis. After expansion and affinity maturation, memory B cells traffic to and take up residence in lymphoid organs where they can immediately respond to reinfection with influenza virus by rapidly differentiating into plasma-secreting B cells. Long-lived memory plasma B cells reside in the bone marrow and continue to produce antibodies with high antigen specificity even in the absence of continued antigenic stimulation.

Neutralization of the influenza virus upon reinfection is mediated by antigens specific to the viral proteins hemagglutinin and neuraminidase. The most effective antibody molecules in preventing reinfection with influenza are hemagglutinin-specific antibodies that prevent either attachment of the virus to the host or intra-endosomal fusion (Knossow *et al.*, 2002). In fact, if hemagglutinin-specific antibody levels are high enough, they may completely prevent the initiation of infection. Antibodies specific to neuraminidase are also generated, but their effect is not as protective as antibodies specific for hemagglutinin; however, they are important in suppressing subsequent viral replication, ultimately reducing morbidity and mortality (Johansson *et al.*, 1989, 1998, 2002).

Vaccines for influenza primarily act through induction of neutralizing antibody responses. Two types of vaccines for influenza virus are currently used: a live, cold-adapted attenuated influenza virus vaccine and a trivalent inactivated vaccine. Both contain the same three influenza strains, A(H3N2), A(H1N1), and B. But because the segmented influenza RNA genome is highly amenable to genetic mutation, from antigenic shift and drift, antigenically distinct subtypes emerge each year, requiring yearly adjustment of the vaccines. The primary mode of protection from the vaccines is the increased plasma cell production of neutralizing IgA and IgG antibodies specific for the hemagglutinin and neuraminidase of the strains used in the vaccines (Cox *et al.*, 1994; Brokstad *et al.*, 1995; el Madhun *et al.*, 1998). If, however, the virus is able to overcome these defenses and infects the host, memory T cells become crucial in controlling and eliminating the virus.

2.6. Memory T-Cell Response to Influenza Virus

Unlike long-lived plasma cells that continually produce antibodies even in the absence of antigen, the effector functions of memory T cells are only induced in the presence of antigen. After primary infection, there is a substantial expansion of the pools of naïve helper and cytotoxic T cells containing T-cell receptors with high specificity for influenza. In comparison with longlasting production of antibodies by plasma cells, clonal expansion and effector T-cell function is short lived, lasting only about 2 weeks. Lasting protection from future infections, therefore, depends on the development of long-lived memory T cells. Although memory T cells do not reside in a state of chronic activation, they are defined by their antigen specificity and their ability to rapidly respond to reinfection by extravasating into sites of infection, producing inflammatory cytokines, and killing virally infected cells more rapidly than naïve T cells (Welsh *et al.*, 2004).

The pathways through which effector T cells become memory T cells have not been completely defined, but it is acknowledged that there are at least three clear stages in the differentiation of memory T cells (i.e., clonal expansion, contraction, and memory). The first phase, clonal expansion, occurs in lymphoid tissues where antigen stimulation drives the differentiation of naïve T cells into effector T cells (i.e., either CD4$^+$ T helper cells or CD8$^+$ cytotoxic T cells) and induces clonal expansion. This burst in antigen-specific effector cells has a significant impact on the development of immunological memory. For example, the size of the influenza-specific memory CD8$^+$ T-cell pool is directly related to the burst size during clonal expansion, which can increase the number of antigen-specific CD8$^+$ T cells by approximately 1000-fold (Hou *et al.*, 1994).

Burst size during clonal expansion is largely dependent on the manner in which T cells are activated. And, as is the case with B cells, professional antigen-presenting cells, such as dendritic cells, are potent activators of T cells. When dendritic cells are present in culture, as little as 2–6 h of antigenic stimulation is needed to drive the expansion of T cells. In their absence, however, T cells need to be exposed to an antigen for more than 24 h (Iezzi *et al.*, 1998; Jelley-Gibbs, 2000). Once stimulated, however, antigen-specific CD8$^+$ T cells appear to undergo a set number of divisions before undergoing apoptosis (Kaech and Ahmed, 2001; Kaech *et al.*, 2002). This finding seems to explain the increased prevalence of antigen-specific T cells in lymphoid tissue days to weeks after the influenza virus has been cleared from host tissues (Flynn *et al.*, 1999).

Given the dramatic increase in the number of antigen-specific effector T cells, it is not surprising that not all of these cells go on to become memory T cells. In fact, after T cells clonally expand and the antigen has been cleared, approximately 90% of the T cells undergo apoptosis, with the remaining cells developing into memory T cells. It is not well understood why some cells remain viable and further differentiate into memory cells, but there are several hypotheses as to the regulation of this contraction phase. The first hypothesis is that as antigen is removed from the system, the production of antiapoptotic cytokines is also reduced, including type I IFNs, IL-2, IL-4, IL-7, and IL-15 (Zhang *et al.*, 1998; Sprent *et al.*, 2000; Tan *et al.*, 2001). It is not known whether these cytokines have longlasting effects on memory cell development, but current research has demonstrated that effector CD8$^+$ T cells with high expression of the IL-7 receptor are more likely to develop into memory CD8$^+$ T cells than are cells with low expression of the IL-7 receptor (Kaech *et al.*, 2003). This finding supports the hypothesis that cytokines in the IL-2 family, which includes IL-7, regulate the development of memory T cells. As antigen exposure wanes, the levels of other effector molecules, such as perforin and IFN-γ, are also reduced. Thus, the differentiation of effector T cells to memory cells may be due to the loss of IFN-γ and perforin gene expression. Data in support of this hypothesis are relatively sparse, but it is known that effector T-cell contraction does not occur normally in perforin or IFN-γ knockout mice (Kagi *et al.*, 1999; Matloubian *et al.*, 1999; Badovinac *et al.*, 2000).

Regardless of which cytokines and effector molecules are involved, it is apparent that the level and duration of the antigen encounter play large roles in shaping the memory response, with a high level of antigenic encounter resulting in a smaller contraction phase but a longer duration driving terminal differentiation and apoptosis. Thus, early control of influenza appears to be important in establishing a robust memory response. Luckily, most strains of influenza, particularly influenza A (H1N1 strains), are easily controlled within 8–12 days (Flynn *et al.*, 1999; Doherty *et al.*, 2000), with contraction of antigen-specific CD8$^+$ T cells occurring shortly thereafter. After the contraction phase in the T-cell response, two types of antigen-specific memory CD8$^+$ T cells emerge: central (T_{CM}) and effector (T_{EM}) memory T cells.

Central memory T cells reside within regional lymphoid tissues and can rapidly respond to reencounter with influenza virus. These T_{CM} cells constitutively express CCR7 and CD62L, two receptors that are also characteristic of naïve T cells and are required for cell extravasation through endothelial venules and migration to T-cell areas of secondary lymphoid organs. Unlike naïve T cells, however, T_{CM} can respond more rapidly to a second infection due to their higher sensitivity to antigenic stimulation and decreased dependency upon costimulation. In addition, T_{CM} can help to regulate the secondary immune response by upregulating CD40L expression, which provides more effective stimulatory feedback to dendritic cells (DCs) and B cells. Upon activation, T_{CM} produce mainly IL-2, with IFN-γ and IL-4 production occurring only after the cells have proliferated (Sallusto et al., 2004).

Effector memory T cells reside within the epithelial tissues of the lungs and are ideally poised to mount the first cytotoxic response to reinfection with influenza virus. T_{EM} no longer express CCR7 but can have low expression of CD62L, which is most likely the reason they reside within the lungs and not secondary lymphoid tissues. Having T_{EM} in the lungs after resolution of an influenza infection makes evolutionary sense because these cells have a more rapid effector function than even T_{CM}, with both T_{CM} and T_{EM} being more rapid than naïve T cells. In fact, within hours of antigenic stimulation, $CD8^+$ T_{EM} can release large amounts of perforin, with both $CD4^+$ and $CD8^+$ T_{EM} producing large amounts of IFN-γ, IL-4, and IL-15 (Geginat et al., 2003a, 2003b; Sallusto et al., 2004). It is because of these T_{CM} and T_{EM} that viral titers in mice reexposed to influenza A are dramatically lower in the lungs 5 days after the infection with little viral replication evident 8 days postinfection (Flynn et al., 1999).

Maintenance of memory T cells can occur through antigen dependent and independent mechanisms. It is unlikely that antigen-dependent mechanisms play a large role in maintaining influenza-specific T cells, because influenza is not a persistent virus that remains in the host for long periods of time. It should be noted, however, that antigen-antibody complexes that are trapped by follicular dendritic cells can be retained for extended periods and can directly stimulate and be acquired by B cells that process the antigen and present it to CD4 T cells in the context of MHC class II molecules (Tew et al., 1984; MacLennan et al., 1994; Qin et al., 2000).

This antigen-dependent maintenance of memory T cells has not been described after influenza infection, making it more likely that homeostatic proliferation of influenza-specific memory T cells is driven by antigen-independent mechanisms. The cytokines IL-15 and IL-7 have been shown to have prominent roles in the maintenance of antigen-specific memory T cells. For example, mice lacking either IL-15 or its receptor lose memory $CD8^+$ T cells (Ma et al., 2000), and exogenous treatment of cells with IL-15 drives the proliferation of $CD8^+$ T cells (Lodolce et al., 2001). Although IL-15 also plays a role in homeostatic proliferation of antigen-specific memory $CD4^+$ T cells, IL-7 has a much larger role in maintaining memory $CD4^+$ cells. For example, IL-7Rα blockade significantly reduced antigen-specific $CD4^+$ T-cell levels in mice previously infected with Lymphocytic Choriomeningitis Virus (LCMV), whereas IL-15 blockade had more moderate effects on memory $CD4^+$ T cells (Lenz et al., 2004).

The immune response to influenza infection is a systemic response that involves the respiratory tract, regional lymph nodes (such as the mediastinal and cervical lymph nodes), and other primary and secondary lymphoid organs. All of these organs are heavily innervated by the nervous system, primarily the autonomic nervous system, and can be affected by the hormones of the hypothalamic-pituitary-adrenal (HPA) axis. In addition, many of the cytokines produced during influenza infection can affect the nervous system and animal behavior, thus creating many points of intersection through which the nervous and immune systems can interact [see Bailey *et al.* (2003) for a review]. These interactions are perhaps most evident when the interactions are disrupted by physical or psychological stressors.

3. Stress and the Immune Response to Influenza Virus

The idea that psychological stress may cause measurable changes in an individual's susceptibility to infectious disease is not new. One of the earliest proposals was put forth by Ishigami (1919) in which he described "the influence of psychic acts on the progress of pulmonary tuberculosis." Now, it is generally acknowledged that humans and animals exposed to chronic psychological or physical stressors have decreased resistance to microbial pathogens. The consequences of stress-induced modulation of the immune system include increased susceptibility and frequency of disease, prolonged healing times, and a greater incidence of secondary health complications associated with infection (Bailey *et al.*, 2003). However, knowledge about how the perception of stress is processed in the central nervous system and then transmitted to peripheral physiological systems is incomplete. Therefore, information about the mechanism(s) by which a stressor, through its effects on the immune system, might increase susceptibility or severity of infectious disease is lacking.

Recently, it has been recognized that the immune system functions as a "sensory system" alerting the central nervous system to the dangers implicit in the invasion of its tissues by pathogenic microorganisms. This recognition has led to intense study of the bidirectional communication among the nervous, endocrine, and immune systems. Many of these studies focus on the HPA axis, which is one of the major pathways activated by stress (Bailey *et al.*, 2003). Stress activates the HPA axis resulting in upregulation of mRNA expression and production of specific hormones including glucocorticoids and endogenous opioids (Hunzeker *et al.*, 2004; Hermann *et al.*, 1995). When examined *in toto*, these findings suggest a rudimentary mechanism by which stress-induced activation of the HPA axis results in increased neuroendocrine hormone production, followed by release of these "stress mediators" into circulation, and subsequent modulatory interactions of these hormones with cells of the innate and adaptive arms of the immune system. The response to stress, however, is not always health aversive, and in fact it may serve to restore homeostasis and thus provide an adaptive response to environmental and internal stressors. This is not a new concept as the literature dating back to the first half of the 20th century contains the beginnings of a quest to understand how organisms successfully respond to environmental challenges (Selye, 1936). The conceptualization of a eustress, or "the good" stress, was articulated in this period and it provided a useful working model to explain the adaptive benefits of the host's response to challenge.

The goal of the studies conducted by our laboratories has been to experimentally model stress in a way that will identify points of intersection among the autonomic nervous system, the HPA axis, and the immune system that lead to meaningful alterations in health. Employing infectious challenge models, we have identified significant changes in the host's ability to generate and/or maintain a protective immune response during stress. These findings are detailed below.

In chronological terms, our work on stress and the immune system began as a collaboration with colleagues in the Ohio State University Medical Center (see Kiecolt-Glaser *et al.*, 1993). Human studies of chronic stress, in which the subjects were caring for spouses with dementia, demonstrated that chronically stressed caregivers responded less well to a number of different commercially prepared vaccines that elicit protective responses to viral and bacterial pathogens (Glaser et al., 1992; Kiecolt-Glaser *et al.*, 1993; Glaser et al., 2000). In an influenza vaccine study, the percentage of caregivers who seroconverted (defined as a fourfold rise in antibody titers to an influenza vaccine by ELISA and HAI) was lower than controls at 1 month postvaccination, and the IL-2 responses of peripheral blood lymphocytes from controls stimulated with influenza vaccine antigen was significantly higher at 30, 90, and 180 days postvaccination than in the caregiver subjects (Kiecolt-Glaser *et al.*, 1993).

Although studies such as these were illuminating and confirmed the negative effect of stress on the generation of protective immunity, it was left to nature to provide the appropriate infectious challenge (e.g., an influenza viral infection). So far, sufficient data have not been available to assess the state of protective immunity in naturally acquired influenza viral infections in these subjects.

Thus, questions about the mechanisms by which stress affects susceptibility and severity of an infection led to the development of animal models of stress and viral infection. The use of well-characterized stressors in rodent models provided an opportunity to conduct infection and immunity studies using microbial challenges with bacteria or viruses. This has been a very useful approach built on detailed knowledge of microbial pathogenesis and antimicrobial immunity in the extant literature.

3.1. Restraint Stress Suppresses the Immune Response to Infection with Influenza Virus

3.1.1. Stress-induced Effects on Innate Immune Responses to Infection with Influenza Virus

3.1.1.1. Stress Effects on Inflammation

In the initial studies of stress and the immune response, a murine model of influenza viral infection was initiated by intranasal challenge of C57BL/6 male mice; physical restraint (RST) was selected as the stressor (Sheridan *et al.*, 1991). Our first observations showed that RST reduced the accumulation of cells in the lungs of influenza-infected mice (Hunzeker *et al.*, 2004). In addition to having effects in the lungs, RST was also responsible for a reduction in lymphadenopathy in the draining lymph nodes (Hermann *et al.*, 1995). The cells whose numbers were suppressed by RST included macrophages, NK cells, and T and B lymphocytes. During the past 15 years or so, our studies have revealed that this effect of RST has a substantial impact on both the innate and adaptive immune response to influenza virus infection.

As mentioned above, the first, innate responses to influenza virus involve the ligation of Toll-like receptors (TLRs) and the subsequent production of the type I interferons (IFN-α and IFN-β), proinflammatory cytokines (IL-1, IL-6, and TNF-α), and chemokines (MCP-1 and MIP-1α). Together these TLR-driven responses are targeted toward creating an antiviral state in the infected tissue and toward recruiting inflammatory cells to the lung and draining lymph nodes.

The antiviral state is created by the type I interferons which play a major role in restricting the spread of virus during the early phase of the infection. As would be predicted, infection with influenza A/PR8 virus induced the expression of both of the type I interferons in the lungs of control mice. However, somewhat unexpectedly, expression of the genes coding for IFN-α and IFN-β was enhanced by RST (Hunzeker *et al.*, 2004). It is not clear whether this reflected an adaptive response to the stressor or whether it was a compensatory response in an attempt to control viral replication while other aspects of the innate response were suppressed.

The development of an activated endothelium, critical to the local inflammatory response during infection, is dependent on the proinflammatory cytokines. Our data have showed that RST has pronounced effects on proinflammatory cytokine responses to influenza infection. For example, lung IL-1α responses in RST/infected mice were suppressed (values were similar to uninfected control responses), and treatment with RU486 resulted in a blockade of the type II steroid receptor and failed to restore the response. In nonstressed control mice, influenza A/PR 8 infection elevated the lung IL-6 response by 24h postinfection (p.i.), with the peak occurring at 48h and levels remaining above background at 72h. Interestingly, RST did not affect the magnitude or the kinetics of the lung IL-6 response during the infection (Konstantinos and Sheridan, 2001). Because two cytokines associated with the early inflammatory responses to infection were differentially regulated in restraint-stressed mice, it appeared that the stress effect on cytokine production was specific for particular cytokines.

The actual transmigration of cells into tissue is dependent on the chemokine signal at the reactive endothelium in the lung. Again, during influenza infection, the β-chemokines are key molecules that aid in the accumulation of mononuclear cells in the lungs of infected mice. In studies in which MIP-1α was knocked out by targeted mutation, infection with influenza virus resulted in reduced pneumonitis and delayed clearance of the virus. Histological analysis of the infected lung tissues showed a significant reduction in the inflammatory infiltrate in the KO mice when compared with the wild type (Cook et al., 1995). In experiments to examine the effects of stress on β-chemokine responses during influenza infection, RST suppressed monocyte chemotactic protein-1 (MCP-1) and macrophage inflammatory protein-1 (MIP-1α) (Hunzeker et al., 2004). Suppression occurred early (before day 3 p.i.) and remained below control levels at day 5 p.i. (Hunzeker et al., 2004). Thus, reduction in the expression of proinflammatory cytokine and β-chemokine responses by RST were likely contributors to the diminished inflammatory response observed in the lungs of infected RST mice.

In sum, the data show that although TLR-mediated IFN-α and IFN-β responses are intact if not elevated, in RST-mice, the proinflammatory cytokine and chemokine responses are both impaired. Presumably, this would impact the recruitment of cells to both the lung and regional lymph nodes. In fact, enumeration of mononuclear cells in the infected lungs confirmed that fewer cells accumulated during A/PR8 infection when the mice were stressed. Subsequent histological studies documented reduced cellularity up to and beyond 7 days postinfection in the infected RST mice.

Of interest to us was the observation that cellularity was restored to RST-stressed animals by treating them with the glucocorticoid-antagonist RU486 (Hermann et al., 1995). Thus, because activation of the HPA axis, and the resultant elevation of plasma corticosterone, was known to affect cell trafficking, studies were performed to examine the effect of RST on circulating plasma corticosterone levels. Samples were obtained before the start of an RST cycle (6 p.m.), at 30 min into the cycle (6:30 p.m.), and again at the end of the cycle (10 a.m.). Three groups were compared including RST/infected, no RST/infected, and no RST/not infected. Both infection and RST individually elevated corticosterone. However, together, RST plus influenza infection had a synergistic effect on corticosterone levels which increased more than sevenfold in comparison with non-stress non-infected controls (Sheridan et al., 1991). Further studies of HPA activation showed that four or more consecutive cycles of restraint broke the circadian rhythm and resulted in persistent high levels of plasma corticosterone throughout the day. Moreover, loss of the HPA rhythm correlated with increased patho-physiology after influenza infection (Hermann *et al.*, 1993).

3.1.1.2. Stress Effects on NK Cell Activity

Natural killer cells play an important role in the early innate defenses to influenza infection as they seek to limit the spread of virus (Leung and Ada, 1981). They respond rapidly in the early phase of the infection to kill virus-infected cells and when activated produce cytokines that initiate and enhance subsequent, specific antiviral immune responses (Biron *et al.*, 1999). It is generally believed that during a viral infection, NK cells limit viral spread until a virus-specific CD8[+] cytotoxic T-cell response can be mounted. In fact, NK cells not only are important in innate resistance to infection but also are required for development of anti-influenza cytotoxic T-cell responses. Mice depleted of NK cells had increased mortality during an influenza viral infection (Stein-Streilein and Guffee, 1986).

Infection of C57BL/6 male mice with influenza A/PR8 virus resulted in an NK response in the lungs that was detectable on day 3 p.i., peaked on day 5 p.i., and was still present in lung tissue 7 days p.i. RST suppressed NK cell cytotoxic activity in the lungs of influenza-infected mice throughout the course of infection. This reduced NK cytotoxic activity was, in part, due to RST-induced suppression of NK cell trafficking to the lungs (Hunzeker *et al.*, 2004). Specifically, RST suppressed IL-1α, MCP-1, and MIP-1α responses at the time that peak NK infiltration was observed in infected control mice. These data suggest that, in stressed animals, NK cells were not accumulating in the lungs to fight infection (Hunzeker *et al.*, 2004).

Restraint-induced reduction in NK cell trafficking was the result of ele-vated corticosterone levels, as evidenced by the finding that NK trafficking to the infected lungs of RST mice was restored by blockade of the type II glucocorticoid receptor with RU486 treatment. Concomitantly, receptor blockade restored expression of MIP-1α and MCP-1 chemokine genes in the lungs of stressed, infected mice. However, blockade of the glucorticoid receptor failed to restore NK cell cytotoxicity (Tseng *et al.*, 2005). Thus, although glucocorticoids induced by stress diminished cell trafficking to the lungs, they were not involved in suppression of NK cytotoxicity. This obser-vation suggested that another "stress mediator" might be involved in regu-lation of natural resistance. Studies conducted by Tseng and colleagues (2005) demonstrated conclusively that suppression of NK cytotoxicity was restored by pharmacologic blockade of the μ-opioid receptor in the stressed animals. Interestingly, NK cells do not appear to have μ-opioid receptors, suggesting that the opioids are acting indirectly rather than directly on the NK cells (Tseng *et al.*, 2005).

When taken together, these studies create a picture in which restraint stress alters three major components of natural resistance to viral infection: the proinflammatory cytokine IL-1α response, the β-chemokine response, and natural killer cell activity. Responses were altered at the site of virus replication and in secondary lymphoid tissues for all three responses. Stress-induced corticosterone reduced lymphadenopathy in draining lymph nodes and diminished mononuclear cell trafficking to the infected lung. In addition, stress-induced corticosterone suppressed cytokine gene expression for some cytokines that were studied (IL-1α, MCP-1, and MIP-1α) while it enhanced, or had no effect on, gene expression of other cytokines induced during infection (e.g., type 1 IFN-α/β and IL-6). Moreover, suppression of NK cell trafficking during stress was corticosterone-dependent, whereas NK cell cytoxicity was suppressed by an opioid response. The question of whether the opioid effect on NK during an influenza infection is mediated centrally or peripherally remains unanswered, but strong evidence to support a centrally mediated mechanism for opioid-associated modulation of immune function has been published (Mellon and Bayer, 1998; Nelson and Lysle, 2001).

3.1.2. Stress-induced Effects on Adaptive Immunity to Infection with Influenza Virus

Previous studies have shown that stress, in addition to affecting natural resistance, also modulates virus-specific T- and B-cell responses during an influenza viral infection (Bonneau et al., 1991; Sheridan et al., 1991; Dobbs et al., 1996). RST reduced both virus-specific CD4+ T-cell cytokine responses and CD8+ cytolytic T-cell responses during infection. Studies of stress-induced neuroendocrine responses showed that sustained, elevated levels of corticosterone-suppressed T-cell cytokine responses reduced the accumulation of T cells in the draining lymph nodes and altered the trafficking of T cells to the lungs of virus-infected animals. Pharmacologic blockade of type II steroid receptors, using RU486 treatment, restored lymphadenopathy, cell trafficking to the lungs, and the expression of T-cell cytokine genes (Dobbs et al., 1996). However, as was the case with suppression of NK cell cytotoxicity, virus-specific cytolytic T-cell responses remained suppressed after RU486 treatment. Subsequent experiments demonstrated that suppression of CD8+ T-cell cytolytic activity was associated with stress-induced activation of the sympathetic nervous system and release of catecholamines; blocking β-adrenergic receptors (with nadolol) restored T-cell cytolytic activity. Thus, RST-induced activation of the sympathetic nervous system results in regulation of virus-specific T-cell cytotoxicity (Dobbs et al., 1993).

3.1.3. Restraint Stress and T-Cell Responses

Elevated serum corticosterone induced by RST suppressed CD4+ T-cell responses after influenza viral infection. RST suppressed the production of IL-2, IFN-γ, and IL-I0 by mononuclear cells from the regional lymph nodes and spleen of A/PR8-infected mice (Dobbs et al., 1996). Because IFN-γ and

IL-10 were both suppressed, this finding suggested that RST suppressed both Th1 and Th2 CD4$^+$ T-cell responses. Thus, during an A/PR8 infection, RST suppressed both CD4$^+$ subsets without biasing the direction of the cytokine response. Diminished cytokine gene expression during infection and stress was restored by RU486 treatment, thus confirming a role for corticosterone in mediating the effect (Dobbs *et al.*, 1996).

To further examine the role of HPA activation in modulating cytokine responses, experiments were conducted using androstenediol (AED; a metabolite of dehydroepiandrosterone; DHEA), which is known to counterregulate glucocorticoid modulation of the immune response (Padgett and Loria, 1994). AED treatment blunted RST-induced HPA activation resulting in lower plasma corticosterone levels during infection. Moreover, the reduced plasma corticosterone levels in AED-treated mice correlated with restoration of lymphadenopathy to draining lymph nodes, enhanced IFN-γ production, and elevated IL-10 gene expression on day 7 p.i. (Padgett and Sheridan, 1999). AED functioned in opposition to corticosterone by regulating T-helper cytokine secretion.

3.1.4. Restraint Stress and B-Cell Responses

The finding of significant suppression of antiviral T-cell responses by RST suggested that T-cell help for antibody production might be limited and therefore that the kinetics and/or magnitude of antibody responses to influenza might also be affected. As hypothesized, RST affected the development of virus-specific B-cell antibody responses by delaying seroconversion in the restrained mice infected with A/PR8 virus. In addition, antibody class switching from IgM to IgG to IgA was delayed. After resolution of the infection, during the memory phase, stressed mice eventually achieved virus-specific antibody titers similar to those found in non-stressed infected control mice (Feng *et al.*, 1991).

3.2. Social Disruption Stress, Glucocorticoid Resistance, and the Immune Response to an Influenza Viral Infection

3.2.1. Social Stress

There are multiple variables that contribute to the stress response. One of the most important is the nature of the stressor. In recent studies, we have employed a social conflict paradigm to investigate how aggressive social interactions, the repeated experience of defeat, and social hierarchy influence the immune response (Avitsur *et al.*, 2001; Stark *et al.*, 2001).

Disruption of a social hierarchy that is established when male mice are caged together is a well-recognized model for social stress in mice (Koolhaas *et al.*, 1997). Introducing an aggressive intruder (social disruption; SDR) elicited aggressive interactions and defeat of the cage residents. In response to the aggressive social interactions, corticosterone levels were markedly elevated and several immune parameters were altered including proinflammatory cytokine responses and splenocyte viability in culture (Avitsur *et al.*, 2001; Stark *et al.*, 2001).

In response to social disruption, mice developed glucocorticoid (GC) resistance. Their splenocytes were less sensitive to the inhibitory effects of corticosterone when stimulated by lipopolysaccharide (LPS). Enhanced cell proliferation and viability was observed in cultures of these splenocytes. *In vivo*, splenomegaly accompanied GC resistance (Avitsur *et al.*, 2001), and there was a significant increase in the number of $CD11b^+$ and $CD62L^+$ monocytes (Avitsur *et al.*, 2002, 2003a). LPS-stimulated splenocytes from SDR mice secreted higher levels of IL-6 and TNF-α in culture than did control mice (Stark *et al.*, 2002; Avitsur *et al.*, 2003a). To date, the mechanisms that drive the development of GC resistance have not been fully elucidated. However, using selective depletion techniques, it was observed that GC resistance was abolished by depletion of $CD11b^+$ cells (monocytes/ macrophages), but not the depletion of $CD19^+$ cells (B lymphocytes), from the cultures. Thus, this observation suggested a role for splenic monocytes/macrophages in GC resistance (Stark *et al.*, 2001). Subsequent analysis of GC resistance demonstrated that it was associated with reduced nuclear translocation of the GC receptor in $CD11b^+$ cells and with the inability of GC to suppress the activity of the inflammatory transcription factor, nuclear factor-κB (Quan *et al.*, 2003).

3.2.2. SDR and LPS Challenge

One of the functional consequences of GC resistance in immune cells was that the ability of glucocorticoids to inhibit the production/activity of proinflammatory cytokines was diminished. Furthermore, the phenotypic changes associated with SDR (i.e., splenomegaly, increased number of $CD11b^+$ monocytes, and elevated proinflammatory cytokine responses) suggested that these mice would have exaggerated inflammatory responses upon microbial challenge. As a surrogate microbial challenge, mice were injected with LPS. This challenge was meant to gauge the level of their proinflammatory cytokine response and determine if SDR altered the sensitivity to endotoxic shock. After the LPS challenge, the expression of IL-1β and TNF-α, was significantly higher in the lung, liver, spleen, and brain of the SDR mice as compared with home cage control animals. SDR increased the mortality of mice challenged with LPS. Histological examination of SDR animals revealed widespread disseminated intravascular coagulation in the brain and lung, extensive meningitis in the brain, severe hemorrhage in the lung, necrosis in the liver, and lymphoid hyperplasia in the spleen, indicating inflammatory organ damage (Quan *et al.*, 2001). Taken together, these results showed that SDR increased the susceptibility to endotoxic shock and suggested that the development of GC resistance and hyper-proinflammatory cytokine responses were the mechanisms for this behavior-induced susceptibility to endotoxic shock.

3.2.3. SDR and Influenza Viral Infection

A significant question to be asked in the SDR GC resistance response was whether these mice, with increased susceptibility to endotoxic shock, displayed an increased susceptibility to viral challenge. Given all the changes in the immune response due to GC resistance, it was likely that the immune response to an influenza A/PR8 viral infection would be modulated in SDR mice. However, a viral challenge is very different compared with an LPS challenge. Influenza A/PR8 viral infection of SDR mice did not result in significant mortality compared with infected home cage controls. Further, SDR mice had an enhanced type I interferon response in the infected lung. The interferon response correlated with enhanced termination of viral replication in SDR mice. Thus, SDR enhanced innate resistance to a viral challenge with influenza A/PR8 virus (Hunzeker, 2004). Additional studies are necessary to determine the impact of SDR on the virus-specific adaptive immune response to viral challenge.

4. General Conclusions

From the early pronouncements by Ishigami (1919) concerning "psychic acts" and progression of pulmonary tuberculosis, to the current assessments of stress-induced molecular regulation of individual immune response genes, the study of stress and health has had a long, productive history. However, we remain on the cusp of understanding the mechanisms of interaction among social, behavioral, and genetic factors that determine susceptibility to disease. Recently, as described by Bruce McEwen, Teresa Seeman, and members of the Allostatic Load Working Group (MacArthur Network on Socioeconomic Status and Health, 1997), "new research has reinforced the fact that the so-called 'stress mediators' have protective and adaptive as well as damaging effects, and the search for biological mechanisms that determine protective versus damaging effects of these mediators is a theme in biobehavioral research (McEwen, 1998)." Thus, one of the major challenges in the 21st century is to identify the mechanisms by which stress-induced activation of neuroendocrine responses affect physiological systems such as the immune system. The use of animal models of infectious disease provides a powerful tool to dissect the interaction between stress mediators and protective host immune responses.

References

Adler, N.E., Boyce, T., Chesney, M.A., Cohen, S., Folkman, S., Kahn, R.L., and Syme, L.S. (1994). Socioeconomic status and health: The challenge of the gradient. *Am. Psychol.* 49:15–24.

Anderson, R.N., and Smith, B.L. (2005). Deaths: Leading causes for 2002. *National Vital Statistics Report* 53:1–89.

Avitsur, R., Stark, J.L., and Sheridan, J.F. (2001). Social stress induces glucocorticoid resistance in subordinate animals. *Horm. Behav.* 39:247–257.

Avitsur, R., Padgett, D.A., Dhabhar, F.S., Stark, J.L., Kramer, K.A., Engler, H., and Sheridan, J.F. (2003a). Expression of glucocorticoid resistance following social stress requires a second signal. *J. Leukoc. Biol.* 74:507–513.

Avitsur, R., Stark, J.L., Dhabhar, F.S., Kramer, K.A., and Sheridan, J.F. (2003b). Social experience alters the response to social stress in mice. *Brain Behav. Immun.* 17:426–437.

Badovinac, V.P., Tvinnereim, A.R., and Harty, J.T. (2000). Regulation of antigen-specific CD8+ T cell homeostasis by perforin and interferon-gamma. *Science* 290: 1354–1358.

Bailey, M., Engler, H., Hunzeker, J., and Sheridan, J.F. (2003). The hypothalamic-pituitary-adrenal axis and viral infection. *Viral Immunol.* 16:141–157.

Barchet, W., Krug, A., Cella, M., Newby, C., Fischer, J.A., Dzionek, A., Pekosz, A., and Colonna, M. (2005). Dendritic cells respond to influenza virus through TLR7- and PKR-independent pathways. *Eur. J. Immunol.* 35:236–242.

Barnes, B., Lubyova, B., and Pitha, P.M. (2002). On the role of IRF in host defense. *J. Interferon Cytokine Res.* 22:59–71.

Biron, C.A., Nguyen, K.B., Pien, G.C., Cousens, L.P., and Salazar-Mather, T.P. (1999). Natural killer cells in antiviral defense: Function and regulation by innate cytokines. *Annu. Rev. Immunol.* 17:189–220.

Bonneau, R.H., Sheridan, J.F., Feng, N., and Glaser, R. (1991). Stress-induced suppression of herpes simplex virus (HSV)-specific cytotoxic T lymphocyte and natural killer cell activity and enhancement of acute pathogenesis following local HSV infection. *Brain Behav. Immun.* 5:170–192.

Brokstad, K.A., Cox, R.J., Major, D., Wood, J.M., and Haaheim, L.R. (1995). Cross-reaction but no avidity change of the serum antibody response after influenza vaccination. *Vaccine* 13:1522–1528.

Cella, M., Facchetti, F., Lanzavecchia, A., and Colonna, M. (2000). Plasmacytoid dendritic cells activated by influenza virus and CD40L drive a potent TH1 polarization. *Nat. Immunol.* 1:305–310.

Chambers, B.J., Wilson, J.L., Salcedo, M., Markovic, K., Bejarano, M.T., and Ljunggren, H.G. (1998). Triggering of natural killer cell mediated cytotoxicity by costimulatory molecules. *Curr. Topics Microbiol. Immunol.* 23:53–61.

Chambers, C.A., and Allison, J.P. (1997). Co-stimulation in T cell responses. *Curr. Opin. Immunol.* 9:396–404.

Conn, C.A., McClellan, J.L., Maassab, H.F., Smitka, C.W., Majde, J.A., and Kluger, M.J. (1995). Cytokines and the acute phase response to influenza virus in mice. *Am. J. Physiol.* 268:R78–R84.

Cook, D., Beck, M.A., Coffman, T.M., Kirby, S.L., Sheridan, J.F., Pragnell, I.B., and Smithies, O. (1995). Requirement of MIP-1α for an inflammatory response to viral infection. *Science* 269:1583–1585.

Cox, R.J., Brokstad, K.A., Zuckerman, M.A., Wood, J.M., Haaheim, L.R., and Oxford, J.S. (1994). An early humoral immune response in peripheral blood following parenteral inactivated influenza vaccination. *Vaccine* 12:993–999.

Darnell, J.E. Jr, Kerr, I.M., and Stark, G.R. (1994). Jak-STAT pathways and transcriptional activation in response to IFNs and other extracellular signaling proteins. *Science* 264:1415–1421.

Debes, G.F., Bonhagen, K., Wolff, T., Kretschmer, U., Krautwald, S., Kamradt, T., and Hamann, A. (2004). CC chemokine receptor 7 expression by effector/memory CD4+ T cells depends on antigen specificity and tissue localization during influenza A virus infection. *J. Virol.* 78:7528–7535.

Deng, Y., Jing, Y., Campbell, A.E., and Gravenstein, S. (2004). Age-related impaired type 1 T cell responses to influenza: reduced activation ex vivo, decreased expansion in CTL culture in vitro, and blunted response to influenza vaccination in vivo in the elderly. *J. Immunol.* 172:3437–3446.

Diebold, S.S., Kaisho, T., Hemmi, H., Akira, S., and Reis e Sousa, C. (2004). Innate antiviral responses by means of TLR7-mediated recognition of single-stranded RNA. *Science* 303:1529–1531.

Dobbs, C.M., Vasquez, M., Glaser, R., and Sheridan, J.F. (1993). Mechanisms of stress-induced modulation of viral pathogenesis and immunity. *J. Neuroimmunol.* 48:151–160.

Dobbs, C.M., Feng, N., Beck, F.M., and Sheridan, J.F. (1996). Neuroendocrine regulation of cytokine production during experimental influenza viral infection: Effects of restraint stress-induced elevation in endogenous corticosterone. *J. Immunol.* 157:1870–1877.

Doherty, P.C., Riberdy, J.M., and Belz, G.T. (2000). Quantitative analysis of the D8+ T-cell response to readily eliminated and persistent viruses. *Philos. Trans. R. Soc. London B Biol. Sci.* 355:1093–1101.

el Madhun, A.S., Cox, R.J., Soreide, A., Olofsson, J., and Haaheim, L.R. (1998). Systemic and mucosal immune responses in young children and adults after parenteral influenza vaccination. *J. Infect. Dis.* 178:933–939.

Feng, N., Pagniano, R., Tovar, C.R., Bonneau, R.H., Glaser, R., and Sheridan, J.F. (1991). The effect of restraint stress on the kinetics, magnitude, and isotype of the humoral immune response to influenza virus infection. *Brain Behav. Immun.* 5: 370–382.

Finberg, R.W., and Kurt-Jones, E.A. (2004). Viruses and Toll-like receptors. *Microbes Infect.* 6:1356–1360.

Flynn, K.J., Riberdy, J.M., Christensen, J.P., Altman, J.D., and Doherty, P.C. (1999). In vivo proliferation of naive and memory influenza-specific CD8(+) T cells. *Proc. Natl. Acad. Sci. U.S.A.* 96:8597–8602.

Fonteneau, J.F., Gilliet, M., Larsson, M., Dasilva, I., Munz, C., Liu, Y.J., and Bhardwaj, N. (2003). Activation of influenza virus-specific CD4+ and CD8+ T cells: A new role for plasmacytoid dendritic cells in adaptive immunity. *Blood* 101:3520–3526.

Forster, R., Mattis, A.E., Kremmer, E., Wolf, E., Brem, G., and Lipp, M. (1996). A putative chemokine receptor, BLR1, directs B cell migration to defined lymphoid organs and specific anatomic compartments of the spleen. *Cell* 87:1037–1047.

Geginat, J., Lanzavecchia, A., and Sallusto, F. (2003a). Proliferation and differentiation potential of human CD8+ memory T-cell subsets in response to antigen or homeostatic cytokines. *Blood* 101:4260–4266.

Geginat, J., Sallusto, F., and Lanzavecchia, A. (2003b). Cytokine-driven proliferation and differentiation of human naive, central memory and effector memory CD4+ T cells. *Pathol. Biol. (Paris)*, 51:64–66.

Gerhard, W., Mozdzanowska, K., Furchner, M., Washko, G., and Maiese, K. (1997). Role of the B-cell response in recovery of mice from primary influenza virus infection. *Immunol. Rev.* 159:95–103.

Glaser, R., Kiecolt-Glaser, J.K., Bonneau, R., Malarkey, W., and Hughes, J. (1992). Stress-induced modulation of the immune response to recombinant hepatitis B vaccine. *Psychosom. Med.* 54:22–29.

Glaser, R., Sheridan, J.F., Malarkey, W.B., MacCallum, R.C., and Kiecolt-Glaser, J.K. (2000). Chronic stress modulates the immune response to a pneumococcal pneumonia vaccine. *Psychosom. Med.* 62:804–807.

Hashimoto, G., Wright, P.F., and Karzon, D.T. (1983). Antibody-dependent cell-mediated cytotoxicity against influenza virus-infected cells. *J. Infect. Dis.* 148: 785–794.

Hermann, G., Tovar, C.A., Beck, F.M., Allen, C., and Sheridan, J.F. (1993). Restraint stress differentially affects the pathogenesis of an experimental influenza viral infection in three inbred strains of mice. *J. Neuroimmunol.* 47:83–94

Hermann, G., Beck, F.M., and Sheridan, J.F. (1995). Stress-induced glucocorticoid response modulates mononuclear cell trafficking during an experimental influenza viral infection. *J. Neuroimmunol.* 56:179–186.

Hordijk, P. (2003). Endothelial signaling in leukocyte transmigration. *Cell Biochem. Biophys.* 38:305–322.

Hornung, V., Schlender, J., Guenthner-Biller, M., Rothenfusser, S., Endres, S., Conzelmann, K.K., and Hartmann, G. (2004). Replication-dependent potent IFN-alpha induction in human plasmacytoid dendritic cells by a single-stranded RNA virus. *J. Immunol.* 173:5935–5943.

Hou, S., Hyland, L., Ryan, K.W., Portner, A., and Doherty, P.C. (1994). Virus-specific CD8+ T-cell memory determined by clonal burst size. *Nature* 369:652–654.

Hovanessian, A.G. (1991). Interferon-induced and double-stranded RNA-activated enzymes: a specific protein kinase and 2′,5′-oligoadenylate synthetases. *J. Interferon Res.* 11:199–205.

Hunzeker, J. (2004). *Differential Effects of Stress on the Immune Response to Influenza A/PR8 Virus Infection in Mice.* Ph.D. dissertation, The Ohio State University, Columbus.

Hunzeker, J., Padgett, D.A., Sheridan, P.A., Dhabhar, F.S., and Sheridan, J.F. (2004). Modulation of natural killer cell activity by restraint stress during an influenza A/PR8 infection in mice. *Brain Behav. Immun.* 18:526–535.

Iezzi, G., Karjalainen, K., and Lanzavecchia, A. (1998). The duration of antigenic stimulation determines the fate of naive and effector T cells. *Immunity* 8:89–95.

Ishigami, T. (1919). The influence of psychic acts on the progress of pulmonary tuberculosis. *Am. Rev. Tuberculosis* 2:470–484.

Jegerlehner, A., Schmitz, N., Storni, T., and Bachmann, M.F. (2004). Influenza A vaccine based on the extracellular domain of M2: weak protection mediated via antibody-dependent NK cell activity. *J. Immunol.* 172:5598–5605.

Jelley-Gibbs, D.M., Lepak, N.M., Yen, M., and Swain, S.L. (2000). Two distinct stages in the transition from naive CD4 T cells to effectors, early antigen-dependent and late cytokine-driven expansion and differentiation. *J. Immunol.* 165:5017–5026.

Johansson, B.E., Bucher, D.J., and Kilbourne, E.D. (1989). Purified influenza virus hemagglutinin and neuraminidase are equivalent in stimulation of antibody response but induce contrasting types of immunity to infection. *J. Virol.* 63: 1239–1246.

Johansson, B.E., Matthews, J.T., and Kilbourne, E.D. (1998). Supplementation of conventional influenza A vaccine with purified viral neuraminidase results in a balanced and broadened immune response. *Vaccine* 16:1009–1015.

Johansson, B.E., Pokorny, B.A., and Tiso, V.A. (2002). Supplementation of conventional trivalent influenza vaccine with purified viral N1 and N2 neuraminidases induces a balanced immune response without antigenic competition. *Vaccine* 20: 1670–1674.

Julkunen, I., Melen, K., Nyqvist, M., Pirhonen, J., Sareneva, T., and Matikainen, S. (2000). Inflammatory responses in influenza A virus infection. *Vaccine* 19(Suppl 1): S32–S37.

Julkunen, I., Sareneva, T., Pirhonen, J., Ronni, T., Melen, K., and Matikainen, S. (2001). Molecular pathogenesis of influenza A virus infection and virus-induced regulation of cytokine gene expression. *Cytokine Growth Factor Rev.* 12:171–180.

Kaech, S.M., and Ahmed, R. (2001). Memory CD8+ T cell differentiation: initial antigen encounter triggers a developmental program in naive cells. *Nat. Immunol.* 2:415–422.

Kaech, S.M., Wherry, E.J., and Ahmed, R. (2002). Effector and memory T-cell differentiation: Implications for vaccine development. *Nat. Rev. Immunol.* 2:251–262.

Kaech, S.M., Tan, J.T., Wherry, E.J., Konieczny, B.T., Surh, C.D., and Ahmed, R. (2003). Selective expression of the interleukin 7 receptor identifies effector CD8 T cells that give rise to long-lived memory cells. *Nat. Immunol.* 4:1191–1198.

Kagi, D., Odermatt, B., and Mak, T.W. (1999). Homeostatic regulation of CD8+ T cells by perforin. *Eur. J. Immunol.* 29:3262–3272.

Kaisho, T., and Akira, S. (2004). Pleiotropic function of Toll-like receptors. *Microbes Infect.* 6:1388–1394.

Katze, M.G., He, Y., and Gale, M. Jr. (2002). Viruses and interferon: A fight for supremacy. *Nat. Rev. Immunol.* 2:675–687.

Kaufmann, A., Salentin, R., Meyer, R.G., Bussfeld, D., Pauligk, C., Fesq, H., Hofmann, P., Nain, M., Gemsa, D., and Sprenger, H. (2001). Defense against influenza A virus interferon: Essential role of the chemokine system. *Immunobiology* 204:603–613.

Kiecolt-Glaser, J., Glaser, R., Gravenstein, S., Malarkey, W.B., and Sheridan, J.F. (1996). Chronic stress alters the immune response to influenza virus vaccine in older adults. *Proc. Natl. Acad. Sci. U.S.A.* 93:3043–3047.

Koolhass, J.M., De Boer, S.F., De Ruiter, A.J.H., Meerlo, P., and Sgoifo, A. (1997). Social stress in rats and mice. *Acta Physiol. Scand.* 161(S640):69–72.

Knossow, M., Gaudier, M., Douglas, A., Barrere, B., Bizebard, T., Barbey, C., Gigant, B., and Skehel, J.J. (2002). Mechanism of neutralization of influenza virus infectivity by antibodies. *Virology* 302:294–298.

Konstantinos, A.P., and Sheridan, J.F. (2001). Stress and influenza viral infection: Modulation of proinflammatory cytokine responses in the lung. *Respir. Physiol.* 128:71–77.

Lanier, L.L. (1998). NK cell receptors. *Annu. Rev. Immunol.* 16:359–393.

Lanzavecchia, A., Lezzi, G., and Viola, A. (1999). From TCR engagement to T cell activation: A kinetic view of T cell behavior. *Cell* 96:1–4.

Lenz, D.C., Kurz, S.K., Lemmens, E., Schoenberger, S.P., Sprent, J., Oldstone, M.B., and Homann, D. (2004). IL-7 regulates basal homeostatic proliferation of antiviral CD4+T cell memory. *Proc. Natl. Acad. Sci. U.S.A.* 101:9357–9362.

Leung, K.N., and Ada, G.L. (1981). Induction of natural killer cells during murine influenza virus infection. *Immunobiology* 160:352–366.

Liu, B., Mori, I., Hossain, M.J., Dong, L., Chen, Z., and Kimura, Y. (2003). Local immune responses to influenza virus infection in mice with a targeted disruption of perforin gene. *Microb. Pathogen.* 34:161–167.

Lodolce, J.P., Burkett, P.R., Boone, D.L., Chien, M., and Ma, A. (2001). T cell-independent interleukin 15Ralpha signals are required for bystander proliferation. *J. Exp. Med.* 194:1187–1194.

Lopez, C.B., Moran, T.M., Schulman, J.L., and Fernandez-Sesma, A. (2002). Antiviral immunity and the role of dendritic cells. *Int. Rev. Immunol.* 21:339–353.

Lund, J.M., Alexopoulou, L., Sato, A., Karow, M., Adams, N.C., Gale, N.W., Iwasaki, A., and Flavell, R.A. (2004). Recognition of single-stranded RNA viruses by Toll-like receptor 7. *Proc. Natl. Acad. Sci. U.S.A.* 101:5598–5603.

Ma, A., Boone, D.L., and Lodolce, J.P. (2000). The pleiotropic functions of interleukin 15: Not so interleukin 2-like after all. *J. Exp. Med.* 191:753–756.

MacArthur Network on Socioeconomic Status and Health. (1997). Revised 6/3/2003. Available at http://www.macses.ucsf.edu/Default.htm.

MacLennan, I.C. (1994). Germinal centers. *Annu. Rev. Immunol.* 12:117–139.

Mailliard, R.B., Egawa, S., Cai, Q., Kalinska, A., Bykovskaya, S.N., Lotze, M.T., Kapsenberg, M.L., Storkus, W.J., and Kalinski, P. (2002). Complementary dendritic cell-activating function of CD8+ and CD4+ T cells: Helper role of CD8+ T cells in the development of T helper type 1 responses. *J. Exp. Med.* 195:473–483.

Matloubian, M., Suresh, M., Glass, A., Galvan, M., Chow, K., Whitmire, J.K., Walsh, C.M., Clark, W.R., and Ahmed, R. (1999). A role for perforin in downregulating T-cell responses during chronic viral infection. *J. Virol.* 73:2527–2536.

Matrosovich, M.N., Matrosovich, T.Y., Gray, T., Roberts, N.A., and Klenk, H.D. (2004). Human and avian influenza viruses target different cell types in cultures of human airway epithelium. *Proc. Natl. Acad. Sci. U.S.A.* 101:4620–4624.

McEwen, B.S. Protective and damaging effects of stress mediators. (1998). *N. Engl. J. Med.* 338:171–179.

McHale, J.F., Harari, O.A., Marshall, D., and Haskard, D.O. (1999). TNF-alpha and IL-1 sequentially induce endothelial ICAM-1 and VCAM-1 expression in MRL/lpr lupus-prone mice. *J. Immunol.* 163:3993–4000.

Mellon, R.D., and Bayer, B.M. (1998). Role of central opioid receptor subtypes in morphine induced alterations in peripheral lymphocyte activity. *Brain Res.* 789: 56–67.

Mills, C.E., Robins, J.M., and Lipsitch, M. (2004). Transmissibility of 1918 pandemic influenza. *Nature* 432:904–906.

Nelson, C.J., and Lysle, D.T. (2001). Involvement of substance P and central opioid receptors in morphine modulation of the CHS response. *J. Neuroimmunol.* 115: 101–110.

Nieto, M., Rodriguez-Fernandez, J.L., Navarro, F., Sancho, D., Frade, J.M., Mellado, M., Martinez-A, C., Cabanas, C., and Sanchez-Madrid, F. (1999). Signaling through CD43 induces natural killer cell activation, chemokine release, and PYK-2 activation. *Blood* 94:2767–2777.

Okada, T., Ngo, V.N., Ekland, E.H., Forster, R., Lipp, M., Littman, D.R., and Cyster, J.G. (2002). Chemokine requirements for B cell entry to lymph nodes and Peyer's patches. *J. Exp. Med.* 196:65–75.

Oran, A.E., and Robinson, H.L. (2004). DNA vaccines: influenza virus challenge of a Th2/Tc2 immune response results in a Th2/Tc1 response in the lung. *J. Virol.* 78:4376–4380.

Oxford, J.S., Lambkin, R., Sefton, A., Daniels, R., Elliot, A., Brown, R., and Gill, D. (2005). A hypothesis: The conjunction of soldiers, gas, pigs, ducks, geese and horses in northern France during the Great War provided the conditions for the emergence of the "Spanish" influenza pandemic of 1918–1919. *Vaccine* 23:940–945.

Padgett, D.A., Loria, R.M. (1994). In vitro potentiation of lymphocyte activation by dehydroepiandrosterone, androstenediol, and androstenetriol. *J. Immunol.* 153: 1544–1552.

Padgett, D.A., and Sheridan, J.F. (1999). Androstenediol (AED) prevents neuroendocrine-mediated suppression of the immune response to an influenza viral infection. *J. Neuroimmunol.* 98:121–129.

Pasare, C., and Medzhitov, R. (2004). Toll-like receptors: linking innate and adaptive immunity. *Microbes Infect.* 6:1382–1387.

Pevzner, V., Wolf, I., Burgstahler, R., Forster, R., and Lipp, M. (1999). Regulation of expression of chemokine receptor BLR1/CXCR5 during B cell maturation. *Curr. Topics Microbiol. Immunol.* 246:79–84.

Proietti, E., Bracci, L., Puzelli, S., Di Pucchio, T., Sestili, P., De Vincenzi, E., Venditti, M., Capone, I., Seif, I., De Maeyer, E., Tough, D., Donatelli, I., and Belardelli, F. (2002). Type I IFN as a natural adjuvant for a protective immune response: Lessons from the influenza vaccine model. *J. Immunol.* 169:375–383.

Qin, D., Wu, J., Vora, K.A., Ravetch, J.V., Szakal, A.K., Manser, T., and Tew, J.G. (2000). Fc gamma receptor IIB on follicular dendritic cells regulates the B cell recall response. *J. Immunol.* 164:6268–6275.

Quan, N., Avitsur, R., Stark, J.L., He, L., Shah, M., Caliguiri, M., Padgett, D.A., Marucha, P.T., and Sheridan, J.F. (2001). Social stress increases the susceptibility to endotoxic shock. *J. Neuroimmunol.* 115:36–45.

Quan, N., Avitsur, R., Stark, J.L., He, L., Lai, W., Dhabhar, F.S., and Sheridan, J.F. (2003). Molecular mechanisms of glucocorticoid resistance in splenocytes of socially stressed male mice. *J. Neuroimmunol.* 137:51–58.

Randolph, G.J., and Furie, M.B. (1995). A soluble gradient of endogenous monocyte chemoattractant protein-1 promotes the transendothelial migration of monocytes in vitro. *J. Immunol.* 155:3610–3618.

Rosseau, S., Selhorst, J., Wiechmann, K., Leissner, K., Maus, U., Mayer, K., Grimminger, F., Seeger, W., and Lohmeyer, J. (2000). Monocyte migration through the alveolar epithelial barrier: Adhesion molecule mechanisms and impact of chemokines. *J. Immunol.* 164:427–435.

Sallusto, F., Geginat, J., and Lanzavecchia, A. (2004). Central memory and effector memory T cell subsets: Function, generation, and maintenance. *Annu. Rev. Immunol.* 22:745–763.

Samuel, C.E. (1991). Antiviral actions of interferon. Interferon-regulated cellular proteins and their surprisingly selective antiviral activities. *Virology* 183:1–11.

Selye, H. (1936). A syndrome produced by diverse nocuous agents. *Nature (London)* 138:32.

Sheridan, J.F., Feng, N.G., Bonneau, R.H., Allen, C.M., Huneycutt, B.S., and Glaser, R. (1991). Restraint stress differentially affects antiviral cellular and humoral immune responses in mice. *J. Neuroimmunol.* 31:245–255.

Sprent, J., Zhang, X., Sun, S., and Tough, D. (2000). T-cell proliferation in vivo and the role of cytokines. *Philos. Trans. R. Soc. London B Biol. Sci.* 355:317–322.

Stark, J., Avitsur, R., Padgett, D.A., and Sheridan, J.F. (2001). Social stress induces glucocorticoid resistance in macrophages. *Am. J. Physiol. Regul. Integr. Comp. Physiol.* 280:R1799–R1805.

Stein-Streilein, J., and Guffee, J. (1986). In vivo treatment of mice and hamsters with antibodies to asialo GM1 increases morbidity and mortality to pulmonary influenza infection. *J. Immunol.* 136:1435–1441.

Stein-Streilein, J., Guffee, J., and Fan, W. (1988). Locally and systemically derived natural killer cells participate in defense against intranasally inoculated influenza virus. *Regional Immunol.* 1:100–105.

Strieter, R.M., Belperio, J.A., and Keane, M.P. (2003). Host innate defenses in the lung: The role of cytokines. *Curr. Opin. Infect. Dis.* 16:193–198.

Tan, J.T., Dudl, E., LeRoy, E., Murray, R., Sprent, J., Weinberg, K.I., and Surh, C.D. (2001). IL-7 is critical for homeostatic proliferation and survival of naive T cells. *Proc. Natl. Acad. Sci. U.S.A.* 98:8732–8737.

Tew, J.G., Mandel, T.E., Phipps, R.P., and Szakal, A.K. (1984). Tissue localization and retention of antigen in relation to the immune response. *Am. J. Anat.* 170:407–420.

Tew, J.G., Wu, J., Fakher, M., Szakal, A.K., and Qin, D. (2001). Follicular dendritic cells: Beyond the necessity of T-cell help. *Trends Immunol.* 22:361–367.

Topham, D.J., Tripp, R.A., and Doherty, P.C. (1997). CD8+ T cells clear influenza virus by perforin or Fas-dependent processes. *J. Immunol.* 159:5197–5200.

Tseng, R., Padgett, D.A., Dhabhar, F.S., Engler, H., and Sheridan, J.F. (2005). Stress-induced modulation of NK activity during influenza viral infection: Role of glucocorticoids and opioids. *Brain Behav. Immun.* 19:153–164.

Walzer, T., Galibert, L., Comeau, M.R., and De Smedt, T. (2005). Plexin C1 engagement on mouse dendritic cells by viral semaphorin A39R induces actin cytoskeleton rearrangement and inhibits integrin-mediated adhesion and chemokine-induced migration. *J. Immunol.* 174:51–59.

Weber, K.S., von Hundelshausen. P., Clark-Lewis, I., Weber, P.C., and Weber, C. (1999). Differential immobilization and hierarchical involvement of chemokines in monocyte arrest and transmigration on inflamed endothelium in shear flow. *Eur. J. Immunol.* 29:700–712.

Zhang, X., Sun, S., Hwang, I., Tough, D.F., and Sprent, J. (1998). Potent and selective stimulation of memory-phenotype CD8+ T cells in vivo by IL-15. *Immunity* 8: 591–599.

9
Autonomic Nervous System Influences on HIV Pathogenesis

ERICA K. SLOAN, ALICIA COLLADO-HIDALGO, AND STEVE W. COLE

1. Introduction

During the past decade, our laboratory has carried out a series of studies analyzing the effects of autonomic nervous system (ANS) activity on HIV-1 pathogenesis (Cole *et al.*, 1998). These studies were motivated by natural history studies showing accelerated HIV disease progression in gay men who had socially inhibited personality characteristics (Cole *et al.*, 1996, 2003). Previous developmental studies have suggested that socially inhibited individuals show elevated levels of ANS activity (Block, 1957; Buck *et al.*, 1974; Cole *et al.*, 1999b; Miller *et al.*, 1999), providing a potential neurobiological basis for differential HIV disease progression. In a subsequent cohort study of 54 HIV-positive gay men with early- to mid-stage infection (no AIDS, and CD4$^+$ T cell levels >200/mm^3), we found that socially inhibited individuals did indeed show elevated levels of ANS activity. ANS activity was measured across a range of end-organ responses including palmar skin conductance, blood pressure, heart rate interbeat interval, finger pulse amplitude, and peripheral pulse transit time (time from heart beat to subsequent finger pulse peak) (Fig. 9.1). Baseline autonomic activity and reactivity to a series of physical, psychological, and social stimuli was found to be stable over time. Individuals showing constitutively high levels of ANS activity also showed elevated plasma viral load (Fig. 9.1) and impaired suppression of viremia and CD4$^+$ T-lymphocyte recovery after the onset of combination antiretroviral therapy (Cole *et al.*, 2001, 2003). Regression-based mediational analyses suggested that individual differences in ANS activity could potentially account for 62–94% of the total association between social inhibition and indicators of HIV pathogenesis (Cole *et al.*, 2003). Although these data represent a cross-sectional analysis, the strong linear correlations between psychosocial characteristics, ANS activity, and virologic parameters suggested that ANS activity could represent an important physiologic influence on HIV-1 pathogenesis. Based on the clinical findings above, we sought to determine whether the biochemical products of autonomic nervous system activity might accelerate HIV-1 replication—the fundamental engine driving HIV disease progression.

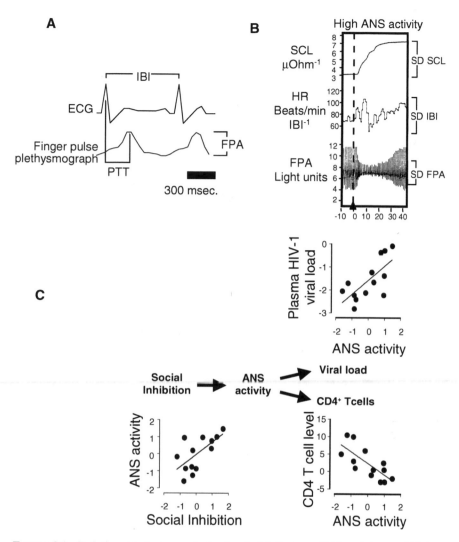

FIGURE 9.1. Relationship between behavioral risk factors, ANS activity, and HIV-1 pathogenesis. ANS activity was measured by skin conductance level (SCL), systolic blood pressure (SBP), electrocardiogram (ECG) interbeat interval (duration between R-spikes; IBI), finger photoplethysmograph pulse peak amplitude (FPA), and peripheral pulse transit time (duration from EKG R-spike to subsequent finger photoplethysmograph peak; PTT) (A). These parameters are altered by ANS activation, as shown in (B), where a participant responds to unexpected inflation of a blood-pressure cuff at the indicated time (dashed line). Constitutive individual differences in ANS activity were quantified by averaging standardized responses on each physiologic indicator over multiple tasks and measurement occasions. (C) Socially inhibited individuals showed elevated ANS activity (units = standard deviation relative to the mean), and high ANS activity was associated with elevated HIV-1 plasma viral load setpoint (not shown) as well as poorer suppression of plasma viral load ($\log_{10}$ copies/ml) and impaired recovery of CD4+ T-lymphocyte levels (% of total lymphocytes) after onset of antiretroviral therapy.

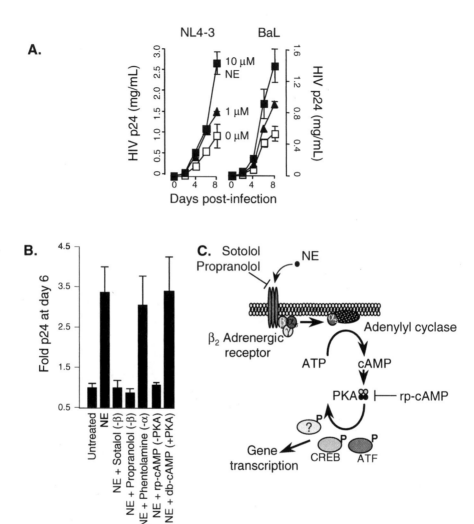

FIGURE 9.2. Norepinephrine accelerates HIV-1 replication in T lymphocytes via β-adrenergic activation of PKA. Physiological concentrations of norepinephrine (NE) accelerated HIV-1 replication as measured by p24 gag levels in supernatants of HIV-1–infected T lymphocytes (A). The β-adrenergic antagonists sotalol or propranalol blocked NE-induced acceleration of HIV-1 replication, but the α-adrenergic antagonist phentolamine did not. The PKA antagonist rp-cAMP also efficiently blocked NE-mediated acceleration of viral replication, and the pharmacologic PKA activator db-cAMP mimicked norepinephrine's effect (B and C).

2. Catecholamine Acceleration of HIV-1 Replication

The sympathetic branch of the ANS mediates fight-or-flight stress responses by activating a network of sympathetic neurons that innervate most organ systems of the body (Weiner, 1992; Sapolsky, 1998). In addition to classical stress-responsive targets such as the cardiovascular, endocrine, and gastrointestinal systems, neurons from the sympathetic nervous system (SNS) also innervate all primary and secondary lymphoid organs (Felten *et al.*, 1984). When activated by stress or other stimuli, these neurons release micromolar concentrations of catecholaminergic neurotransmitters—epinephrine and norepinephrine—into parenchymal tissues, particularly those bearing macrophages and T lymphocytes. We examined norepinephrine's impact on HIV-1 replication in a simple model system of viral replication in primary T lymphocytes. Peripheral blood mononuclear cells (PBMCs) from healthy donors were infected with HIV-1 (CXCR4-tropic NL4-3 or CCR5-tropic BaL strains, MOI ~ 0.05), and T lymphocytes were activated with antibodies to CD3 and the CD28 costimulatory molecule. When these stimuli were supplemented with micromolar concentrations of catecholamines, HIV-1 replication rates increased by up to 11-fold (Fig. 9.2A). Pharmacologic inhibitor studies showed that these effects were mediated by β_2-adrenergic activation of the cyclic adenosine monophosphate/protein kinase A (cAMP/PKA) signaling pathway (Fig. 9.2B). Replication-enhancing effects of norepinephrine could be mimicked by the adenylyl cyclase activator forskolin or the membrane-permeable cAMP analogue db-cAMP. Thus, any stimulus that activates the cellular cAMP/PKA signaling pathway appears to be capable of accelerating HIV-1 replication in primary T lymphocytes.

3. Virologic Mechanisms

To identify specific points of impact on the viral replication cycle, we carried out one-round infection and one-round viral gene expression assays. The HIV-1 replication cycle can be divided into two broad trajectories: (1) a process of *infection* in which the viral RNA genome is introduced into the cell and reverse transcribed into a DNA provirus, which then integrates into the host cell's chromosomes, and (2) a process of viral gene *expression* in which cellular activation induces transcription and translation of viral genes, followed by assembly and budding of new virus particles.

To determine whether norepinephrine might increase cellular vulnerability to infection, we carried out one-round infection assays in which cells were pretreated with either medium or norepinephrine and then exposed to a fixed concentration of HIV-1 virions. Proviral HIV-1 DNA was assayed by PCR 12h later, and cells were cultured in the antiretroviral drug Indinivir to prevent production of infectious virions and subsequent rounds of infection. Exposure to micromolar concentrations of norepinephrine enhanced HIV-1 proviral penetrance by three- to fivefold (Fig. 9.3A). Subsequent studies identified the primary molecular mechanism of this effect in increased expression of the chemokine receptors CCR5 and CXCR4 that collaborate with CD4 to mediate HIV-1 entry into human cells (Cole *et al.*, 1999a) (Fig. 9.3A). Activation of the PKA signaling pathway by norepinephrine or cAMP increased the cell surface density of CXCR4 by 6- to 10-fold on quiescent and activated lymphocytes, and norepinephrine also upregulated cell surface expression of CCR5 (Cole *et al.*, 2001). Chemokine receptors constitutively traffic between the cell surface and endosomal compartments within the cell. Flow cytometric analyses demonstrated that PKA can inhibit CXCR4 internalization and thus enhance the fraction of the total receptor pool localized to the cell surface. HIV-1 binding assays showed that PKA-induced externalization of CXCR4 enhanced virion binding to the cell membrane. PKA-induced externalization of CXCR4 also

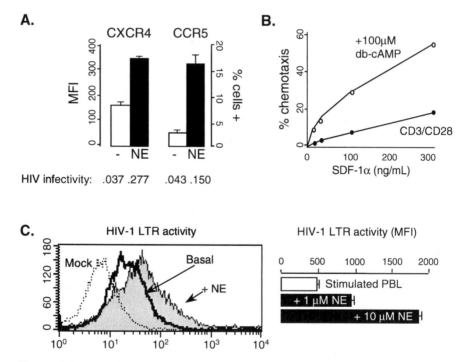

FIGURE 9.3. Norepinephrine increases HIV-1 infectivity and gene expression. Treatment of PBMC with norepinephrine increased cell-surface expression of the CXCR4 and CCR5 chemokine receptors (A, top), resulting in enhanced vulnerability to viral infection as measured by PCR detection of proviral DNA (copies per β-globin) at 12 h after exposure to virus (A, bottom). The PKA activator db-cAMP also enhanced cell-surface CXCR4 density on CD3/CD28 costimulated lymphocytes, resulting in enhanced chemotaxis in response to the chemokine SDF1α (B). Norepinephrine also enhanced activity of the HIV-1 promoter (LTR), as measured by expression of a murine CD24 reporter gene in activated T lymphocytes (C).

enhanced cellular chemotaxis in response to CXCR4's cognate ligand SDF-1α (Bleul *et al.*, 1996; Oberlin *et al.*, 1996) (Fig. 9.3B). Norepinephrine signaling enhanced CXCR4 surface levels in both CD4$^+$ and CD8$^+$ T lymphocytes as well as CD19$^+$ B cells. In CD14$^+$ monocytes, PKA suppressed cell surface expression of CXCR4. These data suggest that catecholamine activation of the PKA signaling pathway might modulate normal cellular trafficking and maturation processes in addition to enhancing cellular vulnerability to HIV-1 infection (Cole *et al.*, 1999a).

To determine whether catecholamines might influence the viral gene expression phase of the life cycle, we also carried out one-round expression assays in which a population of primary PBMCs was infected with an HIV-1 reporter virus bearing the murine CD24 gene (Jamieson and Zack, 1998). Cells were then activated with antibodies to CD3 and CD28 in the presence of Indinivir to prevent subsequent rounds of infection. When these stimulation conditions were supplemented with micromolar concentrations of norepinephrine, flow cytometry showed a three- to fivefold increase in cell surface expression of the reporter gene product (Cole *et al.*, 2001) (Fig. 9.3C). Thus, norepinephrine can enhance expression of genes under the control of the HIV-1 promoter—the viral long terminal repeat (LTR).

In addition to the direct effects on the viral replication cycle, catecholamines can also exert indirect or permissive effects on viral replication by undermining cellular production of antiviral cytokines. Immunoregulatory cytokines such as IL-10 can suppress HIV replication, whereas proinflamatory cytokines enhance HIV replication by activating NF-κB (Finnegan *et al.*, 1996; Badou *et al.*, 2000; Weissman *et al.*, 1995). Analysis of supernatant cytokine concentrations from CD3/CD28 costimulated T lymphocytes showed that norepinephrine could substantially suppress the production of both IL-10 and IFN-γ (Fig. 9.4A). Addition of exogenous IL-10 or IFN-γ to the cultures abrogated the replication-enhancing effects of norepinephrine (Fig. 9.4B), indicating that simultaneous suppression of both cytokines is required for norepinephrine-mediated enhancement of HIV-1 replication. Cytokine modulation might represent an overarching distal influence on more virus-proximal mechanisms such as LTR activity or chemokine receptor trafficking.

In a series of studies on type I interferon responses to HIV-1 or Toll-like receptor (TLR) ligands, we have also found that norepinephrine can suppress production of this crucial innate antiviral cytokine by both myeloid and lymphoid dendritic cells. These effects are mediated via the β-AR/cAMP/PKA signaling cascade, which inhibits expression of both IFN-α and IFN-β genes in response to activation of either TLR-3 or TLR-9 (Fig. 9.4C). Using the IFN-β gene as a model system, we traced PKA's effects to the modulation of transcription control processes in a specific positive regulatory domain (PRD) in the promoter of the IFNB gene (Collado-Hidalgo *et al.*, 2006). Type I interferons are known to suppress HIV-1 replication (Baca-Regen *et al.*, 1994; Pitha, 1994; Agy *et al.*, 1995; Korth *et al.*, 1998), so

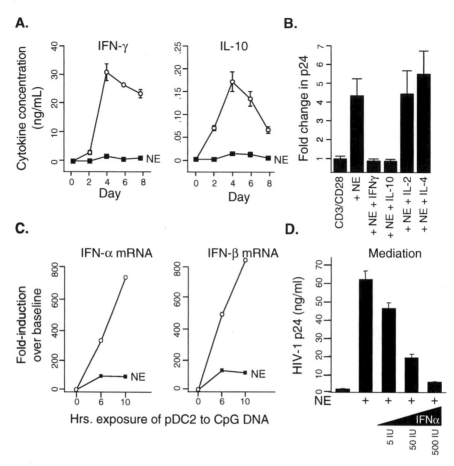

FIGURE 9.4. Norepinephrine inhibits secretion of antiviral cytokines. Norepineph-rine suppressed production of anti-inflammatory cytokines IFN-γ and IL-10 by HIV-1–infected PBMCs (A). Addition of exogenous IFN-γ and IL-10, but not IL-2 or IL-4, inhibited NE acceleration of HIV-1 replication (B). Plasmacytoid dendritic cell expression of type I IFN genes was inhibited by NE (C), and addition of exogenous IFN-α inhibited NE's effect on HIV-1 replication in activated T lymphocytes (D).

we tested their role as potential mediators of norepinephrine's effects on viral replication by supplementing norepinephrine (NE)-treated cultures with exogenous IFN-α. Because these studies focused on the role of cytokines in controlling HIV-1 replication, we sought to isolate the effects of soluble factors from all other impacts of norepinephrine on cellular func-tion (e.g., direct effects on transcription factors activating the LTR or chemokine receptors mediating infection). Uninfected PBMCs were stim-ulated with phytohemagglutinin (PHA) and IL-2 for 2 days in the presence of 0 or 10μM NE to induce cytokine expression. Cells were washed twice in PBS to remove NE and cultured for another 24h in fresh medium to collect soluble factors modulated by prior exposure to norepinephrine. To determine whether the resulting "norepinephrine-conditioned" cytokine profile facilitated HIV-1 replication, filtered supernatants were transferred to fresh PBMCs that had been infected with CXCR4-tropic HIV-1$_{NL4-3}$ and stimulated with phytohemagglutinin (PHA). Over the ensuing 5 days, norepinephrine-conditioned supernatants facilitated a 34-fold increase in HIV-1 replication relative to control supernatants (Fig. 9.4D). To determine whether norepinephrine-induced suppression of type I interferons contributed to this effect, we added graded doses of recombinant IFN-α (0, 5, 50, or 500 IU) to norepinephrine-conditioned cultures in an attempt to reverse that effect. Increasing concentrations of exogenous type I interferon inhibited norepinephrine-mediated support for HIV-1 replication in a dose-dependent manner. Thus norepinephrine-mediated suppression of type I interferon production appears to play a critical role in permitting upregulated HIV-1 replication.

These studies have defined three basic molecular mechanisms by which catecholamine signaling from the ANS can impact the HIV-1 replication cycle. (1) Catecholamines influence the earliest stages of infection by increasing chemokine receptor density on the cell surface and thus enhance viral entry. (2) Catecholamines facilitate subsequent transcription of the viral genome by enhancing transcriptional activity of the HIV-1 LTR. (3) Catecholamine signaling undermines cellular production of antiviral cytokines and thus permits other molecular mechanisms to enhance HIV-1 replication rates. The precise interrelationships among these three processes are not yet known, but their collective effect is sufficient to enhance viral replication by ~10-fold *in vitro*. The magnitude of such differences is commensurate with the 10- to 100-fold variation in plasma viral load setpoints observed in patients with high- versus low levels of constitutive ANS activity (Cole *et al.*, 2001).

4. The Neuroanatomic Basis of ANS Interactions

To determine whether the molecular processes identified *in vitro* contribute to the effects of ANS activity and HIV-1 pathogenesis *in vivo*, we have begun to analyze the relationship between ANS innervation and simian immunodeficiency virus (SIV) replication in lymph nodes from rhesus macaques—the best animal model for human HIV-1 infection. HIV-1 replicates predominantly in lymphoid organs, (Embretson *et al.*, 1993; Pantaleo *et al.*, 1993), and these tissues are known to receive innervation from the sympathetic nervous system (SNS) (Felten *et al.*, 1984, 1987). Catecholaminergic neurons enter lymph nodes in association with the vasculature and radiate out into the parenchyma to terminate in the vicinity of T lymphocytes and macrophages—the cells in which HIV-1 and SIV replicate most efficiently (Stebbing *et al.*, 2004). Sympathetic fibers reach paracortical, medullary, and interfollicular cor-tical regions but avoid nodular regions and germinal centers (Felten and Olschowka, 1987; Felten *et al.*, 1987; Fink

and Weihe, 1988). Several lines of evidence suggest that autonomic innervation of lymphoid tissue may affect immune function (Madden *et al.*,
1995a, 1995b; Downing and Miyan, 2000; Castrillon *et al.*, 2000), but little is
known about its impact on lymphotropic viruses. We mapped the distribution of catecholaminergic neurons in lymph nodes from SIV-infected
macaques using sucrose phosphate glyoxylic acid (SPG) chemofluorescence
and carried out spatial statistical analyses to determine whether SIV replication was increased in the near vicinity. Active SIV replication was mapped
by *in situ* hybridization to *gag, pol*, and *nef* mRNA. Spatial statistical analyses showed that the odds of active SIV replication increased by 3.9-fold in
the vicinity of catecholaminergic neurons (p < 0.0001) (Fig. 9.5) Sloan *et al.*,
2006. Density of SNS innervation and SIV replication differed across cortical, paracortical, and medullary regions of the lymph node, but analyses
controlling for those differences continued to indicate increased SIV replication in the vicinity of catecholaminergic neurons. The density of SNS
innervation was also found to be reduced in lymph nodes showing high concentrations of SIV replication. Such results are consistent with previous
data suggesting that inflammatory processes may deplete sympathetic
neurons (Miller *et al.*, 2000; Kelley *et al.*, 2003) and suggest that SNS influences on lentiviral replication may be most pronounced during the early
stages of infection while innervation is maximal. Such dynamics might
explain why experimental stress manipulations have been shown to alter
viral load setpoints early in infection in SIV-infected macaques (Capitanio
et al., 1998).

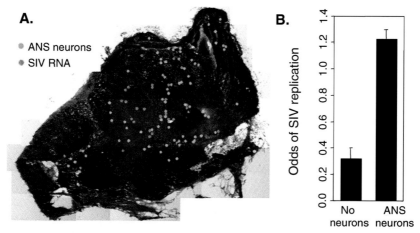

FIGURE 9.5. SIV replication occurs in the vicinity of lymph node neurons. In a study
of 15 rhesus macaques infected with SIVmac251, 22 lymph nodes were biopsied and
mapped for SIV replication using *in situ* hybridization. Catecholaminergic neurons
were mapped by glyoxylic acid chemofluorescence (A), and spatial statistical analysis showed a 3.9-fold increase in the odds of active SIV replication in lymph node
tissue areas containing ANS neurons (B).

5. Conclusion

Neuroanatomic data from the SIV system are consistent with the hypothesis that sympathetic neurons might enhance HIV-1 replication in patients with high levels of ANS activity (Cole *et al.*, 2001, 2003). Several molecular mechanisms that could mediate these effects have been identified, including altered chemokine receptor expression, transcriptional activation of the HIV-1 LTR, and suppression of antiviral cytokine production (Cole *et al.*, 1998, 2001). These findings provide a coherent mechanistic explanation for several studies linking psychosocial characteristics to differential HIV disease progression (Cole and Kemeny, 1997, 2001). It is not yet clear how ANS-targeted interventions might affect the course of HIV-1 infection, but previous studies of correlated psychological risk factors suggest that naturally occurring individual differences in autonomic activity may accelerate the typical 10-year disease trajectory by as much as 2 to 3 years (Cole *et al.*, 1996, 1997). A clearer portrait of clinical significance will emerge from ongoing trials of β-adrenoreceptor antagonists (β-blockers) as adjuvant therapies for early stage HIV infection. As clinical practice shifts toward delayed initiation of antiretroviral therapy (Panel on Clinical Practices for Treatment of HIV Infection, 2004), the availability of safe interventions to slow viral replication during the early stages of disease progression becomes increasingly valuable. If ANS activity exerts a clinically significant impact on HIV replication, long-term disease progression could be significantly slowed by the provision of pharmacologic or behavioral treatments that reduce ANS activity (Antoni, 2003). ANS-targeted interventions may be most effective during early stage infection when lymphoid innervation is most pronounced and behavioral stress can influence long-term viral replication setpoints (Capitanio *et al.*, 1998). Once disease progresses to the point where antiretrovirals are indicated, suppression of ANS activity may also help reduce physiologic support for low-level residual viral replication that continues during antiretroviral treatment (Finzi *et al.*, 1997; Davey *et al.*, 1999; Chun *et al.*, 2000; Ramratnam *et al.*, 2000; Sharkey *et al.*, 2000). To provide a basic virologic context for such interventions, our ongoing work seeks to define the role of previously identified molecular mechanisms *in vivo* and to evaluate the clinical significance of ANS modulation in delaying clinical disease progression and preserving the health of HIV-positive individuals.

Acknowledgements. The authors acknowledge the contributions of John Capitanio, Margaret Kemeny, Bruce Naliboff, Suzanne Stevens, Ross Tarara, and Jerry Zack. This research was supported by the NIAID (AI52737, AI33259, AI36554, AI49135), NIMH (MH15750, MH00820), the Norman Cousins Center, the UCLA AIDS Institute, the University of California AIDS Research Program (K99-LA-030), and the James Pendelton Charitable Trust.

References

Agy, M.B., Acker, R.L., Sherbert, C.H., and Katze, M.G. (1995). Interferon treatment inhibits virus replication in HIV-1- and SIV-infected CD4+ T-cell lines by distinct mechanisms: Evidence for decreased stability and aberrant processing of HIV-1 proteins. *Virology* 214(2):379–386.

Antoni, M.H. (2003). Stress management effects on psychological, endocrinological, and immune functioning in men with HIV infection: Empirical support for a psychoneuroimmunological model. *Stress* 6(3):173–188.

Baca-Regen, L., Heinzinger, N., Stevenson, M., and Gendelman, H.E. (1994). Alpha interferon-induced antiretroviral activities: restriction of viral nucleic acid synthesis and progeny virion production in human immunodeficiency virus type 1-infected monocytes. *J. Virol.* 68(11):7559–7565.

Badou, A., Bennasser, Y., Moreau, M., Leclerc, C., Benkirane, M., and Bahraoui, E. (2000). Tat protein of human immunodeficiency virus type 1 induces interleukin-10 in human peripheral blood monocytes: Implication of protein kinase C-dependent pathway. *J. Virol.* 74(22):10551–10562.

Bleul, C.C., Farzan, M., Choe, H., Parolin, C., Clark-Lewis, I., Sodroski, J., and Springer, T.A. (1996). The lymphocyte chemoattractant SDF-1 is a ligand for LESTR/fusin and blocks HIV-1 entry. *Nature* 382(6594):829–833.

Block, J. (1957). A study of affective responsiveness in a lie-detection situation. *J. Abnorm. Psychol.* 55(1):11–15.

Buck, R., Miller, R.E., and Caul, W.F. (1974). Sex, personality, and physiological variables in the communication of affect via facial expression. *J. Pers. Soc. Psychol.* 30(4):587–596.

Capitanio, J.P., Mendoza, S.P., Lerche, N.W., and Mason, W.A. (1998). Social stress results in altered glucocorticoid regulation and shorter survival in simian acquired immune deficiency syndrome. *Proc. Natl. Acad. Sci. U.S.A.* 95(8):4714–4719.

Castrillon, P.O., Cardinali, D.P., Arce, A., Cutrera, R.A., and Esquifino, A.I. (2000). Interferon-gamma release in sympathetically denervated rat submaxillary lymph nodes. *Neuroimmunomodulation* 8(4):197–202.

Chun, T.W., Davey, R.T., Jr., Ostrowski, M., Shawn Justement, J., Engel, D., Mullins, J.I., and Fauci, A.S. (2000). Relationship between pre-existing viral reservoirs and the re-emergence of plasma viremia after discontinuation of highly active anti-retroviral therapy. *Nat. Med.* 6(7):757–761.

Cole, S.W., and Kemeny, M.E. (1997). Psychobiology of HIV infection. *Crit. Rev. Neurobiol.* 11(4):289–321.

Cole, S., and Kemeny, M.E. (2001). Psychosocial Influences on the Progression of HIV Infection. In R. Ader, D.L. Felten, and N. Cohen (eds.), Psychoneuroimmunology, 3rd ed., Vol. 2. San Diego: Academic Press. 538–612.

Cole, S.W., Kemeny, M.E., Taylor, S.E., Visscher, B.R., and Fahey, J.L. (1996). Accelerated course of human immunodeficiency virus infection in gay men who conceal their homosexual identity. *Psychosom. Med.* 58(3):219–231.

Cole, S.W., Kemeny, M.E., and Taylor, S.E. (1997). Social identity and physical health: accelerated HIV progression in rejection-sensitive gay men. *J. Pers. Soc. Psychol.* 72(2):320–335.

Cole, S.W., Korin, Y.D., Fahey, J.L., and Zack, J.A. (1998). Norepinephrine accelerates HIV replication via protein kinase A-dependent effects on cytokine production. *J. Immunol.* 161(2):610–616.

Cole, S.W., Jamieson, B.D., and Zack, J.A. (1999a). cAMP up-regulates cell surface expression of lymphocyte CXCR4: implications for chemotaxis and HIV-1 infection. *J. Immunol.* 162(3):1392–1400.

Cole, S.W., Kemeny, M.E., Weitzman, O.B., Schoen, M., and Anton, P.A. (1999b). Socially inhibited individuals show heightened DTH response during intense social engagement. *Brain Behav. Immun.* 13(2):187–200.

Cole, S.W., Naliboff, B.D., Kemeny, M.E., Griswold, M.P., Fahey, J.L., and Zack, J.A. (2001). Impaired response to HAART in HIV-infected individuals with high autonomic nervous system activity. *Proc. Natl. Acad. Sci. U.S.A.* 98(22):12695–12700.

Cole, S.W., Kemeny, M.E., Fahey, J.L., Zack, J.A., and Naliboff, B.D. (2003). Psychological risk factors for HIV pathogenesis: mediation by the autonomic nervous system. *Biol. Psychiatry* 54(12):1444–1456.

Davey, R.T., Jr., Bhat, N., Yoder, C., Chun, T.W., Metcalf, J.A., Dewar, R., Natarajan, V., Lempicki, R.A., Adelsberger, J.W., Miller, K.D., Kovacs, J.A., Polis, M.A., Walker, R.E., Falloon, J., Masur, H., Gee, D., Baseler, M., Dimitrov, D.S., Fauci, A.S., and Lane, H.C. (1999). HIV-1 and T cell dynamics after interruption of highly active antiretroviral therapy (HAART) in patients with a history of sustained viral suppression. *Proc. Natl. Acad. Sci. U.S.A.* 96(26):15109–15114.

Downing, J.E., and Miyan, J.A. (2000). Neural immunoregulation: emerging roles for nerves in immune homeostasis and disease. *Immunol. Today* 21(6):281–289.

Embretson, J., Zupancic, M., Ribas, J.L., Burke, A., Racz, P., Tenner-Racz, K., and Haase, A.T. (1993). Massive covert infection of helper T lymphocytes and macrophages by HIV during the incubation period of AIDS. *Nature* 362(6418): 359–362.

Felten, S.Y., and Olschowka, J. (1987). Noradrenergic sympathetic innervation of the spleen: II. Tyrosine hydroxylase (TH)-positive nerve terminals form synaptic-like contacts on lymphocytes in the splenic white pulp. *J. Neurosci. Res.* 18(1): 37–48.

Felten, D.L., Livnat, S., Felten, S.Y., Carlson, S.L., Bellinger, D.L., and Yeh, P. (1984). Sympathetic innervation of lymph nodes in mice. *Brain Res. Bull.* 13(6):693–699.

Felten, D.L., Felten, S.Y., Bellinger, D.L., Carlson, S.L., Ackerman, K.D., Madden, K.S., Olschowki, J.A., and Livnat, S. (1987). Noradrenergic sympathetic neural interactions with the immune system: Structure and function. *Immunol. Rev.* 100: 225–260.

Fink, T., and Weihe, E. (1988). Multiple neuropeptides in nerves supplying mammalian lymph nodes: messenger candidates for sensory and autonomic neuro-immunomodulation? *Neurosci. Lett.* 90(1–2):39–44.

Finnegan, A., Roebuck, K.A., Nakai, B.E., Gu, D.S., Rabbi, M.F., Song, S., and Landay, A.L. (1996). IL-10 cooperates with TNF-alpha to activate HIV-1 from latently and acutely infected cells of monocyte/macrophage lineage. *J. Immunol.* 156(2):841–851.

Finzi, D., Hermankova, M., Pierson, T., Carruth, L.M., Buck, C., Chaisson, R.E., Quinn, T.C., Chadwick, K., Margolick, J., Brookmeyer, R., Gallant, J., Markowitz, M., Ho, D.D., Richman, D.D., and Siliciano, R.F. (1997). Identification of a reservoir for HIV-1 in patients on highly active antiretroviral therapy. *Science* 278(5341):1295–1300.

Jamieson, B.D., and Zack, J.A. (1998). In vivo pathogenesis of a human immuno-deficiency virus type 1 reporter virus. *J. Virol.* 72(8):6520–6526.

Kelley, S.P., Moynihan, J.A., Stevens, S.Y., Grota, L.J., and Felten, D.L. (2003). Sympathetic nerve destruction in spleen in murine AIDS. *Brain Behav. Immun.* 17(2): 94–109.

Korth, M.J., Taylor, M.D., and Katze, M.G. (1998). Interferon inhibits the replication of HIV-1, SIV, and SHIV chimeric viruses by distinct mechanisms. *Virology* 247(2):265–273.

Madden, K.S., Felten, S.Y., Felten, D.L., and Bellinger, D.L. (1995a). Sympathetic nervous system—immune system interactions in young and old Fischer 344 rats. *Ann. N.Y. Acad. Sci.* 771:523–534.

Madden, K.S., Sanders, V.M., and Felten, D.L. (1995b). Catecholamine influences and sympathetic neural modulation of immune responsiveness. *Annu. Rev. Pharmacol. Toxicol.* 35:417–448.

Miller, G.E., Cohen, S., Rabin, B.S., Skoner, D.P., and Doyle, W.J. (1999). Personality and tonic cardiovascular, neuroendocrine, and immune parameters. *Brain Behav. Immun.* 13(2):109–123.

Miller, L.E., Justen, H.P., Scholmerich, J., and Straub, R.H. (2000). The loss of sympathetic nerve fibers in the synovial tissue of patients with rheumatoid arthritis is accompanied by increased norepinephrine release from synovial macrophages. *FASEB J.* 14(13):2097–2107.

Oberlin, E., Amara, A., Bachelerie, F., Bessia, C., Virelizier, J.L., Arenzana-Seisdedos, F., Schwartz, O., Heard, J.M., Clark-Lewis, I., Legler, D.F., Loetscher, M., Baggiolini, M., and Moser, B. (1996). The CXC chemokine SDF-1 is the ligand for LESTR/fusin and prevents infection by T-cell-line-adapted HIV-1. *Nature* 382(6594):833–835.

Panel on Clinical Practices for Treatment of HIV Infection. (2004). Guidelines for the Use of Antiretroviral Agents in HIV-1 Infected Adults and Adolescents. Washington, DC: U.S. Department of Health and Human Services, p. 115.

Pantaleo, G., Graziosi, C., Demarest, J.F., Butini, L., Montroni, M., Fox, C.H., Orenstein, J.M., Kotler, D.P., and Fauci, A.S. (1993). HIV infection is active and progressive in lymphoid tissue during the clinically latent stage of disease. *Nature* 362(6418):355–358.

Pitha, P.M. (1994). Multiple effects of interferon on the replication of human immunodeficiency virus type 1. *Antiviral Res.* 24(2–3):205–219.

Ramratnam, B., Mittler, J.E., Zhang, L., Boden, D., Hurley, A., Fang, F., Macken, C.A., Perelson, A.S., Markowitz, M., and Ho, D.D. (2000). The decay of the latent reservoir of replication-competent HIV-1 is inversely correlated with the extent of residual viral replication during prolonged anti-retroviral therapy. *Nat. Med.* 6(1):82–85.

Sapolsky, R.M. (1998). *Why Zebras Don't Get Ulcers: An Updated Guide To Stress, Stress Related Diseases, and Coping.* New York: W.H. Freeman.

Sharkey, M.E., Teo, I., Greenough, T., Sharova, N., Luzuriaga, K., Sullivan, J.L., Bucy, R.P., Kostrikis, L.G., Haase, A., Veryard, C., Davaro, R.E., Cheeseman, S.H., Daly, J.S., Bova, C., Ellison, R.T., 3rd, Mady, B., Lai, K.K., Moyle, G., Nelson, M., Gazzard, B., Shaunak, S., and Stevenson, M. (2000). Persistence of episomal HIV-1 infection intermediates in patients on highly active anti-retroviral therapy. *Nat. Med.* 6(1):76–81.

Sloan, E.K., Tarara, R.P., Capitanio, J.P., and Cole, S.W. (2006). Enhanced SIV replication adjacent to catechnolaminergic varicosities in primate lymph nodes. *Journal of Virology.* 80(9):4326–4335.

Stebbing, J., Gazzard, B., and Douek, D.C. (2004). Where does HIV live? *N. Engl. J. Med.* 350(18):1872–1880.

Weiner, H. (1992). *Perturbing the Organism: The Biology of Stressful Experience.* Chicago: University of Chicago Press.

Weissman, D., Poli, G., and Fauci, A.S. (1995). IL-10 synergizes with multiple cytokines in enhancing HIV production in cells of monocytic lineage. *J. Acquir. Immune Defic. Syndr. Hum. Retrovirol.* 9(5):442–449.

10

The Effects of Restraint Stress on the Neuropathogenesis of Theiler's Virus-induced Demyelination: A Murine Model for Multiple Sclerosis

C. Jane Welsh, Wentao Mi, Amy Sieve, Andrew Steelman, Robin R. Johnson, Colin R. Young, Thomas Prentice, Ashley Hammons, Ralph Storts, Thomas Welsh, and Mary W. Meagher

1. Introduction

1.1. Stress and the Immune System

Physical and psychosocial stressors have been shown to compromise immune function (Ader et al., 1991; Kielcolt-Glaser and Glaser, 1995). The immune suppressive effects of stress may be more pronounced in individuals that already have limited immune competence, such as infants, individuals with a predisposition to autoimmune disease, and the elderly (Kielcolt-Glaser and Glaser, 1995). An individual's response to a stressor is manifested in physiological, hormonal, behavioral, and immunological changes. These stress-induced responses are initiated by the hypothalamus and translated into action by the hypothalamic-pituitary-adrenal (HPA) axis and the sympathetic nervous system. Products from these two systems (e.g., corticoid hormones and catecholamines) can directly modulate the activity of various immune effector cells (Ader et al., 1991).

Stress has a bidirectional effect on the immune system depending on whether it is acute or chronic. Acute stress enhances antigen-specific cell-mediated immunity (Dhabhar and McEwen, 1996), alters populations of T-cell subsets (Teshima et al., 1987) and modulates mononuclear cell trafficking (Hermann et al., 1995). Acute stressors augment the immune response and result in redistribution of immune cells from the bone marrow into the blood, lymph nodes, and skin (Dhabhar and McEwen, 1996). Redeployment of immune cells into these compartments will allow for heightened responsiveness in the event of a skin wound, a natural consequence of an encounter with a predator as the acute stressor. Likewise, T cell and natural killer cell function are altered by stressful events (Okimura *et al.*, 1986). In contrast, chronic stressful life events are thought to suppress the ability of the immune system to respond to challenge and thus increase susceptibility to infectious diseases and cancers.

Although there is convincing evidence linking stress with the onset and progression of certain infectious diseases (e.g., influenza, herpes), relatively little is known about the role of stress in autoimmune diseases (e.g., multiple sclerosis, rheumatoid arthritis, lupus, insulin-dependent diabetes). However, a few studies indicate that stressful life events and poor social support play a role in the onset and exacerbation of autoimmune diseases such as rheumatoid arthritis (Rimon *et al.*, 1977; Homo-Delarche *et al.*, 1991). Furthermore, intervention studies indicate that cognitive-behavioral stress management decreases the symptomatology of autoimmune disease (Bradley *et al.*, 1987; O'Leary *et al.*, 1988; Radojevic *et al.*, 1992).

1.2. Multiple Sclerosis

Multiple sclerosis (MS) is the most common demyelinating disease of the CNS occurring at a prevalence of 250,000–350,000 in the United States (Anderson *et al.*, 1992). In 1994, the national annual costs of this disease were estimated to be $6.8 billion (Whetten-Goldstein *et al.*, 1998). MS usually affects people between the ages of 15 and 50, and 80% of patients have a relapsing-remitting disease that eventually progresses to a chronic progressive disorder. The MS lesion is characterized by plaques throughout the white matter of the brain and spinal cord. Demyelination is accompanied by inflammatory cell infiltrates consisting of plasma cells, macrophages/microglia, and T and B lymphocytes. In common with other autoimmune diseases, relapsing-remitting MS is more common in women than men, with a ratio of 2:1. Autoimmune responses to myelin components myelin basic protein (MBP) proteolipid protein (PLP), and myelin-oligodendrocyte glycoprotein (MOG) have been detected in MS patients, suggesting an autoimmune etiology for MS (Stinissen *et al.*, 1997).

1.3. Stress and Multiple Sclerosis

Stress was considered to be an important factor in the onset and course of MS in Charcot's original description of the disease (Charcot, 1877). Anecdotal accounts suggest that life stress frequently triggers the development of MS symptoms (Grant, 1993). Recent studies using standardized assessment of life events have begun to shed light on the idea that psychological stress precedes both the onset and recurrence of MS symptoms in 70–80% of cases (Warren *et al.*, 1982). The mechanism involving the role of stress in MS appears to be complex. There is even some evidence of a protective effect of stress under certain conditions (Nisipeanu and Korczyn, 1993). However, in laboratory studies, MS patients and controls had similar immune responses after an acute stressor (as measured by NK cell activity, T-cell proliferation, and changes in cell subsets in the peripheral blood) (Ackerman *et al.*, 1996). More recently, acute life stressors have been shown to be correlated with relapses in MS (Ackerman *et al.*, 2000). Mohr and colleagues conducted a meta-analysis of 14 studies concerning stress and MS and concluded that "there is a consistent association between stressful life events and subsequent exacerbation in multiple sclerosis" (Mohr *et al.*, 2004).

1.4. A Viral Etiology for Multiple Sclerosis

The etiology of MS is unknown, although epidemiological studies have implicated an infective agent as a probable initiating factor (Acheson, 1977; Gilden, 2001). An epidemiological survey reported the increased risk of developing MS was associated with late infection with mumps, measles, and Epstein-Barr virus (Miguel *et al.*, 2001). In addition, exacerbations of MS are frequently preceded by viral infections (Sibley *et al.*, 1985). It is also intriguing that the antiviral agent IFN-β has been reported to have a beneficial effect on relapsing/remitting MS (IFN-β Multiple Sclerosis Study Group, 1993). A number of different viral agents have been isolated from the brains of MS patients, including measles, mumps, parainfluenza type I (Allen and Brankin, 1993), and human herpes simplex type 6 (HHSV6) (Challoner *et al.*, 1995). In common with other autoimmune diseases, stressful life events may precipitate the onset and clinical relapses in MS patients (Whitacre *et al.*, 1994). One mechanism of stress-induced exacerbation might be via increased glucocorticoid levels resulting in immunosuppression and reactivation of latent viruses such as herpes virus.

Viruses are also known to cause demyelination in animals: measles virus in rats; JHM mouse hepatitis virus, Semliki Forest virus, and Theiler's virus in mice; visna in sheep; herpes simplex in rabbits (Dal Canto and Rabinowitz, 1982). Therefore, in order to understand the pathogenesis of MS, it is most appropriate to study an animal model of virus-induced demyelination such as Theiler's virus infection. Theiler's virus infection in mice represents not only an excellent model for the study of the pathogenesis of MS but also a model system for studying disease susceptibility factors, mechanisms of viral persistence within the CNS, and mechanisms of virus-induced autoimmune disease.

1.5. Theiler's Virus-induced Demyelination as a Model for MS

Theiler's murine encephalomyelitis virus (TMEV) is a Picornavirus that causes an asymptomatic gastrointestinal infection and occasionally paralysis (Theiler, 1934). There are two main strains of Theiler's virus, which are classified according to their neurovirulent characteristics. The virulent GDVII strains of Theiler's virus cause fatal encephalitis after intracranial infection (Theiler and Gard, 1940). The persistent TO strains (BeAn, DA, WW, Yale) cause, in susceptible strains of mice, a primary demyelinating disease that is similar to MS (Lipton, 1975; Oleszak et al., 2004). Theiler's virus must establish a persistent infection in the CNS in order to cause later demyelinating disease (Aubert et al., 1987). Strains of mice that are resistant to developing Theiler's virus-induced demyelination (TVID) are able to clear the early viral infection effectively from the CNS. Susceptible strains of mice fail to clear the CNS infection, in part due to inadequate natural killer cell (NK) and cytotoxic T cell (CTL) responses. Persistent viral infection of the CNS is a prerequisite for the development of primary inflammatory demyelination. During the early infection, virus replicates to high levels in the brain and spinal cord (Fig. 10.1) (Welsh et al., 1989). At approximately 1 month postinfection, the viral titers are decreased and this coincides with the development of high neutralizing antibody titers. In this phase of the disease, the virus infects neurons, and mice may develop polio-like disease (i.e., flaccid hind limb paralysis). In the late phase of the disease, the virus infects astrocytes, oligodendrocytes, and macrophage/microglial cells. Autoimmune reactivity to myelin is detected at both the B and T cell level during demyelinating disease.

A number of studies have reported that viral persistence and demyelination in susceptible strains of mice are under multigenic control. Genes coding for major histocompatibility complex (MHC) class I and the T-cell receptor (Melvold et al., 1987) have been implicated in susceptibility to demyelination. Another gene locus on chromosome 6 not linked to the T-cell receptor locus has also been implicated in demyelination (Bureau et al., 1992). Two additional loci, one close to Ifng on chromosome 10 and one near Mbp on chromosome 18, have been associated with viral persistence in some strains of mice (Bureau et al., 1993). Immune recognition of Theiler's virus is clearly an important element in susceptibility to demyelination, as indicated by the genetic association with MHC and the T-cell receptor, although other undefined factors are also involved.

1.6. Interferon and NK Cells in Theiler's Virus Infection

The early events that occur during Theiler's virus infection are crucial in the effective clearance of virus from the CNS. Failure to clear virus results in the establishment of persistent infection of the CNS and subsequent demyelination (Brahic et al., 1981; Rodriguez et al., 1996). The first response to viral infection is the production of type I interferons that are critical in the early clearance of Theiler's virus from the CNS as demonstrated by experimentation with IFN-α/β receptor knockout mice. These mice die within 10 days of infection with severe encephalomyelitis (Fiette et al., 1995).

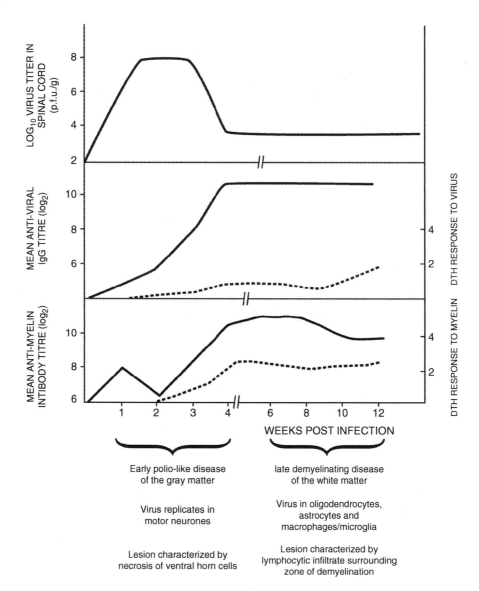

FIGURE 10.1. Disease course of Theiler's virus-induced demyelination. Intracerebral infection of CBA mice with 5×10^4 pfu of the Bean strain of Theiler's virus results in high levels of virus in the spinal cord during the first month of infection (top panel). The CNS viral titers decrease at 4 weeks p.i. when neutralizing antiviral antibodies and viral T-cell responses are detected (middle panel). During late disease, the virus is detected in astrocytes, oligodendrocytes, and macrophage/microglial cells, and autoreactive T- and B-cell responses to myelin are detected in TMEV-infected mice (lower panel).

Natural killer (NK) cells are activated early in viral infections and play an important role in natural resistance to certain viruses, tumor surveillance, and regulation of hematopoiesis. NK cells are active in the CNS as demonstrated in a rat model of quanethidine-killing induced neuronal destruction where they were shown to be the prime mediators of neuronal killing (Hickey et al., 1992). In Theiler's virus infection, susceptible SJL mice were found to have a 50% lower NK cell activity when compared with resistant C57BL/6 mice (Paya et al., 1989). The low activity of NK cells in the SJL mice is due to a differentiation defect in the thymus that impairs the responsiveness of NK cells to stimulation by IFN-β (Kaminsky et al., 1987). When resistant mice were depleted of NK cells by monoclonal antibody to NK 1.1 or anti-asialo-GM1 and then infected with Theiler's virus, they developed severe signs of gray matter disease (Paya et al., 1989). Thus NK cells are critical in the early clearance of Theiler's virus from the CNS.

1.7. Role of CD8+ and CD4+ T Cells in Theiler's Virus Infection

Both CD8+ and CD4+ T cells have been shown to play an important role in early viral clearance (Welsh et al., 1987; Borrow et al., 1992; Murray et al., 1998), but in later disease these T-cell subsets have been implicated in the demyelinating process (Clatch et al., 1987; Rodriguez and Sriram, 1988; Welsh et al., 1989). In early disease, CD4+ T cells are required for B cells to produce antibodies, one of the most important mediators of Picornavirus clearance (Welsh et al., 1987; Borrow et al., 1993). CD4+ T cells also secrete IFN-γ, which has been shown to inhibit the replication of Theiler's virus in vitro (Welsh et al., 1995) and to have a protective role in vivo (Kohanawa et al., 1993; Rodriguez et al., 1995). CD8+ T cells clearly are important in viral clearance as demonstrated by in vivo depletion experiments (Borrow et al., 1992) and studies with gene knockout mice (Fiette et al., 1993; Pullen et al., 1993). CD8+ T-cell-depleted mice fail to clear virus from the CNS and developed more severe demyelinating disease than the immunocompetent controls (Borrow et al., 1992). β2 microglobulin knockout mice were constructed on a TVID-resistant background, and these mice were shown to lack functional cytotoxic T cells (Fiette et al., 1993; Pullen et al., 1993). Histological evidence of demyelination developed in the knockout mice after intracranial infection with Theiler's virus. Introduction of resistant H-2Db (Azoulay et al., 1994) or H-2Dd transgene (Rodriquez and David, 1995) into susceptible strains of mice render these animals resistant to TVID. CD8+ T cells also provide protection against TVID when adoptively transferred to a TVID-susceptible BALB/c substrain, BALB/cAnNCr (Nicolson et al., 1996). Taken together, these investigations clearly implicate CD8+ T cells in viral clearance and resistance to demyelination. Indeed cytotoxic T lymphocyte (CTL) activity has been detected in Theiler's virus-infected SJL/J mice (Lindsley et al., 1991; Rossi et al., 1991) and higher CTL activity in

TVID-resistant C57BL/6 mice (Dethlefs *et al.*, 1997; Lyman *et al.*, 2004). The CTLs may be important either by recognizing viral determinants or by inhibiting delayed type hypersensitivity (DTH) responses (Borrow *et al.*, 1992; Lipton *et al.*, 1995).

1.8. Th1/Th2 Responses in TVID

The relative role of Th1/Th2 cells in susceptibility to TVID is complex. A pathogenic role for Th1 cells during the late demyelinating disease is clear. TVID correlates with DTH responses to TMEV (Clatch *et al.*, 1987). In addition, removal of CD4⁺ T cells during late disease results in amelioration of clinical signs, although this study did not differentiate Th1 and Th2 T cells (Welsh *et al.*, 1987). Furthermore, high levels of proinflammatory Th1 cytokines IFN-γ and TNF-α in late disease correlate with maximal disease activity (Begolka *et al.*, 1998). In addition, the number of TNF-α–producing cells in spinal cord was found to correlate with severity of disease (Inoue *et al.*, 1996).

In early disease, Th1 cytokines are involved in viral clearance. SJL/J mice treated with antibodies to IFN-γ suffered an increase in demyelination (Rodriguez *et al.*, 1995). In addition, IFN-γ knockout mice on a TVID-resistant background suffered increased demyelination and mortality when infected with Theiler's virus (Fiette *et al.*, 1995). These studies suggest the importance of IFN-γ in the resistance to TVID. Administration of the proinflammatory cytokines IL-6 (Rodriguez *et al.*, 1994) and TNF-α (Paya *et al.*, 1990) to TVID-susceptible mice resulted in reduced demyelination. However, another proinflammatory cytokine, IL-1, induced demyelination in TVID-resistant mice (Pullen *et al.*, 1995). The differential effects of these cytokines are probably due to their pleotropic effects. For instance IFN-γ is a potent antiviral agent but also increases inflammation.

Evidence in support of the importance of a Th2 response in protection from TVID comes from Miller and colleagues. They administered ethylene carbonimide–treated splenocytes during early TMEV infection to skew the immune response to TMEV from a predominately Th1 to Th2 response. This procedure proved effective at reducing the later demyelinating disease (Peterson *et al.*, 1993; Karpus *et al.*, 1994, 1995). In contrast, another study by Brahic's group demonstrated that the Th1/Th2 balance did not account for the difference in susceptibility to TVID (Monteyne *et al.*, 1999).

Interestingly, IL-2 secreting tumor cells injected into TVID-susceptible mice increased the frequency of virus-specific precursor CTLs and prevented persistent infection (Larsson-Sciard *et al.*, 1997). This observation supports the notion that a rapid early CTL response is important in early viral clearance and thus protection from demyelinating disease.

TVID-susceptible SJL mice given an immunosuppressive cytokine, TGF-β2, showed a reduction in the number of virus-infected cells and decreased amount of demyelination (Drescher *et al.*, 2000). The mechanism of action was hypothesized to be TGF-β2–dependent reduction in infiltration or activation of virus-infected macrophages into the CNS. Female SJL/J mice infected with the DA strain of Theiler's virus and then given IL-4 or IL-10 or both cytokines in combination showed marked decreases in demyelination and inflammation (Hill et al., 1998). Thus, immunosuppressive cytokines are beneficial in the treatment of TVID.

A number of studies have been conducted into the expression of cytokines and chemokines during infection with Theiler's virus. Investigators have used different time points, different Theiler's virus isolates, and different assay methods for analysis, which makes comparisons difficult. Using the DA strain of TMEV, at 40 days postinfection (p.i.) in SJL/J mice, Sato et al. using RNA protection assays found that the Th1 cytokines IL-5, IL-1, IL-2, and IL-6 were not detectable in the spinal cord, whereas Th2 cytokines, IL-10, and Th1 cytokines TNF-α, IL-12, and IFN-γ were elevated compared with controls. In the brains of the same animals, IL-10, IL-12, TNF-α, IL-1, IL-2, IL-6, and IFN-γ were not detected at 40 days p.i., but IL-4 production was high (Sato et al., 1997). In early disease, susceptible SJL mice were found to express more IL-12p40 mRNA than TVID-resistant mice. In one study (Inoue et al., 1998), IL-12 was shown to play an important exacerbating role in TVID. However, in another study, blocking IL-12 expression did not alter the neuropathogenesis of TVID (Bright et al., 1999).

It has also been reported that in DA-infected SJL/J mice at 60 days p.i., mRNA levels for IFN-γ, IL-1, IL-2, IL-6, IL-12, TNF-α, TGF-β1, IL-4, IL-5, and IL-10 were higher compared with controls. These studies were performed using real-time PCR analysis (Chang et al., 2000). Interestingly, these authors found elevated levels of TGF-β in TVID-susceptible SJL mice, which may account for the low CTLs in this strain of mouse. The Th2 cytokine IL-10 mRNA expression was also observed at particularly high levels in SJL mice; early in disease IL-10 expression may inhibit the CTL response and thus prevent effective viral clearance. Similar results showing increased expression of TNF-α, IL-6, and TGF-β were observed in SJL/J mice infected with DA for 60 days, with an additional result that lymphotoxin-α was also increased (Theil et al., 2000).

In summary, Th1 cytokines are generally pathogenic during late demyelinating disease, and Th2 cytokines are protective. Th1/Th2 cytokine profiles during early disease are more complicated, but clearly the early immunological response to Theiler's virus infection has a profound impact on the later development of demyelinating disease. Strains of mice that are susceptible to TVID produce elevated levels of TGF-β during early disease that are thought to interfere with the recruitment of effective cytotoxic T cells into the CNS. In addition to defective NK cell response in TVID-susceptible mice, the low level of CTLs in the CNS prevents viral clearance, and a persistent infection is established that subsequently leads to demyelinating disease. Additional evidence in support of the importance of CTLs in clearing infections comes from studies with mice depleted of their CD8+ T cells. These mice have increased viral titers in the CNS and more severe later demyelinating disease (Borrow et al., 1992).

1.9. Mechanisms of Theiler's Virus-induced Demyelination

Demyelination in the TVID model is partly mediated by (a) direct viral lysis of oligodendrocytes (Roos and Wollmann, 1984); immune mechanisms including (b) autoimmunity (Welsh *et al.*, 1987, 1989, 1990; Miller *et al.*, 1997; Borrow *et al.*, 1998); (c) bystander demyelination mediated by virus-specific DTH T cells (Clatch *et al.*, 1987; Gerety *et al.*, 1991); (d) cytotoxic T-cell reactivity (Rodriguez and Sriram, 1988) (summarized in Fig. 10.2). Susceptibility to TVID is correlated with (e) increased MHC class II expression *in vitro* on astrocytes (Borrow and Nash, 1992) and (f) cerebrovascular endothelial cells (Welsh *et al.*, 1993) after treatment with IFN-γ. Increased MHC class II expression on cells within the CNS may lead to increased antigen presentation and inflammation.

The autoimmune reactivity seen in TVID may result from viral damage to oligodendrocytes and subsequent activation of autoreactive T cells. Futhermore, these autoimmune T cells have been shown to be pathogenic

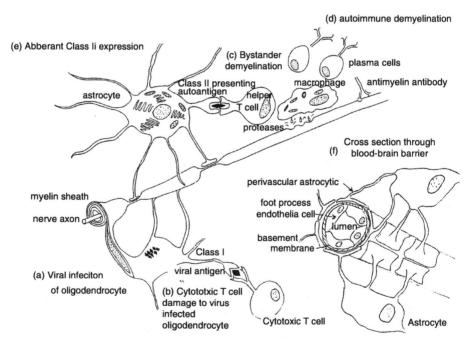

FIGURE 10.2. The mechanisms of demyelination induced by TMEV. Demyelination is partly mediated by (a) direct viral lysis of oligodendrocytes; immune mechanisms including (b) autoimmunity, (c) bystander demyelination mediated by virus-specific DTH T cells, (d) cytotoxic T-cell lysis of virus-infected oligodendrocytes. Susceptibility to TVID is correlated with (e) increased MHC class II expression *in vitro* on astrocytes and (f) cerebrovascular endothelial cells after treatment with IFN-γ.

and are able to demyelinate *in vitro* (Dal Canto *et al.*, 2000). The relative contributions of these mechanisms to the demyelinating process remain to be elucidated.

TVID represents an excellent animal model for MS, and therefore we have been investigating the effects of stress in this model in order to gain a better understanding of how stress impacts the human disease MS. In our first series of experiments, we examined the effect of stress on the early disease induced by Theiler's virus as a model of MS disease onset. Restraint stress was employed as the stressor because there is a great deal of literature in this area.

2. Stress Effects on the Neuropathogenesis of Theiler's Virus Infection

2.1. General Restraint Procedures and Experimental Design

2.1.1. Mice

Three-week-old CBA mice (Harlan Labs, Indianapolis, IN) were used in the initial studies because they are of intermediate susceptibility to the BeAn strain of TMEV, with a disease incidence of 70% (Welsh *et al.*, 1987, 1989). Thus, any alterations in disease incidence due to the effects of stress could be readily detected from this baseline. In addition, the neuropathology, rates of viral clearance, and immune response to Theiler's virus have been previously characterized in this strain (Welsh *et al.*, 1987, 1989; Blakemore *et al.*, 1988). Additional studies were performed with SJL mice, which are highly susceptible to TVID.

2.1.2. Virus

The BeAn strain of Theiler's virus (obtained from Dr. H. L. Lipton, Department of Neurology, Northwestern University, Chicago, IL) was propagated and amplified in BHK-21 cells. The culture supernatant containing infectious virus was aliquoted and stored at −70°C before use (Welsh *et al.*, 1987).

2.1.3. Restraint Stress Protocol

Mice were handled for several minutes each day for 1 week prior to the initiation of restraint stress in order to habituate each mouse to human contact in an attempt to diminish stress due to handling during bleeding, cage changes, and any other contacts that might otherwise have altered stress levels.

Five-week-old mice were randomly assigned to one of three groups, 10 mice per group according to a previously reported protocol (Sheridan *et al.*, 1991, Campbell *et al.*, 2001) and treated as follows: (1) a control group where mice remained undisturbed in their home cages; (2) a group in which food and water (FWD) was withheld for 12 h each of 5 nights per week over a 4-week period; (3) a group in which each mouse was placed in a well-ventilated restraining tube for 12 h each of 5 nights per week. Half the mice in each of the three groups were either infected intracerebrally with Theiler's virus or similarly inoculated with virus-free BHK cell supernatant. Daily food and water deprivation or restraint began 1 day prior to infection and 5 days per week for 1 month postinfection. After the first series of experiments, we did not observe any differences between the food and water–deprived mice and the nonrestrained mice so in the following experiments the design was simplified to four groups noninfected/nonrestrained; noninfected/restrained; infected/nonrestrained; and infected/restrained.

2.2. The Effects of Restraint Stress on Early Theiler's Virus Infection

Restraint stress increased the clinical signs of neurological disease in male CBA mice infected with TMEV (Campbell *et al.*, 2001; Mi *et al.*, 2004). Normally, TMEV infection of CBA mice is asymptomatic for the first 6 weeks of infection. In our first stress study, 80% of the stressed infected mice died during the first 3 weeks of infection. The restraint protocol caused significant weight loss and induced high levels of glucocorticoids (GCs) in the plasma (450 ng/ml after the first 12 h stress session) (Campbell *et al.*, 2001). The restrained mice developed thymic and splenic atrophy (Figs. 10.3a and 10.3b) and reduced numbers of circulating lymphocytes and increased neutrophils (Fig. 10.4). Stressed mice also developed adrenal enlargement (Welsh *et al.*, 2004). In addition, higher viral titers were observed in the brains and spinal cords of infected/restrained mice when compared with infected/nonrestrained mice (Figs. 10.5a and 10.5b). Increased levels of GCs have been implicated in the increased mortality of TMEV-infected mice because these effects could be replicated by simply adding corticosterone to the drinking water of mice infected with TMEV (unpublished observations).

The early lesion of TMEV infection is most prominent in the hippocampus and is characterized by neuronal degeneration, astrocytic hypertrophy/hyperplasia, perivascular cuffing, and microgliosis. In TMEV-infected mice subjected to restraint stress, the lesions were considerably less pronounced

---

FIGURE 10.4. Differential cell analysis was performed on whole blood collected 2 days p.i. from four mice in each of the experimental conditions (noninfected/nonrestrained; noninfected/restrained; infected/nonrestrained; infected/restrained). The lymphocyte and neutrophils are reported as a percentage of total WBC populations. The standard error of the means are shown.

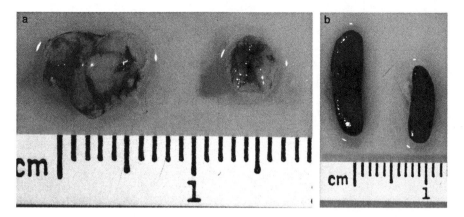

FIGURE 10.3. Thymic and splenic atrophy in restraint stressed CBA mice. (a) The thymus and (b) spleen were removed from a noninfected/nonrestraint stressed mouse (left) and compared with those removed from a noninfected/restrained mouse (right) after 3 consecutive nights of stress.

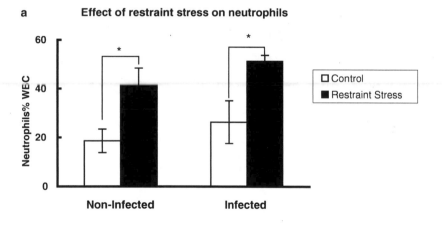

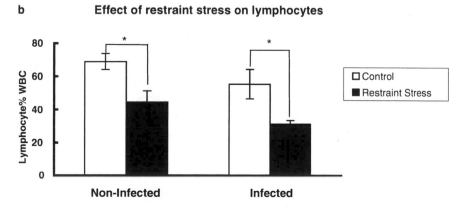

than in the infected nonrestrained mice at day 7 p.i. (Fig. 10.6) (Campbell *et al.*, 2001; Mi *et al.*, 2004). Interestingly, an increase in inflammation was detected at day 24 p.i. in infected/restrained mice (Campbell *et al.*, 2001). This may be due to the persistence of higher viral titers in the CNS that stimulate increased inflammation in the CNS.

Similar results were found in another study with male and female SJL: that chronic restraint stress (8 h per night) administered in the first 4 weeks of TMEV infection decreased body weights, increased clinical symptomatology of infection, and increased plasma GCs levels during the acute viral infection. Although all restraint stressed mice displayed significantly

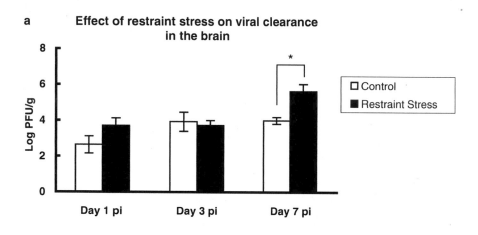

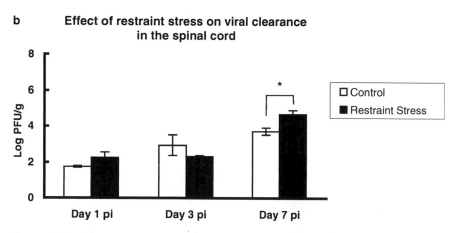

FIGURE 10.5. Restraint stress significantly increases viral titers in the brain and spinal cord of TMEV-infected mice. Mice from each of the experimental conditions (noninfected/nonrestrained; noninfected/restrained; infected/nonrestrained; infected/restrained) were sacrificed at days 1, 3, and 7 p.i and viral levels in the (a) brain and (b) spinal cord were measured by plaque assay on L2 cells.

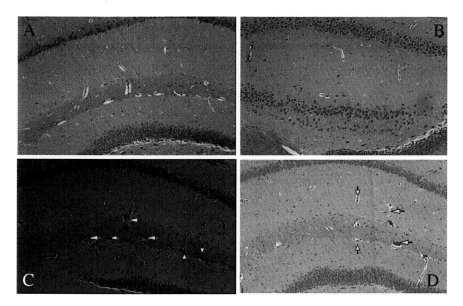

FIGURE 10.6. Restraint stress decreased inflammation in the hippocampus. H&E-stained sections of hippocampus collected at day 7 postinfection (magnification ×10). (A) Noninfected/nonrestrained. (B) Noninfected/restrained. (C) Infected/nonrestrained hippocampus. Mild to moderate microgliosis (as shown in the box) and perivascular cuffing (indicated by arrow). (D) Infected/restrained. Mild perivascular cuffing (indicated by arrow). Reprinted from Mi *et al.* (2004). Alterations in chemokine expression in Theiler's virus infection and restraint stress. *J. Neuroimmunol.* 151:103–115. Copyright 2004 with permission from Elsevier.

increased GCs levels, female SJL mice showed higher basal and stress-induced increases in GCs (Sieve *et al.*, 2004).

The results of these studies suggest that restraint stress increased GCs, which resulted in immunosuppression, reduced inflammatory cell infiltrate into the CNS, and consequently reduced viral clearance. The increased levels of virus replication within the CNS may contribute to the increased mortality observed in the restrained mice.

In more recent studies, the effect of restraint stress on viral dissemination was investigated (Mi *et al.*, 2006a). Stressed mice developed increased levels of virus in the CNS, spleen, lymph nodes, thymus, lungs, and the heart when compared with infected/nonrestrained mice. Interestingly, inflammatory lesions developed in the hearts of the restrained mice. Furthermore, the virus isolated from the hearts of stressed mice had altered and become more cardiotropic when reinjected into normal mice. These findings suggest that stress-induced immunosuppression allows for increased viral replication and spread to sites that would normally remain uninfected. Viral infection of organs that are not normally considered viral targets may then allow for the development of novel diseases.

2.3. Restraint Stress Alters Chemokine/Cytokine mRNA Expression

Experiments were carried out in order to determine the effects of stress on chemokine/cytokine expression in the CNS and spleen (as an example of an immune organ). Groups of male CBA mice were (1) infected/restrained for 7 nights, (2) infected/nonrestrained, (3) noninfected/restrained, or (4) noninfected/nonrestrained. At sacrifice, their brains and spleens were removed and RNA isolated and incorporated in RNase Protection Assay (RPA) to estimate mRNA cytokine and chemokine expression. Infection with TMEV increased the following chemokine expression: lymphotactin (Ltn), interferon-induced protein (IP-10), macrophage inflammatory protein (MIP)-1β, monocyte chemoattractant protein-1 (MCP-1), and thymus derived chemotactic agent (TCA)-3 in the spleen but not the brain at day 2 p.i. The fact that chemokine expression was increased first in the spleen provides evidence that the immune response to TMEV is initiated in the periphery. Ltn, Regulated upon Activation Normal T cell Expressed and Secreted (RANTES), and IP-10 were elevated in both the spleen and the brain at day 7 p.i. and were significantly decreased by restraint in the brain. These chemokines are responsible for the recruitment of CD4$^+$, CD8$^+$ T cells, macrophages, and NK cells and thus may account for the diminished inflammatory cell infiltrate in the CNS of stressed mice and subsequently the reduced viral clearance and increased mortality in virus-infected restraint stressed mice (Mi *et al.*, 2004).

In experiments examining cytokine expression, mice were subjected to the restraint paradigm and, at sacrifice, half the brain taken for viral infectivity assays and the other half for RPA analysis of cytokine RNA levels. TMEV infection elevated IFN-γ, LT-β, IL-12p40, IL-6, and IFN-β in the brain at days 2 and 7. Importantly, restraint attenuated the increases in IFN-γ but elevated IFN-β. RNA levels of IFN-γ, LT-β, and TNF-α were negatively correlated with viral titers in the CNS such that mice with higher cytokine levels had lower virus levels. Thus, these cytokines may play a role in the clearance of virus from the CNS. TNF-α protein levels, as measured by Western blots, gave similar results to the RPA data for this cytokine. Interestingly, stress increased the anti-inflammatory cytokine IL-10 in the spleen, which may contribute to the decrease in proinflammatory cytokine production (Mi *et al.*, 2006b).

The cytokines altered by restraint stress in Theiler's virus infection have pleotropic effects and have vital roles in the neuropathogenesis of this disease. Lymphotoxin-β, a membrane-bound form of lymphotoxin, plays a critical role in the resistance to intracellular pathogens including Theiler's virus (Lin *et al.*, 2003). LT-β induces IFN-β and also increases cytotoxic T-cell activity, which are both important mediators of viral clearance from the CNS. IFN-γ is an important inflammatory mediator produced by NK cells and T cells, which contributes on the one hand to viral clearance and on the other hand to development of demyelination in TVID. The suppressive effect of stress was first detected at day 2 p.i. and attenuated at day 7 p.i. Stress increased the anti-inflammatory cytokine IL-10 and decreased proinflammatory cytokines in the spleen. The increase in IL-10 may have contributed to the decrease in proinflammatory cytokines. Interestingly, stress also caused an increase in IFN-β expression in the brain, which may result from the higher levels of virus within the CNS of these mice, and this may compensate for the impaired viral clearance caused by decreased production of proinflammatory cytokine during stress.

ELISA assays examined the effects of restraint stress on IL-1β and TNF-α levels in serum. No detectable levels of IL-1β were observed in any of the groups of mice, but interestingly, restraint stress induced high levels of TNF-α in the serum of both infected and uninfected mice (Welsh et al., 2004).

To summarize our findings with regard to the effects of restraint stress on the expression of chemokines and cytokines: stress reduced the expression of chemokines responsible for the recruitment of CD4+, CD8+ T cells, macrophages, and NK cells; namely, Ltn, RANTES, and IP-10. Virus-induced IFN-γ expression was also decreased by stress. IFN-γ, TNF-α, and LT-β levels were negatively correlated with viral replication in the brain. These cytokines have important roles in the initiation of immune system activation and also have effective antiviral activities, and therefore lower levels of expression may also result in increased viral replication within the CNS.

Natural killer (NK) cells are known to be important in the early clearance of TMEV as demonstrated by depletion studies (Paya et al., 1989) and are also exquisitely sensitive to stress. Therefore we examined the effect of restraint stress on NK cell activity in CBA mice infected with TMEV. Twenty-four hours postinfection, restraint stress significantly reduced virus-induced NK cell activity in TMEV-infected CBA mice (Welsh et al., 2004) when compared with infected/nonrestrained mice. Decreased NK cell activity may also contribute to the reduced ability to clear virus.

In order to characterize the alterations in spleen cell populations that occur over time after TMEV infection and restraint stress, we conducted flow cytometric analysis experiments on splenocytes using combinations of the following directly labeled antibodies: (1) CD3-FITC, CD19-PE, CD45-PECy5 (leukocyte marker); (2) CD3-FITC, CD8-PE, CD4-PECy7; (3) F4/80-FITC (macrophage marker), DX5-PE (NK cell marker), CD45-PECy5. Preliminary results indicate that at both day 3 and day 7 p.i. in the spleen, restraint stress reduces NK cells and B cells while increasing numbers of T cells overall. No significant differences were seen in macrophages or between CD4+ or CD8+ cells (unpublished observations).

2.4. Restraint Stress Fails to Render TVID-Resistant Mice Susceptible to TVID

Experiments were performed in order to examine whether chronic restraint stress applied during the acute phase of Theiler's virus infection would render the genetically nonsusceptible C57BL/6 mice susceptible to TVID.

Despite the fact that chronic restraint stress has been shown to decrease functions of NK, T and B cells, and these immune functions have been shown to be essential for resistance to TVID, chronic restraint stress failed to render resistant C57BL/6 mice susceptible to the demyelination (Steelman *et al.*, 2006). C57Bl/6 mice have a high basal level of NK cell activity and also a robust CD8[+] T-cell response to TMEV. Although stress may decrease the activity of NK and CD8[+] T cells, it may not completely ablate them, and the residual cells are then still able to effectively clear virus.

2.5. The Effects of Restraint Stress During Acute Infection on the Later Demyelinating Disease

Life stressors precipitate the onset of MS, and we have shown that chronic stress during acute infection with Theiler's virus leads to decreased viral clearance from the CNS. Other studies have shown that increased viral load during acute disease leads to increased demyelinating disease during the late disease (Borrow *et al.*, 1992). Therefore, we hypothesized that stress during the acute viral infection results in higher viral load in the CNS and subsequently increased demyelination in the later disease. Chronic restraint stress, administered during early infection with Theiler's virus, was found to exacerbate the acute CNS viral infection and the subsequent demyelinating phase of disease in SJL male and female mice. During early infection, stressed mice displayed decreased body weights and locomotor activity and increased behavioral signs of illness and plasma GCs levels. During the subsequent demyelinating phase of disease, previously stressed mice had greater behavioral signs of demyelination, worsened rotarod performance, and increased inflammatory demyelinating lesions of the spinal cord, as measured by perivascular cuffing and meningitis (Sieve *et al.*, 2004). Restraint-stressed SJL mice developed higher viral loads in the CNS as compared with nonrestrained TMEV-infected mice (unpublished data).

Correlational analysis of all of the dependent variables found that in the acute phase of disease in SJL mice, plasma corticosterone levels, clinical symptomatology, and loss in body weight were all highly correlated. GCs levels during restraint stress in the acute phase were also highly correlated with histological indications of meningitis, rotarod performance, and clinical symptomatology in the chronic phase of disease. Thus, plasma GCs levels during stress in the acute phase may be a good predictor of disease course in the chronic phase. Acute-phase clinical symptomatology had similar predictive value with chronic-phase clinical symptomatology, rotarod performance, and histological indications of meningitis. We have previously reported increased levels of antibody to myelin membranes during the late demyelinating phase of disease (Welsh *et al.*, 1987). In our more recent study, autoantibodies to myelin basic protein (MBP), proteolipid protein (PLP), or myelin oligodendrocyte glycoprotein (MOG) were detected in virus-infected SJL mice, and this represents the first report of antibodies to specific myelin components and demonstrates the value of TMEV-induced demyelination as a model for MS (Sieve *et al.*, 2004). Female SJL mice had higher antibodies to MOG 33–55 than males at day 69 p.i., and previously stressed female mice had decreased antibody titers to MBP when compared with nonrestrained infected mice. Antibody titers to the Theiler's virus MBP and PLP were no different between the infected/restrained and infected/nonrestrained mice.

In summary, restraint stress during early infection significantly increased both clinical and histological signs of demyelinating disease in SJL mice infected with Theiler's virus. The mice that developed the highest corticosterone levels during the early disease subsequently developed more severe late demyelinating disease. We propose that stress-induced immunosuppression during early infection with TMEV results in increased levels of virus within the CNS and consequently increased disease severity during the late phase of the disease.

2.6. The Effect of Restraint Stress During the Chronic Demyelinating Disease

Stress has been shown to ameliorate experimental allergic encephalomyelitis (EAE), an autoimmune model of MS evoked by injection of spinal cord or myelin components into susceptible strains of mice (Levine *et al.*, 1962; Griffin *et al.*, 1993). The frequency of MBP-specific lymphocytes in the spleen and lymph nodes and both the Th1 and Th2 cytokine responses were suppressed in the stressed mice (Whitacre *et al.*, 1998). Glucocorticoids were implicated as the prime mediators of the disease suppression (Dowdell *et al.*, 1999). Immunosuppressive therapies such as cyclophosphamide or treatment with rabbit anti-thymocyte serum (Lipton and Dal Canto, 1976) or antibody to CD4 T cells (Welsh *et al.*, 1987) have been shown to improve the late demyelinating disease induced by Theiler's virus. Therefore, because restraint stress induces high levels of immunosuppressive glucocorticoids, if this stressor is applied during the late demyelinating disease, we hypothesized that this should result in clinical improvements by reducing inflammatory demyelination. However, experiments with TVID showed that although restraint stress elevated GCs levels, it did not alter the clinical score or histological signs of inflammation. Interestingly, mice infected with Theiler's virus developed high levels of circulating GCs (Welsh *et al.*, 2006). Development of glucocorticoid resistance in restraint stressed mice may factor into these results.

3. Summary and Significance of Research Findings

Chronic restraint stress was shown to increase GCs levels, increase signs of sickness behavior, increase viral titers in the CNS, and increase mortality after infection with TMEV (Campbell *et al.*, 2001). Restraint stress significantly decreased several immune measures including NK cell activity (Welsh *et al.*, 2004), chemokine (Mi *et al.*, 2004) and cytokine expression in the spleen and CNS (Mi *et al.*, 2006b). Reduced chemokine expression may account for the decreased inflammatory cell infiltrates into the CNS. The stress-induced reduction in proinflammatory cytokines may also contribute to the increased levels of virus within the CNS both directly through the reduced cytokine antiviral activity and indirectly by reduced ability of cytokines to induce activation of cytotoxic T cells. As a result of these findings, we propose that restraint stress induces high levels of GCs, which results in immunosuppression, reduced ability to clear virus, and subsequently increased inflammatory demyelinating disease (Sieve *et al.*, 2004). Additionally, restraint stress facilitated the systemic dissemination of TMEV, resulting in increased viral replication in the heart and the development of a cardiotropic variant of TMEV that induced pathology in the heart (Mi *et al.*, 2006a).

Relating our findings to the development of multiple sclerosis and autoimmune diseases in general, stressful events that occur prior to or during infection with an infective agent may result in immunosuppression and failure to eliminate the pathogen. The establishment of persistent infection may then lead to the development of autoimmune disease such as multiple sclerosis. Stress-induced immunosuppression may also facilitate the generation of pathogens with enhanced and altered pathogenecity.

Acknowledgments. The authors acknowledge Lin Bustamante for excellent technical assistance with the preparation of H&E sections. This research was funded by grants to C.J.R.W. and M.W.M. from Texas A&M Interdisciplinary Research Program, the National Multiple Sclerosis Society (RG 3128), and NIH/NINDS (R01 39569). W.M. was a recipient of a Texas A&M University Life Sciences Training Fellowship, and R.R.J. received fellowships from NSF and NIH/NRSA.

References

Acheson, E.D. (1977). Epidemiology of multiple sclerosis. *Br. Med. Bull.* 33:9–14.

Ackerman, K.D., Martino, M., Heyman, R., Moyna, N.M., and Rabin, B.S. (1996). Immunologic response to acute psychological stress in MS patients and controls. *J. Neuroimmunol.* 68(1–2):85–94.

Ackerman, K.D., Stover, A., Heyman, R., Anderson, B.P., Houck, P.R., Frank, E., Rabin B.S., and Baum, A. (2003). Robert Ader New Investigator award. Relationship of cardiovascular reactivity, stressful life events, and multiple sclerosis disease activity. *Brain Behav. Immun.* 17:141–151.

Ader, R., Felten, D.L., and Cohen, N. (1991). *Psychoneuroimmunology*. New York: Academic Press.

Allen, I., and Brankin, B.J. (1993). Pathogenesis of multiple sclerosis—the immune diathesis and the role of viruses. *Neuropathol. Exp. Neurol.* 52:95.

Aubert, C., Chamorro, M., and Brahic, M. (1987). Identification of Theiler's infected cells in the central nervous system of the mouse during demyelinating disease. *Microb. Pathogen.* 3:319–326.

Azoulay, A., Brahic, M., and Bureau, J.F. (1994). FVB mice transgenic for the H-2D[b] gene become resistant to persistent infection by Theiler's virus. *J. Virol.* 68: 4049–4052.

Anderson, D.W., Ellenberg., J.H., Leventhal, C.M., Reingold, S.C., Rodriguez, M., and Silberberg, D.H. (1992). Revised estimate of the prevalence of multiple sclerosis in the United States. *Ann. Neurol.* 31:333–336.

Begolka, W.S., Vanderlugt, C.L., Rahbe, S.H., and Miller, S.D. (1998). Differential expression of inflammatory cytokines parallels progression of central nervous system pathology in two clinically distinct models of multiple sclerosis. *J. Immunol.* 161:4437–4446.

Blakemore, W.F., Welsh, C.J.R., Tonks, P., and Nash, A.A. (1988). Observations on demyelinating lesions induced by Theiler's virus in CBA mice. *Acta. Neuropath.* 76:581–589.

Borrow, P., and Nash, A.A. (1992). Susceptibility to Theiler's virus-induced demyelinating disease correlates with astrocyte class II induction and antigen presentation. *Immunol.* 76:133–139.

Borrow, P., Tonks, P., Welsh, C.J.R., and Nash, A.A. (1992). The role of CD8+ T cells in the acute and chronic phases of Theiler's virus-induced disease in mice. *J. Gen. Virol.* 73:1861–1865.

Borrow, P., Welsh, C.J.R., and Nash, A.A. (1993). Study of the mechanisms by which CD4+ T cells contribute to protection in Theiler's murine encephalomyelitis. *Immunology* 80:502–506.

Borrow, P., Welsh, C.J.R., Dean, D., Tonks, P., Blakemore, W.F., and Nash, A.A. (1998). Investigation of the role of autoimmune responses to myelin in the pathogenesis of TMEV-induced demyelinating disease. *Immunology* 93:478–484.

Bradley, L.A., Young, L.D., Anderson, K.O., Turner, R.A., Agudelo, C.A., McDaniel, L.K., Pisko, E.J., Semble, E.L., and Morgan, T.M. (1987). Effects of psychological therapy on pain behavior of rheumatoid arthritis patients: Treatment outcome and six-month follow-up. *Arthritis Rheum.* 30:1105–1114.

Brahic, M., Stroop, W.G., and Baringer, J.R. (1981). Theiler's virus persists in glial cells during demyelinating disease. *Cell* 26:123–128.

Bright, J.J., Rodriguez, M., and Sriram, S. (1999). Differential influence of Interleukin-IL-12 in the pathogenesis of autoimmune and virus-induced central nervous system demyelination. *J. Virol.* 73:1637–1639.

Bureau, J.F., Montagutelli, X., Lefebvre, S., Guenet, J.L., Pla, M., and Brahic, M. (1992). The interaction of two groups of murine genes determines the persistence of Theiler's virus in the central nervous system. *J. Virol.* 66:4698–4704.

Bureau, J.F., Montagutelli, X., Bihl, F., Lefebvre, S., Guenet, J.L., and Brahic, M. (1993). Mapping loci influencing the persistence of Theiler's virus in the murine central nervous system. *Nat. Genet.* 5:82–91.

Campbell, T., Meagher, M.W., Sieve, A., Scott, B., Storts, R., Welsh, T.H., and Welsh, C.J.R. (2001). The effects of restraint stress on the neuropathogenesis of Theiler's virus-induced demyelination. I. Acute disease. *Brain Behav. Immun.* 15:235–254.

Challoner, P.B., Smith, K.T., Parker, J.D., Macleod, D.L., Coulter, S.N., Rose, T.M., Schultz, E.R., Bennett, J.L., Garber, R.L., Chang, M., Schad, P.A., Sewart, P.M., Nowinski, R.C., Brown, J.P., and Burmer, G.C. (1995). Plaque-associated expression of human herpesvirus 6 in multiple sclerosis. *Proc. Natl. Acad. Sci. U.S.A.* 92: 7440–7444.

Chang, J.R., Zaczynaska, E., Katsetos, C.D., Platsoucas, C.D., and Oleszak, E.L. (2000). Differential expression of TGF-β, IL-2, and other cytokines in the CNS of Theiler's murine encephalomyelitis virus-infected susceptible and resistant strains of mice. *Virology* 278:346–360.

Charcot, J.M. (1877). *Lectures on Diseases on the Nervous System.* G. Sigerson, trans. London: New Sydenham Society.

Clatch, R.J., Lipton, H.L., and Miller, S.D. (1987). Class II-restricted T cell responses in Theiler's murine encephalomyelitis virus (TMEV)-induced demyelinating disease. II. Survey of host immune responses and central nervous system virus titers in inbred mouse strains. *Microb. Pathogen.* 3:327–337.

Clatch, R.J., Miller, S.D., Metzner, R., Dal Canto, M.C., and Lipton, H.L. (1990). Monocytes/macrophages isolated from the mouse central nervous system contain infectious Theiler's murine encephalomyelitis virus (TMEV). *Virology* 176: 244–254.

Dhabhar, F.S., and McEwen, B.S. (1996). Stress-induced enhancement of antigen specific cell-mediated immunity. *J. Immunol.* 156:2608–2615.

Dobbs, C.M., Vasquez, M., Glaser, R., and Sheridan, J.F. (1993). Mechanisms of stress-induced modulation of viral pathogenesis and immunity. *J. Neuroimmunol.* 48:151–160.

Drescher, K.M., Murray, P.D., Lin, X., Carlino, J.A., and Rodriguez, M. (2000). TGF-β2 reduces demyelination, virus antigen expression and macrophage recruitment in a viral model of multiple sclerosis. *J. Immunol.* 164:3207–3213.

Dal Canto, M.C., and Rabinowitz, S.G. (1982). Experimental models of virus-induced demyelination of the central nervous system. *Ann. Neurol.* 11:109–127.

Dal Canto, M.C., Calenoff, M.A., Miller, S.D., and Vanderlugt, C.L. (2000). Lymphocytes from mice chronically infected with Theiler's murine encephalomyelitis virus produce demyelination of organotypic cultures after stimulation with the major encephalitic epitope of myelin proteolipid protein. Epitope spreading in TMEV infection has functional activity. *J. Neuroimmunol.* 104:79–84.

Dethlefs, S., Brahic, M., and Larsson-Sciard, E.L. (1997). An early abundant cytotoxic T-lymphocyte response against Theiler's virus is critical for preventing for viral persistence. *J. Virol.* 71:8875–8878.

Dhabhar, F.S., and McEwen, B.S. (1996). Stress-induced enhancement of antigen specific cell-mediated immunity. *J. Immunol.* 156:2608–2615.

Dhabhar, F.S., and McEwen, B.S. (1997). Acute stress enhances while chronic stress suppresses cell-mediated immunity in vivo: a potential role for leukocyte trafficking. *Brain Behav. Immun.* 11:286–306.

Dhabhar, F.S., Miller, A.H., McEwen, B.S., and Spencer, R.L. (1995). Effects of stress on immune cell distribution: dynamics and hormonal mechanisms. *J. Immuonl.* 154:5511–5527.

Dobbs, C.M., Vasquez, M., Glaser, R., and Sheridan, J.F. (1993). Mechanisms of stress-induced modulation of viral pathogenesis and immunity. *J. Neuroimmunol.* 48:151–160.

Dowdell, K.C., Gienapp, I.E., Stuckman, S., Wardrop, R.M., and Whitacre, C.C. (1999). Neuroendocrine modulation of chronic relapsing experimental autoimmune encephalomyeltis: a critical role for the hypothalamic-pituitary-adrenal axis. *J. Neuroimmunol.* 100:243–251.

Fiette, L., Aubert, C., Brahic, M., and Ross, C.P. (1993). Theiler's virus infection of β2-microglobulin deficient mice. *J. Virol.* 67:589–592.

Fiette, L., Aubert, C., Ulrike, M., Huang, S., Aguet, M., Brahic, M., and Bureau, J.F. (1995). Theiler's virus infection of 129Sv mice that lack the interferon α/β or IFN-γ receptors. *J. Exp. Med.* 181:2069–2076.

Gerety, S.J., Clatch, R.J., Lipton, H.L., Goswami, R.G., Rundell, M.K., and Miller, S.D. (1991). Class II-restricted T cell responses in Theiler's murine encephalomyelitis virus-induced demyelinating disease. IV Identification of an immunodominant T cell determinant, on the N-terminal end of the VP-2 capsid protein in susceptible SJL/J mice. *J. Immunol.* 146:4322–4326.

Gilden, D.H. (2001). Viruses and multiple sclerosis. *JAMA* 286:3127–3129.

Grant, I. (1993). Psychosomatic-somatopsychic aspects of multiple sclerosis In U. Hailbreich (ed.), *Multiple Sclerosis: A Neuropsychiatric Disorder.* Washington, DC: American Psychiatric Press, pp. 119–136.

Griffen, A., Lo, W., Wolny, A., and Whitacre, C. (1993). Suppression of experimental allergic encephalomyelitis by restraint stress: sex differences. *J. Neuroimmunol.* 44:103–116.

Hermann, G., Beck, F.M., and Sheridan, J.F. (1995). Stress-induced glucocorticoid response modulates mononuclear cell trafficking during an experimental influenza viral infection. *J. Neuroimmunol.* 56:179–186.

Hickey, W.F., Ueno, K., Hiserodt, J.C., and Schmidt, R. (1992). Exogenously-induced, natural killer cell-mediated neuronal killing: a novel pathogenetic mechanism. *J. Exp. Med.* 176:811–817.

Hilakivi-Clarke, L.A., Turkka, J., Lister, R.G., and Linnoila, M. (1991). Effects of early postnatal handling on brain beta-adrenoceptors and behavior in tests related to stress. *Brain Res.* 542:286–292.

Hill, K.E., Pigmans, M., Fujinami, R.S., and Rose, J.W. (1998). Gender variations in early Theiler's virus-induced demyelinating disease: differential susceptibility and effects of IL-4, IL-10 and combined IL-4 with IL-10. *J. Neuroimmunol.* 85: 44–51.

Homo-Delarche, F., Fitzpatrick, F., Christeff, N., Nunez, E.A., Bach, J.F., and Dardenne, M. (1991). Sex steroids, glucocorticoids, stress, and autoimmunity. *J. Steroid Biochem. Mol. Biol.* 40:619–637.

IFN-β Multiple Sclerosis Study Group. (1993). IFN-β1-b is effective in relapsing remitting multiple sclerosis. I. Clinical results of a multi-center, randomized, double blind, placebo controlled trial. *Neurology* 43:655–661.

Inoue, A., Koh, C.S., Yahikozawa, H., Yanagisawa, N., Yagita, H., Ishihara, Y., and Kim, B.S. (1996). The level of tumor necrosis factor-alpha producing cells in the spinal cord correlates with the degree of Theiler's murine encephalomyelitis virus-induced demyelinating disease. *Int. Immunol.* 8:1001–1008.

Inoue, A., Koh, C-S., Yamazaki, M., Yahikozawa, H., Ichikawa, M., Yagita, H., and Kim, B.S. (1998). Suppressive effect on Theiler's murine encephalomyelitis virus-induced demyelinating disease by the administration of anti-IL12 antibody. *J. Immunol.* 161:5586–5593.

Kaminsky, S.G., Nakamura, I., and Cudkowicz, G. (1987). Defective differentiation of natural killer cells in SJL mice. Role of the thymus. *J. Immunol.* 138:1020–1025.

Kappel, C.A., Melvold, R.W., and Kim, B.S. (1990). Influence of sex on susceptibility in the Theiler's murine encephalomyelitis virus model for multiple sclerosis. *J. Neuroimmunol.* 29:15–19.

Kappel, C.A., Dal Canto, M.C., Melvold, R.W., and Kim, B.S. (1991). Hierarchy of effects of the MHC and T cell receptor beta-chains in susceptibility to Theiler's murine encephalomyelitis virus-induced demyelinating disease. *J. Immunol.* 147: 4322–4326.

Karpus, W.J., Clatch, R.J., and Miller, S.D. (1993). Split tolerance of Th1 and Th2 cells in tolerance to Theiler's murine encephalomyelitis virus. *Eur. J. Immunol.* 23:46–55.

Karpus, W.J., Peterson, J.D., and Miller, S.D. (1994). Anergy in vivo: Down regulation of antigen specific CD4+ Th1 but not Th2 cytokine responses. *Int. Immunol.* 6:721–730.

Karpus, W.J., Pope, J.G., Peterson, J.D., Dal Canto, M.C., and Miller, S.D. (1995). Inhibition of Theiler's virus mediated demyelination by peripheral immune tolerance induction. *J. Immuol.* 155:947–957.

Keith, L.D., Winslow, J.R., and Reynolds, R.W. (1978). A general procedure for estimation of corticosteroid response in individual rats. *Steriods* 31:523–531.

Kiecolt-Glaser, J.K., and Glaser, R. (1995). Psychoneuroimmunology and health consequences: Data and shared mechanisms. *Psychosom. Med.* 57:269–274.

Kohanawa, M., Nakane, A., and Minagawa, T. (1993). Endogenous gamma interferon produced in central nervous system by systemic infection with Theiler's virus in mice. *J. Neuroimmunol.* 48:205–211.

Larsson-Sciard, E.L., Dethlefs, S., and Brahic, M. (1997). In vivo administration of interleukin-2 protects susceptible mice from Theiler's virus persistence. *J. Virol.* 71:797–799.

Levine, S., Strebel, R., Wenk, E., and Harman, P. (1962). Suppression of experimental autoimmune encephalomyelitis by stress. *Exp. Biol. Med.* 109:294–298.

Lindsley, M.D., Thiemann, R., and Rodriguez, M. (1991). Cytotoxic T cells isolated from the central nervous system of mice infected with Theiler's virus. *J. Virol.* 65: 6612–6620.

Lipton, H.L. (1975). Theiler's virus infection in mice: an unusual biphasic disease process leading to demyelination. *Infect. Immun.* 11:1147–1155.

Lipton, H.L., and Dal Canto, M.C. (1976). Theiler's virus-induced demyelination: prevention by immunosuppression. *Science* 192:62–64.

Lipton, H.L., Twaddle, G., and Jelachich, M.L. (1995). The predominant virus antigen burden is present in macrophages in Theiler's murine encephalomyelitis virus-induced demyelinating disease. *J. Virol.* 69:2525–2533.

Lyman, M.A., Myoung, J., Mohindrum, M., and Kim, B.S. (2004). Quantitative, not qualitative, differences in CD8+ T cell responses to Theiler's murine encephalomyelitis virus between resistant C57BL/6 and susceptible SJL/J mice. *Eur. J. Immunol.* 34:2730–2739.

Melvold, R.W., Jokinen, D.M., Knobler, R.L., and Lipton, H.L. (1987). Variations in genetic control of susceptibility to Theiler's virus (TMEV)-induced demyelinating disease. *J. Immunol.* 138:1429–1433.

Mi, W., Belyavskyi, M., Johnson, R.R., Sieve, A.N., Storts, R., Meagher, M.W., and Welsh, C.J.R. (2004). Alterations in chemokine expression in Theiler's virus infection and restraint stress. *J. Neuroimmunol.* 151:103–115.

Mi, W., Young, C.R., Storts, R., Steelman, A., Meagher, M.W., and Welsh, C.J.R. (2006a). Stress alters pathogenecity and facilitates systemic dissemination of Theiler's virus. Microbial Pathogenesis in press.

Mi, W., Prentice T.W., Young, C.R., Johnson, R.R., Sieve, A.N., Meagher, M.W., Welsh, C.J.R. (2006). Restraint stress decreases virus-induced pro-inflammatory cytokine expression during acute Theiler's virus infection. (In revision for J. Neuroimmounol.)

Miguel, H., Zhang, S.M., Lipworth, L., Olek, M.J., and Ascherio, A. (2001). Multiple sclerosis and age at infection with common viruses. *Epidemiology* 12:301–306.

Miller, S.D., VanDerlugt, C.L., Begolka, W.S., Pao, W., Yauch, R.L., Neville, K.L., Katz-Levy, Y., Carrizosa, A., and Kim, B.S. (1997). Persistent infection with Theiler's virus leads to CNS autoimmunity via epitope spreading. *Nat. Med.* 3:1133–1136.

Mohr, D.C., Hart, S.L., Julian, L., Cox, D., and Pelletier, D. (2004). Association between stressful life events and exacerbation in Multiple Sclerosis: a meta analysis. *B.M.J.* 328(7442):731–735.

Monteyne, P., Bihl, F., Levillayer, F., Brahic, M., and Bureau, J.F. (1999). The Th1/Th2 balance does not account for the difference of susceptibility of mouse strains to Theiler's virus persistent infection. *J. Immunol.* 162:7330–7334.

Murray, P.D., Pavelko, K.D., Leibowitz, J., Lin, X., and Rodriguez, M. (1998). CD4 (+) and CD8 (+) T cells make discrete contributions to demyelination and neurologic disease in a viral model of multiple sclerosis. *J. Virol.* 72:7320–7329.

Nicholson, S.M., Peterson, J.D., Miller, S.D., Wang, K., Dal Canto, M.C., and Melvold, R.W. (1994) BALB/c substrain difference in susceptibility to Theiler's murine encephalomyelitis virus-induced demyelinating disease. *J. Neuroimmunol.* 52: 19–24.

Nisipeanu, P., and Korczyn, A.D. (1993). Psychological stress as risk factor for exacerbations in multiple sclerosis. *J. Auton. Nerv. Syst.* 26:77–84.

Okimura, T., Ogawa, M., and Yamauchi, T. (1986). Stress and immune responses III. Effect of restraint stress on delayed type hypersensitivity (DTH) response, natural killer (NK) activity and phagocytosis in mice. *Jpn. J. Pharmacol.* 41:229–235.

O'Leary, A., Shoor, S., Lorig, K., and Holman, H.R. (1988). A cognitive behavioral treatment for rheumatoid arthritis. *Health Psychol.* 7:527–544.

Oleszak, E.L., Chang, J.R., Friedman, H., Katestos, C.D., and Platsoucas, C.D. (2004). Theiler's virus infection: a model for multiple sclerosis. *Clin. Microbiol. Rev.* 17: 174–207.

Paya, C.V., Patick, A.K., Leibson, P.J., and Rodriguez, M. (1989). Role of natural killer cells as immune effectors in encephalitis and demyelination induced by Theiler's virus. *J. Immunol.* 143:95–102.

Paya, C.V., Leibson, P.J., Patick, A.K., and Rodriguez, M. (1990). Inhibition of Theiler's virus-induced demyelination in vivo by tumor necrosis factor-α. *Int. Immunol.* 2:909.

Peterson, J.D., Waltenbaugh, C., and Miller, S.D. (1992). IgG subclass responses to Theiler's murine encephalomyelitis virus infection and immunization suggest a dominant role for Th1 cells in susceptible mouse strains. *Immunology* 75:652–658.

Pullen, L.C., Miller, S.D., Dal Canto, M.C., and Kim, B.S. (1993). Class I-deficient resistant mice intracerebrally inoculated with Theiler's virus show an increased T cell response to viral antigens and susceptibility to demyelination. *Eur. J. Immunol.* 23:2287–2293.

Pullen, L.C., Miller, S.D., Dal Canto, M.C., Van der Meide, P.H., and Kim, B.S. (1994). Alteration in the level of interferon-γ results in acceleration of Theiler's virus-induced demyelinating disease. *J. Neuroimmunol.* 55:143–152.

Pullen, L.C., Park, S.H., Miller, S.D., Dal Canto, M.C., and Kim, B.S. (1995). Treatment with bacterial LPS renders genetically resistant C57BL/6 mice susceptible to Theiler's virus-induced demyelinating disease. *J. Immunol.* 155:4497.

Radojevic, V., Nicassio, P.M., and Weisman, M.H. (1992). Behavioral intervention with and without family support for rheumatoid arthritis. *Behav. Ther.* 23:13–30.

Rimon, R., and Laakso, R.L. (1985). Life stress and rheumatoid arthritis: A 15-year follow-up study. *Psychother. Psychosom.* 43:38–43.

Rodriguez, M., and David, C.S. (1995). H-2Dd transgene suppresses Theiler's virus-induced demyelination in susceptible strains of mice. *J. Neurovirol.* 1:111–117.

Rodriguez, M., and Sriram, S. (1988). Successful therapy of Theiler's virus-induced demyelination (DA strain) with monoclonal anti-Lyt2 antibody. *J. Immunol.* 140: 2950–2955.

Rodriguez, M., Pavelko, K.D., McKinney, C.W., and Leibowitz, J.L. (1994). Recombinant human IL-6 suppresses demyelination in a viral model of multiple sclerosis. *J. Immunol.* 153:3811–3821.

Rodriguez, M., Pavelko, K., and Coffman, R.L. (1995). Gamma interferon is critical for resistance to Theiler's virus-induced demyelination. *J. Virol.* 69:7286–7290.

Rodriguez, M., Pavelko, K.D., Njenga, M.K., Logan, W.C., and Wettstein, P.J. (1996). The balance between persistent virus infection and immune cells determines demyelination. *J. Immunol.* 157:5699–5709.

Roos, R.P., and Wollmann, R. (1984). DA strain of Theiler's murine encephalomyelitis virus induces demyelination in nude mice. *Ann. Neurol.* 15:494–499.

Rossi, P.C., McAllister, A., Fiette, L., and Brahic, M. (1991). Theiler's virus infection induces a specific cytotoxic T lymphocyte response. *Cell. Immunol.* 138:341–348.

Rossi, C.P., Delcroix, M., Huitinga, I., McAllister, A., Rooijen, N., Claassen, E., and Brahic, M. (1997). Role of macrophages during Theiler's virus infection. *J. Virol.* 71:3336–3340.

Sato, S., Reiner, S.L., Jensen, M.A., and Roos, R.P. (1997). Central nervous system cytokine mRNA expression following Theiler's murine encephalomyelitis virus infection. *J. Neuroimmunol.* 76:213–223.

Selmaj, K.W., and Raine, C.S. (1987). Tumor necrosis factor mediates myelin and oligodendrocyte damage in vitro. *Ann. Neurol.* 23:339–346.

Sheridan, J.F. (1998). Stress-induced modulation of anti-viral immunity. *Brain Behav. Immun.* 12:1–6.

Sheridan, J.F., Feng, N., Bonneau, R.H., and Allen, C.M. (1991). Restraint stress differentially affects anti-viral cellular and humoral immune responses in mice. *J. Neuroimmunol.* 31:245–255.

Sibley, W.A., Bamford, C.R., and Clark, K. (1985). Clinical viral infections and multiple sclerosis. *Lancet* 1:1313–1315.

Sieve, A.N., Steelman, A.J., Young, C.R., Storts, R., Welsh, T.H., Welsh, C.J.R., and Meagher, M.W. (2004). Chronic restraint stress during early Theiler's virus infection exacerbates the subsequent demyelinating disease in SJL mice. *J. Neuroimmunol.* 155:103–118.

Steelman, A., Mi, W., Alford, E., Young, C.R., Meagher, M.W., and Welsh, C.J.R. (2006). Restraint stress fails to render resistant C57Bl/6 mice susceptible to Theiler's virus-induced demyelination. (Manuscript in preparation.)

Stininssen, P., Raus, J., and Zhang, J. (1997). Autoimmune pathogenesis of multiple sclerosis: role of autoreactive T lymphcytes and new immunotherapeutic strategies. *Crit. Rev. Immunol.* 17:33–75.

Teshima, H., Sogawa, H., Kihara, H., Magata, S., Ago, Y., and Nakagawa, T. (1987). Changes in populations of T-cell subsets due to stress. *Ann. N.Y. Acad. Sci.* 496: 459–466.

Theil, D.J., Tsunoda, I., Libbey, J.E., Derfuss, T.J., and Fujinami, R.S. (2000). Alterations in cytokine but not chemokine mRNA expression during three distinct Theiler's virus infections. *J. Neuroimmunol.* 104:22–30.

Theiler, M. (1934). Spontaneous encephalomyelitis of mice—a new virus. *Science* 80:122.

Theiler, M., and Gard, S. (1940). Encephalomyelitis of mice. I. Characteristics and pathogenesis. *J. Exp. Med.* 72:79.

Warren, S., Greenhill, S., and Warren, K.G. (1982). Emotional stress and the development of multiple sclerosis: case-control evidence of a relationship. *J. Chronic Dis.* 35:821–831.

Welsh, C.J.R., Tonks, P., Nash, A.A., and Blakemore, W.F. (1987). The effect of L3T4 T cell depletion on the pathogenesis of Theiler's murine encephalomyelitis virus infection in CBA mice. *J. Gen. Virol.* 68:1659–1667.

Welsh, C.J.R., Blakemore, W.F., Tonks, P., Borrow, P., and Nash, A.A. (1989). Theiler's murine encephalomyelitis virus infection in mice: A persistent viral infection of the central nervous system which induces demyelination. In N. Dimmock (ed.), *Immune Responses, Virus Infection and Disease.* Oxford: Oxford University Press, pp. 125–147.

Welsh, C.J.R., Tonks, P., Borrow, P., and Nash, A.A. (1990). Theiler's virus: An experimental model of virus-induced demyelination. *Autoimmunity* 6:105–112.

Welsh, C.J.R., Sapatino, B., Rosenbaum, B., Smith. R., and Linthicum, D.S. (1993). Correlation between susceptibility to demyelination and interferon gamma induction of major histocompatibility complex class II antigens on murine cerebrovascular endothelial cells. *J. Neuroimmunol.* 48:91–98.

Welsh, C.J.R., Sapatino, B.V., Rosenbaum, B., and Smith, R. (1995). Characteristics of cloned cerebrovascular endothelial cells following infection with Theiler's virus. I. Acute infection. *J. Neuroimmunol.* 62:119–125.

Welsh, C.J.R., Bustamante, L., Nayak, M., Welsh, T.H., Dean, D.D., and Meagher, M.W. (2004). The effects of restraint stress on the neuropathogenesis of Theiler's virus infection II: NK cell function and cytokine levels in acute disease. *Brain Behav. Immun.* 18:166–174.

Welsh, C.J.R., Sieve, A., Johnson, R., Storts, R., Welsh, T.H., and Meagher, M.W. (2006). The effects of restraint stress on the neuropathogenesis of Theiler's virus infection. II. Chronic disease. (Manuscript in preparation.)

Whetten-Goldstein, K., Sloan, F.A., Goldstein, L.B., and Kulas, E.D. (1998). A comprehensive assessment of the cost of multiple sclerosis in the United States. *Mult. Scler.* 4:419–425.

Whitacre, C.G., Cummings, S.O., and Griffin, A.C. (1994). The effects of stress on autoimmune disease. In R. Glaser, and J.K. Kiecolt-Glaser (eds.), *Handbook of Human Stress and Immunity.* San Diego: Academic Press, pp. 77–100.

Yoon, J.W., Austin, M., Onodera, T., and Notkins, A.L. (1979). Virus induced diabetes mellitus. Isolation of a virus from the pancreas of a child with diabetic ketoacidosis. *N. Engl. J. Med.* 300:113.

Zurbriggen, A., and Fujinami, R.S. (1988). Theiler's virus infection in nude mice: viral RNA in vascular endothelial cells. *J. Virol.* 62:3589–3596.

Zwilling, B.S., Brown, D., Feng, N., Sheridan, J., and Pearl, D. (1993). The effect of adrenalectomy on the restraint stress induced suppression of MHC class II expression by murine peritoneal macrophages. *Brain Behav. Immun.* 7:29–35.

11
Social Stress Alters the Severity of an Animal Model of Multiple Sclerosis

MARY W. MEAGHER, ROBIN R. JOHNSON, ELISABETH GOOD, AND C. JANE WELSH

1. Introduction

Multiple sclerosis (MS) is a demyelinating disease of the central nervous system (CNS) and a leading cause of disability among young adults (Anderson *et al.*, 1992; Jacobson *et al.*, 1997; Noonan *et al.*, 2002; Sorpedra and Martin, 2005). Common symptoms include loss of motor control or sensation in the limbs, loss of bowel or bladder control, neuropathic pain, optic neuritis, sexual dysfunction, and cognitive dysfunction. The etiology of MS remains uncertain; however, considerable evidence suggests that environmental factors interact with genetic factors to cause disease (Kurtzke and Hyllested, 1987; Noseworthy *et al.*, 2000; Sospedra and Martin, 2005). Suspected environmental factors include viral infection and stress.

In the sections that follow, we will review evidence suggesting that viral infection and stress contribute to the pathogenesis of MS in humans and in animal models of MS. First, the pathophysiology of MS is briefly reviewed along with evidence for a viral etiology in humans and in animal models of MS (see Welsh *et al.* in this volume for an extensive review). This section is followed by a discussion of the effects of stress on MS and research indicating that social stress can alter immune function. Finally, we review evidence indicating that social stress alters the course of an animal model for MS, Theiler's murine encephalomyelitis virus (TMEV) infection.

2. Multiple Sclerosis

As discussed in the preceding chapter 10 (Welsh *et al.*, 2006), MS is a cell-mediated autoimmune disease in which immune responses are directed against the myelin sheath surrounding the axons of neurons in the CNS. Demyelination is associated with the infiltration of inflammatory cells into the brain and spinal cord, including macrophages/microglia, plasma cells, and T and B lymphocytes. An autoimmune etiology is suggested because MS patients show autoimmune responses to myelin components myelin basic protein (MBP), proteolipid protein (PLP), and myelin-oligodendrocyte glycoprotein (MOG; Stinissen *et al.*, 1997).

MS is a heterogeneous disease that can follow a relapsing-remitting, secondary progressive, or chronic progressive disease course (Sorpedra and Martin, 2005). As the disease progresses, plaques are formed throughout the white matter in the brain and spinal cord, resulting in significant loss of motor, sensory, autonomic, and neurological function. Although the determinants of individual differences in disease course and susceptibility remain poorly understood, complex genetic traits and exposure to environmental stressors may result in differential susceptibility to viral infections and dysregulation of the immune response to CNS inflammatory insults. A growing body of evidence suggests that MS develops in genetically susceptible individuals exposed to an environmental trigger (Dyment *et al.*, 2004, Sorpedra and Martin, 2005). Family and twin studies indicate that genetic susceptibility is necessary but not sufficient for disease vulnerability. While rates of disease are higher among relatives of MS patients, concordance rates for identical twins (approximately 25%) are modest, indicating that environmental factors must be involved.

2.1. A Viral Etiology for MS

Both human and animal research suggests that viral infection is a likely environmental trigger of MS (Gilden, 2005; Sospedra and Martin, 2005). Postmortem analyses of the brains of MS patients have identified a number of viral agents, including mumps, measles, parainfluenza type I (Allen and Brankin, 1993), and human herpes virus simplex type 6 (Challoner *et al.*, 1995). Although no single viral candidate has been implicated in all cases, it is possible that several viruses may serve as an environmental trigger of MS (Gilden, 2005; Sospedra and Martin, 2005). Indeed, several viruses have been found to initiate demyelination in humans (Johnson, 1994; Soldan and Jacobson, 2001) and animals (Dal Canto and Rabinowitz, 1982; Johnson, 1994; Soldan and Jacobson, 2001). Two common human herpes viruses that induce persistent infections are likely candidates in human MS: human herpes virus 6 (HHV-6) and Epstein-Barr virus (EBV). In the general population, both viruses are widespread, with seroprevalence rates of 80% and 90%, respectively (Martyn *et al.*, 1993; Soldan *et al.*, 2000; Wandinger *et al.*, 2000; Moore *et al.*, 2002). With both viruses, seroconversion tends to occur during adolescence, which corresponds with epidemiological studies suggesting that viral exposure during adolescence is a linked to the subsequent development of MS.

2.2. Animal Models of Virus-induced Demyelination

Animal models provide experimental evidence indicating that viruses can induce demyelination. Although it is generally agreed that TMEV infection provides the best characterized model of virus-induced CNS demyelination in mice, other viruses such as mouse hepatitis virus, Semliki Forest virus, canine distemper virus, visna virus, caprine arthritis-encephalitis virsus, and measles virus induce demyelination in animals. After establishing a persistent infection, these viruses trigger virus- and/or autoimmune-mediated demyelination. TMEV studies indicate that autoimmune demyelination is initiated through mechanisms of molecular mimicry and/or bystander activation (Olson *et al.*, 2001, 2002, 2004; Sospedra and Martin, 2005). Molecular mimicry involves T and B cells that cross-react with virus, antigenic determinants, or peptides of self-antigens, such as MBP, MOG, and PLP. In contrast, bystander activation involves the activation of autoreactive T cells by nonspecific inflammatory responses triggered by infection (e.g., cytokines activation). Studies of animal models of virus-induced demyelination are important because they have increased our understanding of potential mechanisms contributing to the pathogenesis of MS in humans, including environmental susceptibility factors such as exposure to viral infection and stress.

2.3. MS and Stress

MS patients frequently report elevated levels of stress prior to initial diagnosis and/or disease exacerbation (Warren *et al.*, 1982; Grant *et al.*, 1989; Ackerman *et al.*, 2000; Mohr *et al.*, 2000; Mohr *et al.*, 2004). A recent meta-analysis was conducted on 14 studies investigating the impact of stressful life events on MS exacerbation. The results indicated that stressful life events significantly increased the risk of subsequent disease exacerbation (Mohr *et al.*, 2004). Importantly, 13 of the 14 studies measured common stressful life events, mostly interpersonal stressors at family and work. However, one of the 14 studies did not observe a negative effect of stress on MS. This study examined the impact of a traumatic stressor, missile attacks during the Gulf War, finding reduced relapse rates and no change in lesion development (Nisipeanu and Korczyn, 1993). Thus, the characteristics of the stressor may alter the impact of stress on initial susceptibility and disease course in MS. Chronic social stressors appear to exacerbate disease, whereas acute traumatic stressors may have a beneficial effect.

The majority of studies examining the relationship between stress and MS have used patient self-reports of symptoms and/or neurologist ratings of symptoms that are based in part on patient reports. Therefore, it is possible that reports of disease exacerbations during periods of stress may be biased by the patient's mood or some other spurious subjective variable. Objective markers of disease exacerbation are clearly needed to resolve this issue, however only one human study meets this criterion. Using a prospective design, Mohr and colleagues (2000) examined the association between stressful life events and subsequent lesion development in MS patients. To objectively measure changes in blood-brain barrier (BBB) disruption that are linked to CNS inflammation in MS, patients received a monthly gadolinium (Gd+) MRI. Gd+ is injected peripherally and then crosses the BBB at sites of MS inflammation, allowing the quantification of active lesions during the MRI scan. Patient reports of stressful life events were found to predict the development of new Gd+ brain lesions. Importantly, chronic social stressors and disruptions in daily routine were found to have the strongest relationship with lesion development.

Although human studies suggest that social stress is correlated with later MS disease exacerbation, animal studies are critical to determine whether there is a *causal* relationship between stress and disease exacerbation. Animal experiments are advantageous because the stressor can be experimentally manipulated and its impact on disease course can be quantified using objective behavioral and physiological markers of disease. Animal studies are also important because they can readily identify the underlying immune and endocrine *mechanisms* mediating the effects of social stress on disease.

3. Theiler's Virus Infection as a Model for MS

After intracerebral infection with either the BeAn or DA strains of TMEV, susceptible strains of mice develop a primary inflammatory demyelinating disease that is remarkably similar to MS (Lipton, 1975; Oleszak *et al.*, 2004). TMEV must establish a persistent infection of the CNS in order to cause demyelination (Aubert *et al.*, 1987). Resistant strains of mice, which are able to clear the virus from the CNS during the first month of infection, do not develop the demyelination.

3.1. The Immune Response in TMEV Infection

As discussed in the preceding chapter, TMEV induces a biphasic disease. The acute phase of disease (first month) is primarily a gray matter disease similar to poliomyelitis. The later chronic disease occurs several months postinfection and is characterized by primary inflammatory demyelination. Immunosuppressive regimens during the acute disease are detrimental because an intact immune system is required to control infection. In contrast, immunosuppression during chronic disease is beneficial because it ameliorates inflammation as the immune response becomes pathogenic (Welsh *et al.*, 1987).

The early immune events that occur during TMEV infection are crucial in the effective clearance of the virus from the CNS. Failure to clear virus results in persistent infection of the CNS and subsequent demyelination (Brahic *et al.*, 1981; Rodriguez *et al.*, 1996). As noted in the preceding chapter 10 (Welsh *et al.*, 2006), innate cytokine responses to infection play an important role in shaping downstream innate and adaptive immune responses (Biron, 1998). For example, the cytokine IFN-β plays a pivotal role in the early immune response to TMEV (Fiette *et al.*, 1995). Natural killer (NK) cells and CD8⁺ and CD4⁺ T cells are also activated early in infection and play an important role in viral clearance (Kaminsky *et al.*, 1987; Welsh *et al.*, 1987; Borrow *et al.*, 1992; Dethlefs *et al.*, 1997; Murray *et al.*, 1998). However, during chronic disease, CD8⁺ and CD4⁺ T-cell subsets contribute to demyelination (Clatch *et al.*, 1987; Rodriguez *et al.*, 1988; Welsh *et al.*, 1989).

3.2. *Theiler's Virus-induced Demyelination (TVID)*

TMEV-induced demyelination is partly mediated by direct viral lysis of oligodendrocytes (Roos and Wollmann, 1984). In addition, CD4⁺ T cells also play a major role in mediating CNS demyelination by generating both antivirus and antimyelin autoimmune responses (Welsh *et al.*, 1987, 1989, 1990; Miller *et al.*, 1997; Borrow *et al.*, 1998; Dal Canto *et al.*, 2000). Susceptibility to inflammation and demyelination of the spinal cord are strongly related to the ability of the mouse to mount a delayed-type hypersensitivity (DTH) response to viral antigen. Other evidence suggests that demyelination is initiated by virus-specific T cells that target the virus presented by persistently infected macrophages/microglia in the CNS. This leads to initial myelin destruction due to proinflammatory cytokines produced by TMEV-specific Th1 cells that target CNS-persistent virus. As the disease progresses, myelin destruction induces epitope spreading, characterized by the development of myelin-epitope specific autoreactive Th1 cells (Miller *et al.*, 2001; Croxford *et al.*, 2002; Olson *et al.*, 2004). The autoimmune aspects of this disease enhance bystander demyelination mediated by virus-specific DTH T cells (Clatch *et al.*, 1987; Gerety *et al.*, 1991) and cytotoxic T-cell reactivity (Rodriguez and Sriram, 1988).

In summary, TMEV infection provides a well-characterized animal model of MS in which to study the effects of social stress on disease course. Using this model, we have previously examined the impact of restraint stress on disease course to gain a better understanding of how stress impacts the human disease (for a recent review, see Welsh *et al.*, this volume). In the following sections, we will review evidence that social stress can alter immune function and TMEV disease vulnerability.

4. Social Stress and Immunity

The nature of a stressful event determines its biobehavioral consequences. For example, social stress, but not restraint stress, can lead to the reactivation of a latent herpes virus infection (Padgett *et al.*, 1998), increased inflammatory responses to a lipolysaccharide (LPS) challenge (Quan *et al.*, 2001), and increased CNS inflammation during acute TMEV infection (Johnson *et al.*, 2004a). Virtually all mammals experience social stress, thus there is a strong rationale for investigating the effects of social conflict on immune function and disease vulnerability in both humans and animals (Blanchard *et al.*, 2001).

Human research suggests that chronic social stress is immunosuppressive and associated with increased risk for infectious illnesses. Studies investigating the chronic social stress of caring for a spouse with Alzheimer disease indicate that caregiving is linked to suppressed antibody responses to vaccination challenge (Kiecolt-Glaser *et al.*, 1996; Glaser *et al.*, 2000). Similarly, chronic social stress has been shown to increase the risk for subsequently developing upper respiratory infections after an influenza viral challenge (Cohen *et al.*, 1991, 1993). Although these correlational studies suggest that social stress is associated with adverse immunological and health effects in humans, it is also possible that this association may be due to a spurious third variable, such as a personality variable that is linked to both stress and illness. To determine causality, animal experiments have been conducted to examine the impact of social stress on immunity and the underlying mechanisms that mediate these effects. (Cohen *et al.*, 1997)

4.1. Social Stress, Immunity, and Glucocorticoid Resistance in Mice

Animal studies investigating the effects of social stress on immune function have tended to use male rodents exposed to social conflict. Several models of social conflict have been developed that involve disruption of a preexisting social hierarchy and social defeat. The social disruption (SDR) model is one example in which an aggressive male intruder is introduced into the residence of a group of younger male mice. The introduction of the dominant intruder elicits observable aggressive confrontations (e.g., posturing, fighting, and wounding) and defeat of the younger resident mice.

Chronic social stress has profound effects on endocrine and immune function in mice (Padgett *et al.*, 1998; Avitsur *et al.*, 2001; Quan *et al.*, 2001; Johnson *et al.*, 2004a). For instance, SDR produces significant increases in circulating levels of corticosterone and glucocorticoid (GC) resistance (Avitsur *et al.*, 2001). GC resistance refers to a decrease in the immune system's capacity to respond to the inhibitory effects of corticosterone in terminating inflammatory responses. When mice are exposed to social stress, their splenocytes become less sensitive to corticosterone when stimulated with lipopolysaccharide (LPS), showing increased proliferation and viability relative to control splenocytes (Avitsur *et al.*, 2001). In addition, these mice exhibit splenomegaly, an increased number of CD11b$^+$ monocytes, and elevated levels of the proinflammatory cytokines TNF-a, IL-6, and IL 1β (Quan *et al.*, 2001; Avitsur *et al.*, 2002, 2003b; Stark *et al.*, 2002). SDR has also been shown to increase susceptibility to endotoxic shock after an LPS *in vivo* challenge. Compared with controls, SDR mice subjected to LPS challenge show pronounced inflammatory organ damage and exaggerated proinflammatory cytokine responses in lung, spleen, liver, and brain compared with controls (Quan *et al.*, 2001). Together, these results suggest that GC resistance and elevated cytokine expression mediate the increase in susceptibility to endotoxic shock observed after social stress.

Humans exposed to chronic stress also exhibit alterations in hypothalamic pituitary adrenal axis function (McEwen, 1998) and the development of GC resistance (Lowy *et al.*, 1984; Stratakis *et al.*, 1994; Miller *et al.*, 2002). For instance, the parents of children undergoing treatment for cancer report higher levels of psychological distress, flatter diurnal cortisol rhythms, and GC resistance relative to the parents of healthy children (Miller *et al.*, 2002). To assess GC resistance, these investigators examined whether the parents' immune responses showed decreased sensitivity to the anti-inflammatory effects of dexamethasone, a synthetic GC. Compared with parents of healthy children, the ability of a dexamethasone to inhibit the *in vitro* production of the proinflammatory cytokine IL-6 was suppressed among the parents of cancer patients.

4.2. Social Stress and Glucocorticoid Resistance in MS

Mohr and colleagues (2006) have provided evidence that social stress is associated with increased exacerbations and new lesions in multiple sclerosis patients. One potential mechanism explaining the link between stress and exacerbation of MS is the development of GC resistance. Indeed, a few clinical studies provide evidence that the immune cells of MS patients are less sensitive to the regulatory effects of GC when compared with healthy controls (Stefferl *et al.*, 2001; DeRijk *et al.*, 2004; van Winsen *et al.*, 2005). GC resistance has also been observed in other inflammatory and autoimmune diseases (i.e., lupus, rheumatoid arthritis, asthma, Crohn disease, ulcerative colitis) and is thought to contribute to disease pathogenesis. For example, reduced sensitivity to *in vitro* cortisol is correlated with poor clinical response to steroid therapy in lupus patients (Tanaka *et al.*, 1992). This finding suggests that GC resistance may be one factor associated with susceptibility to autoimmune disease. Although the mechanism mediating GC resistance in autoimmune disease remains unknown, it has been suggested that GC resistance may develop due to frequent administration of exogenous GC. However, a recent study of MS patients demonstrated that GC resistance was unrelated to the frequency or interval of prior treatment with exogenous GC (van Winsen *et al.*, 2005). In light of the critical role that endogenous GCs play in immune regulation, a reduction in tissue sensitivity to GCs induced by chronic social stress could be one mechanism that increases vulnerability to infectious and autoimmune diseases. To test this hypothesis, our laboratory has examined the impact of SDR-induced GC resistance on acute and chronic TMEV infection.

5. The Effects of Social Stress on TMEV Infection

5.1. Effect of Social Stress During Acute TMEV Infection

To investigate the effects of social stress on MS disease onset, our first set of experiments examined the effects of SDR during the acute phase of TMEV infection. We hypothesized that exposure to SDR prior to infection (PRE-SDR) would induce GC resistance, resulting in increased CNS inflammation and increased behavioral signs of acute infection compared with infected/non-stressed mice. This hypothesis was derived from prior studies indicating that SDR-induced GC resistance increased inflammation after LPS challenge, whereas restraint stress did not (Avisur et al., 2001; Quan et al., 2001). The effects of SDR applied concurrent with infection (CON-SDR) were also evaluated. Similar to our previous studies of restraint stress, the CON-SDR procedure parallels the timing of the application of restraint (Campbell et al., 2001; Mi et al., 2004; Sieve et al., 2004; Welsh et al., 2004; Mi et al., in press). When restraint stress was applied concurrent with TMEV infection, we observed decreased CNS inflammation at day 7 postinfection. In light of this finding, it was possible that CON-SDR might reduce CNS inflammation and associated behavioral signs of acute infection. However, an alternative possibility is that the impact of a concurrent stressor may vary depending upon the nature of the stress paradigm; restraint may reduce CNS inflammation, whereas SDR may enhance inflammation.

Our first experiment was designed to resolve this issue by evaluating the impact of SDR applied either before or concurrent with TMEV infection. The timeline of the experimental procedures is presented in panel A of Figure 11.1. PRE-SDR mice received six sessions of SDR 1 week prior to infection, CON-SDR mice received six sessions of SDR concurrent with infection, and NON-SDR mice remained undisturbed. Intruders for the SDR manipulation were older aggressive male breeders that were introduced into the cage of resident mice at the beginning of their dark cycle for 2-h sessions. After the last SDR session, the PRE-SDR mice were infected with TMEV, whereas the CON-SDR mice were infected after their first session. Thereafter, the mice were periodically assessed for the development of behavioral signs of motor impairment and illness behavior before being sacrificed at days 7 or 21 postinfection. In comparison with other susceptible mouse strains, Balbc/J mice are unique because of polio-like motor impairment during acute TMEV infection. Using a behavioral scoring system that measured the natural progression of motor impairment during TMEV infection, trained observers blind to experimental condition scored the degree of hind limb impairment. Footprint stride length and grid hang latencies were also measured to assess TMEV induced changes in gait and strength, respectively.

A.   Experimental Design

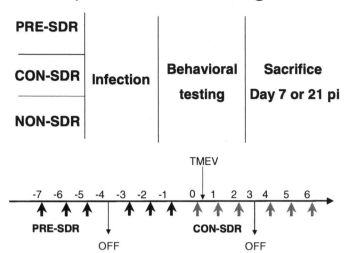

B. Motor Impairment

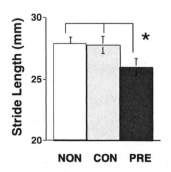

C. Histology

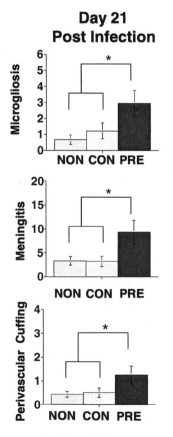

The physiological impact of SDR on acute TMEV infection was assessed by examining CNS inflammation and CNS viral clearance. In half of the mice in each condition, brains and spinal cords were sectioned and stained with hematoxylin and eosin (H&E). Raters who were blind to experimental condition scored sections for CNS inflammatory markers for acute TMEV infection, including perivascular cuffing (accumulation of lymphocytes and macrophages around blood vessels), meningitis (accumulation of lymphocytes and macrophages in the meninges), and microgliosis (presence of increased microglia/macrophages within the parenchyma of the brain and spinal cord). Previous research indicates that CNS viral titers peak at 1–2 weeks postinfection, but by 3–4 weeks postinfection the virus is cleared to nondetectable levels (Welsh *et al.*, 1987, 1989). Therefore, to determine the effect of SDR on CNS viral clearance, half of the brains and spinal cords were used to determine viral titer at days 7 and 21 postinfection using plaque assays.

Social stress presented prior to infection (PRE-SDR) resulted exacerbated disease course, whereas concurrent presentation of social stress with infection (CON-SDR) resulted attenuated disease severity (Fig. 11.1, panels B–D). Greater levels of motor impairment were observed in the PRE-SDR mice (Fig. 11.1, panel B). PRE-SDR mice showed hind limb impairment, reduced stride length, and reduced grid hang time compared with the non-stressed controls and CON-SDR. Histological analyses of H&E-stained sections of spinal cord and brain revealed that mice exposed to SDR prior to infection exhibited increased levels of inflammation in spinal cord and brain compared with the no-stress control and CON-SDR mice. Although significant increases in microgliosis, perivascular cuffing, and meningitis were observed in both spinal cord and brain, these effects were most pronounced in spinal cord and most prominent at day 21 (Fig. 11.1, panel C). SDR was also found to alter viral load on days 7 and 21 postinfection. As anticipated, a reduction in viral load was observed in the non-stressed animals over time, suggesting that these mice were clearing the virus to low levels. In contrast, the CON-SDR group showed even greater clearance over time, whereas the PRE-SDR animals showed no reduction over time. This is significant because disruption of the viral clearance process during

FIGURE 11.1. The experimental design and timing of procedures and data collection for the *acute* Theiler's virus infection studies (A). Stressed mice received three consecutive SDR sessions, with one night off, followed by an additional three consecutive SDR sessions. PRE-SDR mice received SDR 1 week prior to infection, CON-SDR mice received SDR concurrent with infection, and NON-SDR mice remained undisturbed in their home cage. The figures show effects of SDR on behavioral measures of motor impairment (B), including hind limb impairment (top panel) and stride length (bottom panel), and histological signs of spinal cord inflammatory markers at day 21 postinfection (C).

the acute phase of TMEV increases the risk of developing the chronic demyelinating phase. Collectively, these results suggest that the timing of SDR application in relation to infection determines the impact of SDR on acute TMEV infection, with PRE-SDR exacerbating disease course and CON-SDR attenuating disease severity.

5.2. Social Stress, GC Resistance, and Exacerbation of Acute TMEV

We propose that the development of GC resistance prior to infection may account for the increase in CNS inflammation and motor impairment observed in the infected PRE-SDR mice. Similar to previous SDR studies conducted using C57BL/6 mice (Avitsur, 2001; Stark *et al.*, 2001), we found that exposure to SDR increased circulating levels of corticosterone and induced GC resistance in the uninfected Balbc/J mice. Normally, SDR-induced increases in corticosterone would be expected to decrease inflammation. However, the development of GC resistance in the PRE-SDR mice before infection would be expected to amplify the CNS inflammatory response during TMEV infection, potentially altering the timing and nature of the innate and specific immune response to early infection. Consistent with this view, PRE-SDR resulted in increased CNS inflammation and more severe behavioral signs of infection. In contrast, GC resistance was not observed in the CON-SDR mice, and the inflammatory response to infection was not amplified. Rather, it appears that SDR-induced increases in corticosterone in the CON-SDR mice attenuated the CNS inflammatory response to infection relative to the NON-SDR infected control mice. These findings suggest that the induction of GC resistance by social stress may be one factor that increases the severity of CNS inflammatory responses to TMEV immune challenge.

5.3. The Role of IL-6 in Mediating the Adverse Effects of Social Stress

We have recently proposed that the induction of cytokine expression by social stress *prior* to infection may mediate the adverse effects of SDR on acute TMEV. Consistent with this hypothesis, our laboratory has previously observed elevated circulating levels of IL-6 in PRE-SDR mice after infection (Johnson *et al.*, 2006a). IL-6 is a proinflammatory cytokine that centrally regulates acute phase, immune, fever, and HPA-axis responses (Akira *et al.*, 1990). Prior research suggests that IL-6 levels are elevated after SDR (Avitsur *et al.*, 2001; Stark *et al.*, 2002). For example, SDR has been show to enhance IL-6 responses in plasma and in LPS-stimulated splenocytes. These observations are consistent with prior research indicating that circulating levels of IL-6 are increased in animal exposed to other stressor

(Zhou *et al.*, 1993; Nukina *et al.*, 1998) and in humans suffering from major depression (Maes *et al.*,1997). Importantly, other work indicates that IL-6 levels are elevated in the CNS during Theiler's virus infection (Theil *et al.*, 2000; Palma *et al.*, 2003; Mi *et al.*, in press). IL-6 has also been implicated in the pathogenesis of MS. Support for this view comes from studies indicating that IL-6–deficient mice do not develop experimental allergic encephalomyelitis (EAE), an autoimmune model of MS mediated by Th1 cells (Eugster *et al.*,1998; Mendel *et al.*,1998; Samoilova *et al.*, 2000; DeRijk *et al.*, 2004). Moreover, clinical studies indicate that IL-6 is expressed in the lesions and CSF of human MS patients (Padberg *et al.*, 1999; Schonrock *et al.*, 2000).

Based on this set of findings, it seems plausible that SDR-induced increases in central IL-6 prior to infection could enhance the IL-6 response to TMEV infection, thereby exacerbating TMEV-induced inflammation and behavioral signs of acute disease. This would represent an example of cross-sensitization of proinflammatory cytokine expression. This phenomenon has been demonstrated in prior studies where prior exposure to stress was found to enhance LPS-induced cytokine expression (Chancellor-Freeland *et al.*, 1995; Persoons *et al.*, 1995; Zhu *et al.*, 1995; Tilder and Schmidt, 1999; Johnson *et al.*, 2002). From this perspective, the increase in histological signs of CNS inflammation observed in mice exposed to social stress prior to TMEV infection may be attributable to the cross-sensitization of central proinflammatory cytokine activity. If SDR-induced increases in central IL-6 mediate the adverse effects of PRE-SDR, then blocking its effects during the stress exposure period should prevent the exacerbation of acute disease.

This hypothesis was tested by determining whether intracranial administration of a neutralizing antibody to IL-6 could reverse the adverse effects of SDR during acute TMEV infection (Johnson *et al.*, 2005a, 2006b). Before each SDR session, mice in the PRE-SDR or no-stress groups received either an intracranial injection of a neutralizing antibody to IL-6 or the vehicle. After their last SDR session, the mice were infected with TMEV and monitored for the development of illness behaviors and motor impairment. As anticipated, exposure to SDR prior to infection led to a loss of sucrose preference, allodynia, decreased locomotor activity, a loss of body weight, reduced stride length, and greater hind limb impairment in the vehicle control group. In contrast, pretreatment with the IL-6 neutralizing antibody blocked the effects of SDR on illness behavior and motor function. In addition, administration of the neutralizing antibody to IL-6 reversed PRE-SDR–induced increases in meningitis, perivascular cuffing, and microgliosis in spinal cord. Together, these findings suggest that SDR-induced increases in central IL-6 contribute to the adverse effects of social stress during acute TMEV infection. It must be acknowledged, however, that only a partial reversal was observed on some behavioral measures in the PRE-SDR mice treated with the IL-6 antibody. Partial reversal suggests that social stress may increase the central expression of other proinflammatory cytokines. Therefore, we are currently investigating whether the adverse effects of social stress on acute TMEV infection also depend on the release of TNF-a and IL-1β in CNS.

5.4. Significance of Acute Social Stress Effects

The finding that PRE-SDR exacerbates acute TMEV infection in adolescent mice is significant in light of epidemiological evidence suggesting that MS may be triggered by viral infection in adolescence (Acheson, 1977; Kurtzke, 1993). We propose that systemic conditions at the time of infection play an important role in modulating immune processes leading to the development of MS. Specifically, we suggest that the induction of GC resistance and/or the release of proinflammatory cytokines by social stress exacerbates acute infection, which in turn would be expected to have cascading adverse effects during the demyelinating phase of disease.

5.5. Impact of Social Stress on Chronic TMEV Infection

Using a longitudinal design, we have recently examined whether the timing of SDR during early infection alters the development of late disease (Johnson *et al.*, 2006a). In this experiment, mice received SDR either immediately prior to (PRE-SDR) infection, concurrent with (CON-SDR) infection, or remained undisturbed before and during infection (NON-SDR). During the acute phase of disease, mice were examined regularly for behavioral signs of acute disease, and circulating levels of the proinflammatory cytokine IL-6 were measured on day 9 postinfection. After the resolution of the acute phase (4 weeks), mice were monitored monthly until they exhibited behavioral signs of the chronic disease. Thereafter, weekly behavioral measures were taken to examine the development of chronic-phase disease. In addition, blood samples were taken on a monthly basis to determine whether SDR altered the development of circulating levels of antibodies to TMEV and several myelin components [myelin basic protein (MBP), myelin oligodendrocyte glycoprotein peptide (MOG), and proteolipid protein peptide (PLP)].

Replicating and extending our prior research, PRE-SDR was found to exacerbate both the *acute* and *chronic* phases of TMEV infection. Consistent with previous studies, the PRE-SDR mice developed more severe behavioral signs of *acute* infection, including hind limb impairment, reduced stride length, and reduced spontaneous locomotor activity, compared with both CON-SDR and NON-SDR mice. The PRE-SDR mice also developed earlier and more severe behavioral signs of *chronic* disease (Fig. 11.2), including hind limb impairment (Fig. 11.2A), impaired motor coordination on the rotarod (Fig. 11.2B), reduced stride length (Fig. 11.2C), reduced spontaneous locomotor activity (Fig. 11.2D), impaired inclined plane

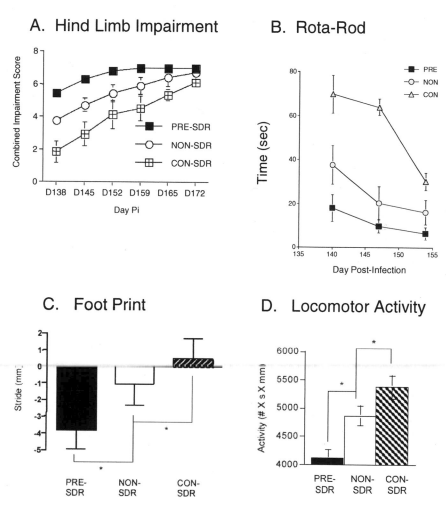

FIGURE 11.2. The effect of SDR on behavioral measures of motor impairment during *chronic* Theiler's virus infection. PRE-SDR animals had earlier onset (day 138 postinfection) and significantly greater hind limb impairment over time (A). Motor coordination on the rotarod test (B) was significantly impaired in the PRE-SDR mice compared with the NON-SDR and CON-SDR groups. In addition, stride length (C) and horizontal locomotor activity (D) were significantly reduced in the PRE-SDR mice. In contrast, the CON-SDR mice showed less impairment over time and they did not exhibit significant hind limb impairment until day 165 postinfection. The CON-SDR mice also showed increased improved rotarod performance, increased stride length, and increased horizontal locomotor activity, relative to the infected NON-SDR controls and PRE-SDR mice. Average age-matched control data was subtracted from all data points prior to analysis for stride length. Asterisks indicate significant differences.

performance, and reduced sensitivity to von Frey mechanical stimulation. In contrast, the CON-SDR during the first week of TMEV infection reduced the severity of functional impairments during both the acute and chronic phase of disease.

The severity of the chronic phase of disease was predicted by several acute-phase behavioral and immunological measures. Behavioral measures collected on day 7 postinfection (hind limb impairment and spontaneous locomotor activity) predicted the level of behavioral impairment during the chronic phase (hind limb impairment, footprint stride length, and spontaneous locomotor activity at day 136 p.i., significant *r* values ranged from 0.33 to 0.72). Moreover, several physiological measures collected early in infection (IL-6 at day 9 p.i., antibody to TMEV, MOG, and MBP at day 42 p.i., and body weights) predicted the level of behavioral impairment during the chronic phase. Correlations between the acute- and chronic-phase measures were found for both the onset of the chronic phase and the later time points. These findings suggest that social stress applied prior to infection leads to more severe acute-phase symptomology, increasing the risk for developing more severe behavioral symptoms during the chronic phase of disease.

5.6. Neonatal Experience Alters the Impact of Social Stress

Our laboratory has recently investigated whether early life events can induce longlasting effects on TMEV vulnerability. These studies also examined whether neonatal experience alters the subsequent impact of SDR on TMEV infection. In these experiments, mouse pups undergo brief daily maternal separation during the first 2 weeks of life. Previous research by other investigators had suggested that exposure to brief maternal separation leads to a hypoactive HPA-axis, with long-term protective effects (for a review, see Meaney, 2001). In contrast, we found that brief maternal separation may not be protective when paired with later social stress (Johnson *et al.*, 2004b, 2005c).

In this experiment, pups were exposed daily to either brief maternal separation for 15 min or to an undisturbed control condition during the first 2 weeks of life. Mice were later infected with TMEV during adolescence and monitored for behavioral and physiological indicators of acute and chronic disease. Normally, when SDR occurs prior to TMEV infection, the outcome is detrimental, whereas SDR concurrent with infection reduces disease severity (Johnson *et al.*, 2004a, 2005a). However, when the brief maternal separation mice were socially stressed later in life (either prior to or concurrent with infection), they developed more severe hind limb impairment compared with mice that were not separated as neonates. In contrast, the brief maternal separation mice that were not socially stressed showed the classic pattern of protection, with less severe levels of impairment. These effects were observed during both the acute and chronic phase of infection on several functional measures, including hind limb impairment, body weight, open field vertical activity measures, and open field horizontal measures. Additionally, the brief maternal separation mice that were not socially stressed exhibited lower circulating levels of antibody to virus during late disease, suggesting that neonatal experience influenced disease activity. Blunted corticosterone (CORT) responses were also observed in the brief maternally separated mice exposed to SDR before or concurrent with infection. It is tempting to speculate that blunted HPA function creates a permissive environment that increases acute CNS inflammation, which in turn alters the immune response to the virus and the development of GC resistance after SDR. These findings suggest that whereas brief maternal separation may be protective against some types of later stressors (*e.g.*, TMEV infection alone), it may increase vulnerability to the adverse effects of later social stressors on TMEV infection (Laban *et al.*, 1995; Shank and Lightman, 2001).

6. General Conclusions

A growing body of evidence suggests that two environmental factors, viral infection and stress, may contribute to the development of MS in genetically susceptible humans and animals. Epidemiological studies indicate that MS is triggered by viral infection. Evidence for a relationship between stressful life events and MS exacerbation is also mounting, including longitudinal studies indicating that naturally occurring stressors predict the occurrence of new lesions. However, these human studies are limited because the association between stress and disease exacerbation may be attributable to a spurious third variable that influences both stress and disease. To determine whether there is a causal relationship between chronic stress and viral infection in MS, animal experiments are essential. Using this approach, we have shown that social stress interacts with TMEV infection to determine the severity of behavioral impairment and CNS inflammation. The timing of the stressor in relation to infection has been shown to play a critical role in determining disease course. Moreover, neonatal experiences have been shown to influence the response to infection as well as the impact of later social stress on disease course. Finally, we have begun to examine the underlying endocrine and immune mechanisms whereby social stress exacerbates CNS inflammation and disease course, including the induction of GC resistance and cross-sensitization of central proinflammatory cytokine activity (IL-6).

Exposure to social stress prior to TMEV infection resulted in more severe manifestations of acute and chronic disease in adolescent mice. This finding may have important implications for understanding disease vulnerability in humans. We propose that exposure to social stress prior to infection may result in increased CNS inflammation and dysregulation of the early immune response to infection. The induction of GC resistance and the sensitization of central proinflammatory cytokine expression by social stress may contribute to the dysregulation of central inflammatory responses during early infection. Because early immune responses shape the specific immune response to infection, dysregulation of this response may contribute to the failure to eliminate the pathogen and exacerbation of acute infection. We hypothesize that the establishment of a persistent infection, combined with a heightened inflammatory environment in the CNS, may contribute to the development of autoimmune diseases such as multiple sclerosis. To test this hypothesis, future research will need to examine whether social stress–induced GC resistance and central sensitization of IL-6 expression will have cascading effects that increase the severity of the demyelinating phase of disease.

Acknowledgments. This research was supported by an NSF fellowship and NIH/NRSA to R.R.J. and funded by grants to C.J.R.W. and M.W.M. from the National Multiple Sclerosis Society (RG 3128) and NIH/NINDS (R01 39569).

References

Acheson, E.D. (1977). Epidemiology of multiple sclerosis. *Br. Med. Bull.* 33:9–14.

Ackerman, K.D., Rabin, B.S., Heyman, R., and Baum, A. (2000). Stressful life events and disease activity in multiple sclerosis. *Brain Behav. Immun.* 14:77.

Akira, S., Hirano, T. Taga, T., and Kishimoto, T. (1990). Biology of multifunctional cytokines: IL-6 and related molecules (IL1 and TNF). *FASEB J.* 4:2860–2867.

Allen, I., and Brankin, B.J. (1993). Pathogenesis of multiple sclerosis—the immune diathesis and the role of viruses. *Neuropathol. Exp. Neurol.* 52:95.

Anderson, D.W., Ellenberg., J.H., Leventhal, C.M., Reingold, S.C., Rodriguez, M., and Silberberg, D.H. (1992). Revised estimate of the prevalence of multiple sclerosis in the United States. *Ann. Neurol.* 31:333–336.

Aubert, C., Chamorro, M., and Brahic, M. (1987). Identification of Theiler's infected cells in the central nervous system of the mouse during demyelinating disease. *Microb. Pathogen.* 3:319–326.

Avitsur, R., Stark, J.L., and Sheridan, J.F. (2001). Social stress induces glucocorticoid resistance in subordinate animals. *Hormones Behav.* 39:247–257.

Avitsur, R., Stark, J.L., Dhabhar, F.S., Padgett, D.A., and Sheridan, J.F. (2002). Social disruption-induced glucocorticoid resistance: Kinetics and site specificity. *J. Neuroimmunol.* 124(1–2):54–61.

Avitsur, R., Padgett, D.A., Dhabhar, F.S., Stark, J.L., Kramer, K.A., Engler, H., and Sheridan, J.F. (2003a). Expression of glucocorticoid resistance following social stress requires a second signal. *J. Leukoc. Biol.* 74:507–513.

Avitsur, R., Stark, J.L., Dhabhar, F.S., Kramer, K.A., and Sheridan, J.F. (2003b). Social experience alters the response to social stress in mice. *Brain Behav. Immun.* 17:426–437.

Barak, O., Goshen, I., Ben-Hur, T., Weidenfeld, J., Taylor, A.N., and Yirmiya, R. (2002a). Involvement of brain cytokines in the neurobehavioral disturbances induced by HIV-1 glycoprotein 120. *Brain Res.* 933:98–108.

Barak, O., Weidenfeld, J., Goshen, I., Ben-Hur, T., Taylor, A.N., and Yirmiya, R. (2002b). Intracerebral HIV-1 glycoprotein 120 produces sickness behavior and pituitary-adrenal activation in rats: role of prostaglandins. *Brain Behav. Immun.* 16:720–735.

Biron, C.A. (1998). Role of early cytokines, including alpha and beta interferons ((IFN-a/b), in innate and adaptive immune responses to viral infections. *Semin. Immunol.* 10:383–390.

Blanchard, R.J., McKittrick, C.R., and Blanchard, D.C. (2001). Animal models of social stress: Effects on behavior and brain neurochemical systems. *Neurosci. Biobehav. Rev.* 73:261–271.

Borrow, P., Tonks, P., Welsh, C.J.R., and Nash, A.A. (1992). The role of CD8+ T cells in the acute and chronic phases of Theiler's virus-induced disease in mice. *J. Gen. Virol.* 73:1861–1865.

Borrow, P., Welsh, C.J.R., and Nash A.A. (1993). Study of the mechanisms by which CD4+ T cells contribute to protection in Theiler's murine encephalomyelitis. *Immunology* 80:502–506.

Borrow, P., Welsh, C.J.R., Dean, D., Tonks, P., Blakemore, W.F., and Nash, A.A. (1998). Investigation of the role of autoimmune responses to myelin in the pathogenesis of TMEV-induced demyelinating disease. *Immunology* 93:478–484.

Brahic, M., Stroop, W.G., and Baringer, J.R. (1981). Theiler's virus persists in glial cells during demyelinating disease. *Cell* 26:123–128.

Campbell, T., Meagher, M.W., Sieve, A., Scott, B., Storts, R., Welsh, T.H., and Welsh, C.J. (2001). The effects of restraint stress on the neuropathogenesis of Theiler's virus infection: I. Acute disease. *Brain Behav. Immun.* 15:235–254.

Challoner, P.B., Smith, K.T., Parker, J.D., MacLeod, D.L., Coulter, S.N., Rose, T.M., Schultz, E.R., Bennett, J.L., Garber, R.L., Chang, M., Schad, P.A., Stewart, P.M., Nowinski, R.C., Brown, J.P., and Burmer, G.C. (1995). Plaque-associated expression of human herpesvirus 6 in multiple sclerosis. *Proc. Natl. Acad. Sci. U.S.A.* 92:7440–7444.

Chancellor-Freeland, C., Zhu, G., Kage, R., Beller, D., Leeman, S., and Black, P. (1995). Substance P and stress-induced changes in macrophages. *Ann. N.Y. Acad. Sci.* 771:472–484.

Chrousos, G.P. (1995). The hypothalamic-pituitary-adrenal axis and immune-mediated inflammation. *N. Engl. J. Med.* 332(20):1351–1362.

Clatch, R.J., Lipton, H.L., and Miller, S.D. (1987). Class II-restricted T cell responses in Theiler's murine encephalomyelitis virus (TMEV)-induced demyelinating disease. II. Survey of host immune responses and central nervous system virus titers in inbred mouse strains. *Microb. Pathogen.* 3:327–337.

Clatch, R.J., Miller, S.D., Metzner, R., Dal Canto, M.C., and Lipton, H.L. (1990). Monocytes/macrophages isolated from the mouse central nervous system contain infectious Theiler's murine encephalomyelitis virus (TMEV). *Virolosy.* 176:244–254.

Cohen, S., Tyrrell, D.A.J., and Smith, A.P. (1991). Psychological stress in humans and susceptibility to the common cold. *N. Engl. J. Med.* 325:606–612.

Cohen, S., Kaplan, J.R., Cunnick, J.E., Manuck, S.B., and Rabin, B.S. (1992). Chronic social stress, affiliation and cellular immune response in nonhuman primates. *Psychol. Sci.* 3:301–304.

Cohen, S., Tyrrell, D.A.J., and Smith, A.P. (1993). Life events, perceived stress, negative affect and susceptibility to the common cold. *J. Pers. Social Psychol.* 64:131–140.

Cohen, S., Line, S., Manuck, S.B., Rabin, B.S., Heise, E., and Kaplan, J.R. (1997). Chronic social stress, social status and susceptibility to upper respiratory infections in nonhuman primates. *Psychosom. Med.* 59:213–221.

Croxford, J.L., Olson, J.K., and Miller, S.D. (2002). Epitope spreading and molecular mimicry as triggers of autoimmunity in the Theiler's virus-induced demyelinating disease model of multiple sclerosis. *Autoimmun. Rev.* 1:251–260.

Dal Canto, M.C., and Rabinowitz, S.G. (1982). Experimental models of virus-induced demyelination of the central nervous system. *Ann. Neurol.* 11:109–127.

Dal Canto, M.C., Calenoff, M.A., Miller, S.D., and Vanderlugt, C.L. (2000). Lymphocytes from mice chronically infected with Theiler's murine encephalomyelitis virus produce demyelination of organotypic cultures after stimulation with the major encephalitic epitope of myelin proteolipid protein. Epitope spreading in TMEV infection has functional activity. *J. Neuroimmunol.* 104:79–84.

DeRijk, R.H., Eskandari, F., and Sternberg, E.M. (2004). Corticosteroid resistance in a subpopulation of multiple sclerosis patients as measured by ex vivo dexamethasone inhibition of LPS induced IL-6 production. *J. Neuroimmunol.* 151(1–2):180–188.

Dethlefs, S., Brahic, M., and Larsson-Sciard, E.L. (1997). An early abundant cytotoxic T-lymphocyte response against Theiler's virus is critical for preventing for viral persistence. *J. Virol.* 71:8875–8878.

Dyment, D.A., Ebers, G.C., and Sadovnick, A.D. (2004). Genetics of multiple sclerosis. *Lancet Neurol.* 3:104–110.

Elenkov, I.J., and Chrousos, G.P. (2002). Stress hormones, proinflammatory and anti-inflammatory cytokines, and autoimmunity. *Ann. N.Y. Acad. Sci.* 966:290–303.

Eugster, H.P., Frei, K., Kopf, M., Lassmann, H., and Fontana, A. (1998). IL-6-deficient mice resist myelin oligodendrocyte glycoprotein-induced autoimmune encephalomyelitis. *Eur. J. Immunol.* 28:2178–2187.

Fiette, L., Aubert, C., Brahic, M., and Ross, C.P. (1993). Theiler's virus infection of b2-microglobulin deficient mice. *J. Virol.* 67:589–592.

Fiette, L., Aubert, C., Ulrike, M., Huang, S., Aguet, M., Brahic, M., and Bureau, J.F. (1995). Theiler's virus infection of 129Sv mice that lack the interferon a/b or IFN-g receptors. *J. Exp. Med.* 181:2069–2076.

Filippini, G., Munari, L., Incorvaia, B., Ebers, G.C., Polman, C., D'Amico, R., and Rice, G.P. (2003). Interferons in relapsing remitting multiple sclerosis: a systematic review. *Lancet* 361(9357):545–552.

Gerety, S.J., Clatch, R.J., Lipton, H.L., Goswami, R.G., Rundell, M.K., and Miller, S.D. (1991). Class II-restricted T cell responses in Theiler's murine encephalomyelitis virus-induced demyelinating disease. IV Identification of an immunodominant T cell determinant, on the N-terminal end of the VP-2 capsid protein in susceptible SJL/J mice. *J. Immunol.* 146:4322–4326.

Gilden, D.H. (2005). Infectious causes of multiple sclerosis. *Lancet Neurol.* 4:195–202.

Glaser, R., Kiecolt-Glaser, J.K., Bonneau, R., Malarkey, W., and Hughes, J. (1992). Stress-induced modulation of the immune response to recombinant hepatitis B vaccine. *Psychosom. Med.* 54:22–29.

Glaser, R., Sheridan, J.F., Malarkey, W.B., MacCallum, R.C., and Kiecolt-Glaser, J.K. (2000). Chronic stress modulates the immune response to a pneumococcal pneumonia vaccine. *Psychosom. Med.* 62:804–807.

Grant, I., Brown, G.W., Harris, T., McDonald, W.I., Patterson, T., and Trimble, M.R. (1989). Severely threatening events and marked life difficulties preceding onset or exacerbation of multiple sclerosis. *J. Neurol. Neurosurg. Psychiatry* 52:8–13.

Gutierrez, J., Vergara, M.J., Guerrero, M., Fernandez, O., Piedrola, G., Morales, P., and Maroto, M.C. (2002). Multiple sclerosis and human herpesvirus 6. *Infection* 30:145–149.

Hernan, M.A., Zhang, S.M., Lipworth, L., Olek, M.J., and Ascherio, A. (2001). Multiple sclerosis and age at infection with common viruses. *Epidemiology* 12:301–306.

Jacobson, D.L., Gange, S.J., Rose, N.R., and Graham, N.M. (1997). Epidemiology and estimated population burden of selected autoimmune diseases in the United States. *Clin. Immunol. Immunopathol.* 84(3):223–243.

Johnson, R.T. (1994). The virology of demyelinating diseases. *Ann. Neurol.* 36(Suppl.):S54–60.

Johnson, J.D., O'Connor, K.A., Deak, T., Stark, M., Watkins, L.R., and Maier, S.F. (2002). Prior stressor exposure sensitizes LPS-induced cytokine production. *Brain Behav. Immunol.* 16:461–476.

Johnson, R.R., Storts, R., Welsh, T.H., Jr., Welsh, C.J., and Meagher, M.W. (2004a). Social stress alters the severity of acute Theiler's virus infection. *J. Neuroimmunol.* 148:74–85.

Johnson, R.R., Bridegam, P., Prentice, T.W., Welsh, T.H., Welsh, C.J.R., and Meagher, M.W. (2004b). Early life experience interacts with later social stress in the development of Theiler's virus infection. *Society for Neuroscience Abstracts.* 30, 728.5.

Johnson, R.R., Good, E.A., Hardin, E.A., Connoer, M.A., Prentice, T.W., Welsh, C.J.R., and Meagher, M.W. (2005a). Necessity of IL-6 in effects of social stress in acute Theiler's virus infection. *Soc. Neurosci. Abstr.* 31:1012–1020.

Johnson, R.R., Prentice, T.W., Bridegam, P., Young, C.R., Steelman, A.J., Welsh, T.H., Welsh, C.J.R., and Meagher, M.W. (2006a). Social stress alters the severity and onset of the chronic phase of Theiler's virus infection (in press Journal of Neuroimmunol).

Johnson, R.R., Bridegam, P., Prentice, T.W., Welsh, T.H., Welsh, C.J.R., and Meagher, M.W. (2006b). Early life experience interacts with later social stress in the development of acute and chronic Theiler's virus infection (in preparation).

Kaminsky, S.G., Nakamura, I., and Cudkowicz, G. (1987). Defective differentiation of natural killer cells in SJL mice. Role of the thymus. *J. Immunol.* 138:1020–1025.

Kiecolt-Glaser, J., Glaser, R., Gravenstein, S., Malarkey, W.B., and Sheridan, J.F. (1996). Chronic stress alters the immune response to influenza virus vaccine in older adults. *Proc. Natl. Acad. Sci. U.S.A.* 93:3043–3047.

Kohanawa, M., Nakane, A., and Minagawa, T. (1993). Endogenous gamma interferon produced in central nervous system by systemic infection with Theiler's virus in mice. *J. Neuroimmunol.* 48:205–211.

Kurtzke, J.F. (1993). Epidemiologic evidence for multiple sclerosis as an infection. *Clin. Microbiol. Rev.* 6:382–427.

Kurtzke, J.F., and Hyllested, K. (1987). MS epidemiology in Faroe Islands. *Rev. Neurol.* 57:77–87.

Laban, O., Dimitrijevic, M., von Hoersten, S., Markovic, B.M., and Jankovic, B.D. (1995). Experimental allergic encephalomyelitis in adult DA rats subjected to neonatal handling or gentling. *Brain Res.* 676:133–140.

Levin, L.I., Munger, K.L., Rubertone, M.V., Peck, C.A., Lennette, E.T., Spiegelman, D., and Ascherio, A. (2003). Multiple sclerosis and Epstein-Barr virus. *JAMA* 289:1533–1536.

Lipton, H.L. (1975). Theiler's virus infection in mice: An unusual biphasic disease process leading to demyelination. *Infect. Immun.* 11:1147–1155.

Lipton, H.L., Twaddle, G., and Jelachich, M.L. (1995). The predominant virus antigen burden is present in macrophages in Theiler's murine encephalomyelitis virus-induced demyelinating disease. *J. Virol.* 69:2525–2533.

Lindsley, M.D., Thiemann R., and Rodriguez, M. (1991). Cytotoxic T cells isolated from the central nervous system of mice infected with Theiler's virus. *J. Virol.* 65: 6612–6620.

Lowy M.T., Reder A.T., Antel J.P., and Meltzer, H.Y. (1984). Glucocorticoid resistance in depression: the dexamethasone suppression test and lymphocyte sensitivity to dexamethasone. *Am. J. Psychiatry* 141:1365–1370.

Mack, C.L., Vanderlugt-Castaneda, C.L., Neville, K.L., and Miller, S.D. (2003). Microglia are activated to become competent antigen presenting and effector cells in the inflammatory environment of the Theiler's virus model of multiple sclerosis. *J. Neuroimmunol.* 144:68–79.

Maes, M., Bosmans, E., De Jongh, R., Kenis, G., Vandoolaeghe, E., and Neels, H. (1997). Increased serum IL-6 and IL-1 receptor antagonist concentrations in major depression and treatment resistant depression. *Cytokine* 9:853–858.

Majid A., Galetta S.L., Sweeney C.J., Robinson, C., Mahalingam, R., Smith, J., Forghani, B., and Gilden D.H. (2002). Epstein-Barr virus myeloradiculitis and encephalomyeloradiculitis. *Brain* 125:1–7.

Martyn, C.N., Cruddas, M., and Compston, D.A. (1993). Symptomatic Epstein-Barr virus infection and multiple sclerosis. *J. Neurol. Neurosurg. Psychiatry* 56:167–68.

McEwen, B.S. (1998). Protective and damaging effects of stress mediators. *N. Engl. J. Med.* 338:171–179.

McGavern, D.B., Zoecklein, L., Drescher, K.M., and Rodriguez, M. (1999). Quantitative assessment of neurologic deficits in a chronic progressive murine model of CNS demyelination. *Exp. Neurol.* 158:171–181.

McGavern, D.B., Zoecklein, L., Sathornsumetee, S., and Rodriguez, M. (2000). Assessment of hindlimb gait as a powerful indicator of axonal loss in a murine model of progressive CNS demyelination. *Brain Res.* 877:396–400.

Meaney, M.J. (2001). Maternal care, gene expression, and the transmission of individual differences in stress reactivity across generations. *Annu. Rev. Neurosci.* 24: 1161–1192.

Mendel, I., Katz, A., Kozak, N., Ben-Nun, A., and Revel, M. (1998). Interleukin-6 functions in autoimmune encephalomyelitis: a study in gene-targeted mice. *Eur. J. Immunol.* 28:1727–1737.

Mi, W., Prentice, T.W., Young, C.R., Johnson, R.R., Sieve, A.N., Meagher, M.W., Welsh, C.J.R. (2006). Restraint Stress Decreases Virus-induced Pro-inflammatory Cytokine Expression during Acute Theiler's Virus Infection. (in press J. Neuroimmunol.)

Miller, S.D., VanDerlugt, C.L., Begolka, W.S., Pao, W., Yauch, R.L., Neville, K.L., Katz-Levy, Y., Carrizosa, A., and Kim, B.S. (1997). Persistent infection with

Theiler's virus leads to CNS autoimmunity via epitope spreading. *Nat. Med.* 3: 1133–1136.

Miller, S.D., Olson, J.K., and Croxford J.L. (2001). Multiple pathways to induction of virus-induced autoimmune demyelination: Lessons from Theiler's virus infection. *J. Autoimmunol.* 16:219–227.

Miller, G.E., Cohen, S., and Ritchey, A.K. (2002). Chronic psychological stress and the regulation of pro-inflammatory cytokines: A glucocorticoid-resistance model. *Health Psychol.* 21(6):531–541.

Mohr, D.C., Goodkin, D.E., Bacchetti, P., Boudewyn, A.C., Huang, L., Marrietta, P., Cheuk, W., and Dee, B. (2000). Psychological stress and the subsequent appearance of new brain MRI lesions in MS. *Neurology* 55:55–61.

Mohr, D.C., Goodkin, D.E., Nelson, S., Cox, D., and Weiner, M. (2002). Moderating effects of coping on the relationship between stress and the development of new brain lesions in multiple sclerosis. *Psychosom. Med.* 64(5):803–809.

Mohr, D.C., Hart, S.L., Julian, L., Cox, D., and Pelletier, D. (2004). Association between stressful life events and exacerbation in multiple sclerosis: a meta-analysis. *Br. Med. J.* 328(7442):731–735.

Mohr, D.C. (2006). The relationship between stressful life events and inflammation among patients with multiple sclerosis. In C. Jane Welsh, Mary W. Meagher and Esther Sternberg (eds.), *Neural and neuroendocrine mechanisms in host defense and autoimmunity.* New York: Springer, pp. 225–273.

Mohr, D.C., and Pelletier, D. (2006). A temporal framework for understanding the effects of stressful life events on inflammation in patients with multiple sclerosis. *Brain Behav. Immun.* 20:27–36.

Moore, F.G., and Wolfson, C. (2002). Human herpes virus 6 and multiple sclerosis. *Acta Neurol. Scand.* 106:63–83.

Murray, P.D., Pavelko, K.D., Leibowitz, J., Lin, X., and Rodriguez, M. (1998). CD4 (+) and CD8 (+) T cells make discrete contributions to demyelination and neurologic disease in a viral model of multiple sclerosis. *J. Virol.* 72:7320–7329.

Nicholson, S.M., Peterson, J.D., Miller, S.D., Wang, K., Dal Canto, M.C., and Melvold, R.W. (1994). BALB/c substrain differences in susceptibility to Theiler's murine encephalomyelitis virus-induced demyelinating disease. *J. Neuroimmunol.* 52: 19–24.

Nisipeanu, P., and Korczyn, A.D. (1993). Psychological stress as risk factor for exacerbations in multiple sclerosis. *Neurology* 43:1311–1312.

Noonan, C.W., Kathman, S.J., and White, M.C. (2002). Prevalence estimates for MS in the United States and evidence of an increasing trend for women. *Neurology* 58(1):136–138.

Noseworthy, J.H., Lucchinetti, C., Rodriguez, M., and Weinshenker, B.G. (2000). Multiple sclerosis. *N. Engl. J. Med.* 343:938–952.

Oleszak, E.L., Chang, J.R., Friedman, H., Katsetos, C.D., and Platsoucas, C.D. (2004). Theiler's virus infection: a model for multiple sclerosis. *Clin. Microbiol. Rev.* 17: 174–207.

Olson, J.K., Girvin, A.M., and Miller, S.D. (2001). Direct activation of innate and antigen-presenting functions of microglia following infection with Theiler's virus. *J. Virol.* 75:9780–9789.

Olson, J.K., Eagar, T.N., and Miller, S.D. (2002). Functional activation of myelin specific T cells by virus-induced molecular mimicry. *J. Immunol.* 169:2719–2726.

Olson, J.K., Ludovic Croxford, J., and Miller, S.D. (2004). Innate and adaptive immune requirements for induction of autoimmune demyelinating disease by molecular mimicry. *Mol. Immunol.* 40:1103–1108.

Padberg, F., Feneberg, W., Schmidt, S., Schwarz, M.J., Korschenhausen, D., Greenberg, B.D., Nolde, T., Muller, N., Trapmann, H., Konig, N., Moller, H.J., and Hampel, H. (1999). CSF and serum levels of soluble interleukin-6 receptors (sIL-6R and sgp130), but not of interleukin-6 are altered in multiple sclerosis. *J. Neuroimmunol.* 99:218–223.

Padgett, D.A., Sheridan, J.F., Dorne, J., Berntson, G.G., Candelora, J., and Glaser, R. (1998). Social stress and the reactivation of latent herpes simplex virus type 1. *Proc. Natl. Acad. Sci. U.S.A.* 95:7231–7235.

Palma, J.P., Kwon, D., Clipstone, N.A., and Kim, B.S. (2003). Infection with Theiler's murine encephalomyelitis virus directly induces proinflammatory cytokines in primary astrocytes via NF-kappaB activation: Potential role for the initiation of demyelinating disease. *J. Virol.* 77:6322–6331.

Persoons, J.H., Schornagel, K., Breve, J., Berkenbosch, F., and Kaal, G. (1995). Acute stress affects cytokines and nitric oxide production by alveolar macrophages differently. *Am. J. Respir. Crit. Care Med.* 152:619–624.

Pollak, Y., Ovadia, H., Goshen, I., Gurevich, R., Monsa, K., Avitsur, R., and Yirmiya, R. (2000). Behavioral aspects of experimental autoimmune encephalomyelitis. *J. Neuroimmunol.* 104:31–36.

Pullen, L.C., Miller, S.D., Dal Canto, M.C., and Kim, B.S. (1993). Class I-deficient resistant mice intracerebrally inoculated with Theiler's virus show an increased T cell response to viral antigens and susceptibility to demyelination. *Eur. J. Immunol.* 23:2287–2293.

Quan, N., Avitsur, R., Stark, J.L., He, L., Shah, M., Caliguiri, M., Padgett, D.A., Marucha, P.T., and Sheridan, J.F. (2001). Social stress increases the susceptibility to endotoxic shock. *J. Neuroimmunol.* 115:36–45.

Quan, N., Avitsur, R., Stark, J.L., He, L., Lai, W., Dhabhar, F.S., and Sheridan, J.F. (2003). Molecular mechanisms of glucocorticoid resistance in splenocytes of socially stressed male mice. *J. Neuroimmunol.* 137:51–58.

Rodriguez, M., Pavelko, K., and Coffman, R.L. (1995). Gamma interferon is critical for resistance to Theiler's virus-induced demyelination. *J. Virol.* 69:7286–7290.

Rodriguez, M., Pavelko, K.D., Njenga, M.K., Logan., W.C., and Wettstein, P.J. (1996). The balance between persistent virus infection and immune cells determines demyelination. *J. Immunol.* 157:5699–5709.

Rodriguez, M., and Sriram, S. (1988). Successful therapy of Theiler's virus-induced demyelination (DA strain) with monoclonal anti-Lyt2 antibody. *J. Immunol.* 140: 2950–2955.

Roos, R.P., and Wollmann, R. (1984). DA strain of Theiler's murine encephalomyelitis virus induces demyelination in nude mice. *Ann. Neurol.* 15:494–499.

Rossi, P.C., McAllister, A., Fiette, L., and Brahic, M. (1991). Theiler's virus infection induces a specific cytotoxic T lymphocyte response. *Cell. Immunol.* 138:341–348.

Samoilova, E.B., Horton, J.L., Hilliard, B., Liu, T.S., and Chen, Y. (1998). IL-6-deficient mice are resistant to experimental autoimmune encephalomyelitis: Roles of IL-6 in the activation and differentiation of autoreactive T cells. *J. Immunol.* 161:6480–6486.

Schonrock, L.M., Gawlowski, G., and Bruck, W. (2000). Interleukin-6 expression in human multiple sclerosis lesions. *Neurosci. Lett.* 294:45–48.

Shanks, N., and Lightman, S.L. (2001). The maternal-neonatal neuro-immune inter-face: Are there long-term implications for inflammatory or stress-related disease? *J. Clin. Invest.* 108:1567–1573.

Sibley, W.A., Bamford, C.R., and Clark, K. (1985). Clinical viral infections and mul-tiple sclerosis. *Lancet* 1:1313–1315.

Sieve, A.N., Steelman, A.J., Young, C.R., Storts, R., Welsh, T.H., Welsh, C.J., and Meagher, M.W. (2004). Chronic restraint stress during early Theiler's virus infection exacerbates the subsequent demyelinating disease in SJL mice. *J. Neuroimmunol.* 155:103–118.

Soldan, S.S., Leist, T.P., Juhng, K.N., McFarland, H.F., and Jacobson, S. (2000). Increased lymphoproliferative response to human herpesvirus type 6A variant in multiple sclerosis patients. *Ann. Neurol.* 47:306–313.

Soldan, S.S., and Jacobson S. (2001). Role of viruses in etiology and pathogenesis of multiple sclerosis. *Adv. Virus Res.* 56:517–555.

Sospedra, M., and Martin, R. (2005). Immunology of multiple sclerosis. *Annu. Rev. Immunol.* 23:683–747.

Stark, J.L., Avitsur, R., Padgett, D.A., Campbell, K.A., Beck, F.M., and Sheridan, J.F. (2001). Social stress induces glucocorticoid resistance in macrophages. *Am. J. Physiol. Regul. Integr. Comp. Physiol.* 280(6):R1799–1805.

Stark, J.L., Avitsur, R., Hunzeker, J., Padgett, D.A., and Sheridan, J.F. (2002). Inter-leukin-6 and the development of social disruption-induced glucocorticoid resis-tance. *J. Neuroimmunol.* 124(1–2):9–15.

Stefferl, A., Storch, M.K., Linington, C., Stadelmann, C., Lassmann, H., Pohl, T., Holsboer, F., Tilders, F.J., and Reul, J.M. (2001). Disease progression in chronic relapsing experimental allergic encephalomyelitis is associated with reduced inflammation-driven production of corticosterone. *Endocrinology* 142(8):3616–3624.

Stininssen, P., Raus, J., and Zhang, J. (1997). Autoimmune pathogenesis of multiple sclerosis: Role of autoreactive T lymphocytes and new immunotherapeutic strate-gies. *Crit. Rev. Immunol.* 17:33–75.

Stratakis, C.A., Karl, M., Schulte, H.M., and Chrousos, G.P. (1994). Glucocorticos-teroid resistance in humans. Elucidation of the molecular mechanisms and impli-cations for pathophysiology. *Ann. N.Y. Acad. Sci.* 746:362–374.

Tanaka, H., Akama, H., Ichikawa, Y., Makino, I., and Homma, M. (1992). Gluco-corticoid receptor in patients with lupus nephritis: relationship between receptor levels in mononuclear leukocytes and effect of glucocorticoid therapy. *J. Rheumatol.* 19:878–883.

Theil, D.J., Tsunoda, I., Libbey, J.E., Derfuss, T.J., and Fujinami, R.S. (2000). Alterations in cytokine but not chemokine mRNA expression during three distinct Theiler's virus infections. *J. Neuroimmunol.* 104:22–30.

Theiler, M. (1934). Spontaneous encephalomyelitis of mice—a new virus. *Science* 80:122.

Tilder, F.J., and Schmidt, E.D. (1999). Cross-sensitization between immune and non-immune stressors. A role in depression? *Adv. Exp. Med. Biol.* 461:179–197.

van Winsen, L.M., Muris, D.F., Polman, C.H., Dijkstra, C.D., van den Berg, T.K., and Uitdehaag, B.M. (2005). Sensitivity to glucocorticoids is decreased in relapsing remitting multiple sclerosis. *J. Clin. Endocrinol. Metab.* 90(2):734–740.

Wandinger, K.P., Jabs, W., Siekhaus, A., Bubel, S., Trillenberg, P., Wagner, H., Wessel, K., Kirchner, H., and Hennig, H. (2000). Association between clinical disease activity and Epstein-Barr virus reactivation in MS. *Neurology* 55:178–184.

Warren, S., Greenhill, S., and Warren, K.G. (1982). Emotional stress and the development of multiple sclerosis: case-control evidence of a relationship. *J. Chronic Dis.* 35:821–831.

Welsh, C.J.R., Tonks, P., Nash, A.A., and Blakemore, W.F. (1987). The effect of L3T4 T cell depletion on the pathogenesis of Theiler's murine encephalomyelitis virus infection in CBA mice. *J. Gen. Virol.* 68:1659–1667.

Welsh, C.J.R., Blakemore, W.F., Tonks, P., Borrow, P., and Nash, A.A. (1989). Theiler's murine encephalomyelitis virus infection in mice: A persistent viral infection of the central nervous system which induces demyelination. In N. Dimmock (ed.), *Immune Responses, Virus Infection and Disease.* Oxford: Oxford University Press, pp. 125–147.

Welsh, C.J.R., Tonks, P., Borrow, P., and Nash, A.A. (1990). Theiler's virus: An experimental model of virus-induced demyelination. *Autoimmunity* 6:105–112.

Welsh, C.J.R., Sapatino, B.V., Rosenbaum, B., and Smith, R. (1995). Characteristics of cloned cerebrovascular endothelial cells following infection with Theiler's virus. I. Acute Infection. *J. Neuroimmunol.* 62:119–125.

Welsh, C.J., Mi, W., Sieve, A., Steelman, A.J., Johnson, R.R., Young, C.R., Prentice, T., Hammons, A., Storts, R., Welsh T., and Meagher, M.M. (2006). The effect of restraint stress on the neuropathogenesis of Theiler's virus-induced demyelination, a murine model for multiple sclerosis. In C. Jane Welsh, Mary, W. Meagher and Esther Sternberg (eds.), *Neural and neuroendocrine mechanisms in host defense and autoimmunity.* New York: Springer, pp. 190–225.

Zhou, D., Kusnecov, A.W., Shurin, M.R., DePaoli, M., and Rabin, B.S. (1993). Exposure to physical and psychological stressors elevates plasma interleukin 6: relationship to the activation of hypothalamic-pituitary-adrenal axis. *Endocrinology* 133:2523–2530.

Zhu, G., Chancellor-Freeland, C., Berman, A., Kage, R., Leeman, S., Beller, D., and Black, P. (1995). Endogenous substance P mediates cold-water-stress induced increase in interleukin-6 section from peritoneal macrophages. *J. Neurosci.* 16: 3745–3752.

12

Early Postnatal Nongenetic Factors Modulate Disease Susceptibility in Adulthood: Examples from Disease Models of Multiple Sclerosis, Periodontitis, and Asthma

MICHAEL STEPHAN, THOMAS SKRIPULETZ, and
STEPHAN VON HÖRSTEN

1. The Development of Nongenetic Acquired Individual Differences

1.1. Introduction

The genesis of individuality serves as an adaptive mechanism and includes the nongenetic modification of several developmental dimensions such as growth, maturation, and learning. As a consequence, modified stress responsiveness, coping strategies, and susceptibility for diseases develop interdependently. Until today, knowledge on the neurobiology of individuality has only marginally been integrated into the understanding of the pathophysiology, prevention, and therapy of diseases.

To explore these interdependent links, we investigated the effects of postnatal stimulation or deprivation on adult behavioral responses and the clinical course of disease models that are thought to be associated with alterations of stress responsiveness.

1.2. Stress and Adaptation

Regulation of adaptive responses is liable to lifelong modifications. The outcome of this developmental perspective of an individual is due to many factors including genetic background and aging per se but is also due to repeated environmental challenges (Fig. 12.1). Dealing with the stresses and strains of life, each individual reacts with a specific activation pattern of regulatory systems. In turn, this activation causes a secondary modification via a feedback reaction, thereby shaping specific response patterns depending on the point of time, frequency, duration, and power of the environmental

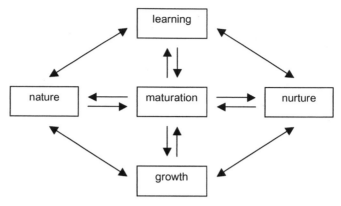

FIGURE 12.1. Possible interdependent links programming individual differences.

challenge. Therefore, every new environmental challenge causes a reaction, which depends on former experiences. In 1993, McEwen and Stellar defined *allostasis* as the process for adapting to environmental challenges to maintain homeostasis. Systems that vary according to demand, like the hypo- thalamic-pituitary-adrenal (HPA) axis and the autonomic nervous system (ANS), actually help protect and maintain those systems that are truly homeostatic, (e.g., the blood pH or body temperature). Large variations in the HPA axis do not lead directly to death, as would large deviations in truly homeostatic systems. Therefore, McEwen and Stellar proposed that *allostasis* is the adequate term for physiological coping mechanisms. Exam- ples of allostasis go to broader aspects of individual survival. In the immune system, an acute stress–induced release of catecholamines and glucocorti- coids facilitates the circulation and migration of immune cells to parts of the body where they are needed to fight an infection or to produce other immune responses (Dhabhar and McEwen, 1999). Furthermore, glucocor- ticoids and catecholamines act concertedly to promote the formation of memories of events of potentially dangerous situations so that the individ- ual can avoid them in the future (Roozendaal, 2000). Yet, each of these adaptive processes has a potential cost to the body when allostasis is either called upon too often or is inefficiently managed, and that cost is referred to as *allostatic load* (McEwen and Stellar, 1993).

1.3. The Early Environment and Maternal Behavior in Rodents

The early postnatal period of life is an exceptionally sensitive time in which the nervous system as well as hormonal and immunological regulatory systems develop and mature within the interaction of the individual and its environment. Early severe stress or disturbance of the postnatal environment produces a cascade of neurobiological modifications that have the potential to cause enduring changes in brain development, physiology, and behavior (Anisman *et al.*, 1998; Pryce and Feldon, 2003). Therefore, the quality of the early family environment can serve as a major source of vulnerability in later life.

In mammals, the mother is the major source of an infant's nutrition and, beside the littermates, the most important social stimulus. Under the influence of maternal hormones, the new dam is maternally responsive to pups as soon as they are born. After retrieving pups that have strayed from the nest, the dam adopts a nursing posture over them. By 24 h after parturition, nearly all dams have established maternal behavior and are routinely engaging in nursing, retrieving, and licking of pups. Being undisturbed, the dam suckles them almost continuously for the first few days and then gradually takes longer and longer absences (Ader and Grota, 1970). During this time, pups regulate their own body temperature poorly and can neither hear nor see. By the end of the first week, the dam is absent for longer time periods. During the second week of life, pups have gained the rudiments of competent thermoregulation and are beginning to show more physiologic autonomy. Once the pups can hear and see, they explore their own environment more frequently, and by weaning, pups are fully able to live on their own, largely independently.

For the maintenance of maternal behavior, it is crucial that dams receive stimulation from their newborns during the first few days. The dam appears to respond to a complex of stimuli from pups and a complex mother-infant relationship has evolved out of the nutritional need, such that the mother provides essential thermal, somatosensory, and auditory stimulation (Hofer, 1987). Hofer has been pioneering in describing how observable behavioral interactions of parent and offspring mediate nonobservable events that have important and widespread effects (Hofer, 1994). As such, the maternal environment constitutes one of the most significant environments that any mammal will encounter throughout its entire life span.

Even slight modifications of the way a dam nurses her pups can result in a different maturation and development of the young. Usually, there are two types of licking observable: pup body licking and pup anogenital licking. Body licking can be observed during various circumstances in the maternal cage (e.g., before retrieval, between retrievals, during nursing), whereas anogenital licking tends to be observed while pups are nursing and are on their backs. This type of licking stimulation plays an important role in development of the pups. The quality of pups grooming and the amount of pups licking affects the development of offspring's emotionality and cognition (Caldji *et al.*, 1998; Liu *et al.*, 2000). The offspring of high- and low-licking mothers differed in behavioral responses to novelty. As adults, the offspring of dams that licked their pups seldom showed increased anxiety-like behavior patterns, decreased novelty-induced exploration, and longer latencies to eat food provided in a novel environment. These animals also showed altered brain receptor levels that might contribute to the more anxious phenotype (Caldji *et al.*, 1998). Furthermore, Liu *et al.* (2000) reported that variations in maternal behavior are related to differential expression of genes encoding N-methyl d-aspartate (NMDA) receptor sub-units, which enhance hippocampal sensitivity to glutamate, and increase Brain derived growth factor (BDNF) gene expression and thus hippocampal synaptic development. These effects form the basis for the development of stable, individual differences in stress reactivity and certain forms of cognition. In conclusion, the quality of pup nursing is an important source of stimulation that has strong effects on the maturation and development of the offspring.

Disturbances of the rats' postnatal environment have been demonstrated to adjourn and somehow to modulate maternal behavior. For example, a prolonged separation of the dam from her pups results in a lower amount of licking and grooming pups after reunion compared with a short separation (Skripuletz *et al.*, submitted). Therefore, beside separation-induced effects, the altered mother-pup interaction might be one factor contributing to the observed robust and marked effects on reported neurobiological, physiological, and behavioral phenotypes in adulthood after specific postnatal manipulations.

2. Modulation of Disease Susceptibility by Early Experiences

2.1. Maternal Deprivation and Postnatal Handling Stimulation (Prolonged vs. Short Separation): Oppositional Paradigms of Early Deprivation and Stimulation

In laboratory rats, the chronic effects of postnatal manipulation of the dam-pup relationship have been studied experimentally for nearly 50 years. From the standpoint of neurobiology, exposure to early stress programs the individual to display enhanced stress responsiveness, especially through a modification of the HPA axis resulting in the production of glucocorticoids and the sympathetic nervous systems resulting in the production of catecholamines and neuropeptides. For example, postnatal handling (HA; i.e., brief maternal separation and exposure to novelty) stimulates pups as well as dams, thereby intensifying maternal behavior. In contrast with HA, daily maternal deprivation (MD; i.e., maternal separations for hours) can be considered an early life stressor that deprives maternal care and leads to an understimulation of pups. MD and HA are often used postnatal manipulations that represent oppositional models of early deprivation and stimulation (for review, see Pryce and Feldon, 2003) but are seldom investigated within a single test design. The group of Ellenbroek and Cools recently reported that MD leads to a reduction in acoustic startle habituation and auditory sensory gating in adult rats (Ellenbroek *et al.*, 2004). Because a number of deficits (e.g., a disturbance in prepulse inhibition) are similar to abnormalities observed in schizophrenic patients, Ellenbroek and Cools went to such lengths as to propose MD as a model for schizophrenia.

However, MD and HA are powerful enough to induce modifications of behavioral patterns, even in adulthood. We recently found that MD as well as HA do not alter home-cage activity pattern but novelty-induced anxiety and exploration behavior depending on gender and genetics (Skripuletz *et al.*, submitted). HA increased the activity and exploration behavior in the open-field as well as in the holeboard test, whereas anxiety-like behavior was decreased. In partial contrast, maternally deprived rats showed less exploration behavior in the open-field test. These results fit the observation that HA decreases and MD increases the magnitude of behavioral and endocrine responses to stress in adulthood (Francis and Meaney, 1999; Pryce and Feldon, 2003).

Beside these modifications of anxiety-like behavior, activity, and exploration, postnatal manipulations have also longlasting effects on pain sensitivity. We have reported that postnatal handling prolongs, whereas MD shortens, the time for adult rats to show a pain reaction (Stephan *et al.*, 2002a). Although hot-plate and tail-flick testing are not exclusively measuring supra- or infra-spinal pain processing, because polysynaptic nociceptive responses often involve supra-spinal loops and descending modulation, primarily supra-spinal pain processing seemed to be affected by the postnatal environments, as hot-plate but not tail-flick testing showed acquired differences. Because HA is known to decrease the central CRH/HPA/stress responsiveness whereas MD does the opposite (Ladd *et al.*, 2000), and central application of corticotropin reteasing hormone (CRH) produces analgesia (Lariviere, 2000), we had expected oppositional directed effects in handling and MD than we actually observed. Therefore, we concluded that the observed effects of postnatal manipulations on adult pain sensitivity could not be mainly attributed to changes in the central CRH/HPA axis systems.

Furthermore, we found that both postnatal manipulations have oppositional effects on plasticity and motor function in adulthood. MD decreases the ability to run on an accelerating rod, whereas HA improved the performance. These effects were due to a selective reduction of cerebellar granule cells in maternally deprived rats, while the volume of white matter, molecular cell layer, granular cell layer, and numbers of Purkinje cells remained unchanged. These examples clarify the complex and large-scale changes after specific postnatal challenges.

Here, we report on the different effects of the introduced postnatal manipulations on disease susceptibility of different diseases in adulthood as well as on possible treatment strategies.

2.2. Individual Differences in Response to Ligature-induced Periodontitis

2.2.1. The Animal Model of Periodontitis

Ligature-induced periodontitis is a disease model that fulfils several important criteria: first, it is sensitive to individual differences of HPA axis reactivity. Second, it should involve innate immune functions being most sensitive to alterations of the stress responsiveness. Third, it should be relevant to human disease. The disease is a locally restricted destructive inflammatory process triggered by Gram-negative oral microorganisms that colonize tooth surfaces (dental plaque) in the gingival sulci. The disease is characterized by breakdown of the tooth-supporting tissues (the periodontium), which may lead to tooth loss in the most severe cases. Periodontal disease represents a major health problem with an incompletely understood pathomechanism. A high incidence of periodontal disease is associated with genetics, increased age, negative life events and depression, heavy smoking, and poorly controlled diabetes mellitus. Until now, it is unknown whether there might be a common factor predisposing individuals for the severe forms of the disease. Because all these epidemiological risk factors are associated with changes in stress regulation, we hypothesized that adaptation to stress may play a key role in the pathogenesis of the disease (Breivik *et al.*, 1996). Recently, we were able to show that a genetically determined high HPA responsiveness increases disease susceptibility and that increased disease activity of periodontal disease further stimulates the HPA axis (Breivik *et al.*, 2001). Thus, early disturbance stress may lead to increased stress reactivity thereby predisposing individuals to periodontal pathology in adulthood.

2.2.2. Different Effects of HA and MD Depend on Genetics

The Lewis (LEW) and Fischer (F344) rat strains provide a comparative model of HPA function in which LEW rats are relatively hyporeactive to environmental challenge (Stöhr *et al.*, 2000). Comparing the control groups of both strains, we found that F344 showed a dramatic increase of bone and tissue loss compared with LEW. These effects were due to higher ACTH and corticosterone levels in F344 upon challenge. Beyond doubt, the susceptibility to periodontitis is partly mediated via the genetic background, but interestingly these effects correlate with the intriguing observation that LEW pups of the control group were groomed three times longer than F344 pups. The different HPA axis activity observed between both strains might therefore be partly due to different childhood experiences and rearing conditions.

Investigating the effects of postnatal stimulation or deprivation in both strains, in F344 neither HA nor MD modulated the progress of disease in a significant manner compared with the control group. These effects might

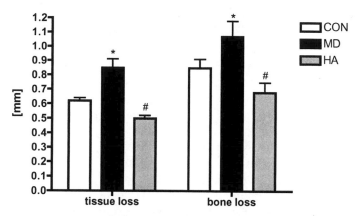

FIGURE 12.2. Periodontal disease susceptibility in adult rats subjected to different postnatal treatments. Panels provide means in millimeters of periodontal attachment fiber loss and bone loss on the experimental side. Data represent means ± SEM; significant differences versus controls are indicated by asterisks (p < 0.05), and significant differences of HA versus MD are indicated by number signs (p < 0.05).

be due to high tissue breakdown in all three groups due to the genetic background, which prevented us from finding significant differences.

On the other side, MD significantly increased both tissue and bone loss in LEW, whereas HA exerted protective effects (Fig. 12.2). These effects were due to lower plasma corticosterone levels but elevated interferon-gamma levels in handled rats compared with the controls. In maternally deprived rats, we observed an oppositional directed trend that does not suffice to explain the observed effects.

2.2.3. Amelioration of Disease Progress by Antidepressant Treatment

As a treatment option in adulthood, we started a chronic imipramine treatment via the drinking water at the age of 6 weeks, accomplished according to the method of Reul and colleagues (1993). Because imipramine is light sensitive, black drinking bottles were used, and drinking solutions were renewed every day. Weighing the drinking bottles, we calculated the daily intake of imipramine aiming at therapeutic doses of 2.5 mg/rat per 24 h (about 10 mg/kg per day) via fluid intake.

Again, MD increased the bone loss compared with the control group. Interestingly, the antidepressant treatment was powerful enough to rescue from periodontal breakdown. Furthermore, imipramine treatment significantly decreased bone loss compared with the control group. Research on possible mechanisms is still ongoing, but again cytokine and hormone levels in the peripheral blood seem to be insufficient to indicate what could explain the observed effects.

2.3. Asthma

2.3.1. Disease Model of Experimentally Induced Asthma Bronchiale

Allergic asthma is an immunological common disease that has increased dramatically in prevalence in industrialized countries during the past 20 years (Umetsu *et al.*, 2002).

It has become well established that allergic asthma is characterized by chronic airway inflammation associated with goblet cell hyperplasia and mucus plugging of airways, subepithelial fibrosis, focal desquamation of bronchial epithelium, and airway smooth muscle cell hypertrophy. Development of this pulmonary inflammatory process is thought to be dependent on inflammatory and immune cells such as T lymphocytes and eosinophil granulocytes by producing a variety of inflammatory mediators, toxic oxygen radicals, and cytokines (Larche *et al.*, 2003). These cells are present in lung parenchyma and bronchoalveolar lavage (BAL) fluid during asthmatic reactions. During the past few years, the hygiene hypothesis has gained strong support (for review, see Eder and von Mutius, 2004). Following this hypothesis, exposure with specific endotoxins in the postnatal period is potentially protective against the development of asthma and allergies in children because of its inflammatory effects that lead to a strong T helper 1 (Th1)-type immune response (Heumann and Roger, 2002).

Because probably the postnatal antigenic environment is not the only factor affecting adult immune responsiveness, here, we want to draw attention to other nongenetic, noninfectious aspects of the postnatal environment (i.e., the impact of postnatal stressors). We investigated the impact of MD and HA on the onset and course of the ovalbumin (OVA)-induced airway inflammation in rats. Animal models have been valuable for the investigation of the underlying pathology of allergic pulmonary diseases, and especially the OVA-induced airway inflammation has become a well-investigated asthma model in rats (Schneider *et al.*, 1997; Kruschinski *et al.*, 2005). Therefore, we used that model and provoked an asthma-like response by sensitizing and challenging rats by application of OVA as previously described (Schuster *et al.*, 2000).

2.3.2 Differential Effects of HA and MD

HA and MD showed oppositional effects concerning the acute reactions to an OVA-challenge. In brief, maternally deprived rats showed more asthma-like changes in immune parameters; for example, the increase of absolute eosinophil numbers in the BAL fluid (Fig. 12.3).

Interestingly, handled rats showed significantly lower adrenocorticotropic hormone (ACTH) levels in the peripheral blood than controls, but as a reaction to the OVA-challenge, the plasma levels increased dramatically (Fig. 12.4). This increase might be one factor that rescued handled rats from stronger asthma-like reactions, because glucocorticoids represent still

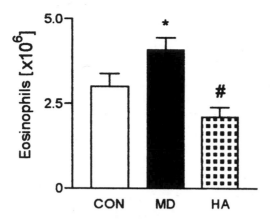

FIGURE 12.3. Total recruitment of eosinophil granulocytes to the lungs after challenge with ovalbumin: determination of cells in the BAL fluid of rats 22 h after challenge. Data represent means ± SEM; significant differences versus controls are indicated by asterisk ($p < 0.05$), and significant differences of HA versus MD are indicated by rhomb ($p < 0.05$).

a major supporting pillar in the treatment of the disease. In partial contrast, MD rats showed higher basal ACTH levels than controls but increased symptoms associated with a blunted stress-induced increase of ACTH. These data hint for the first time to a noninfectious and nongenetic modulation of disease susceptibility for asthma via an experience-induced modulation of the psycho-neuro-immunological network.

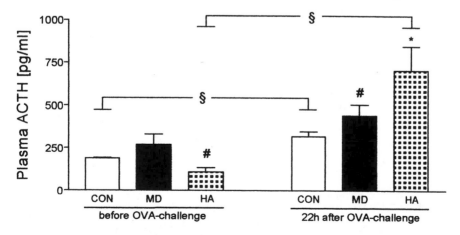

FIGURE 12.4. Plasma concentration levels of ACTH prechallenge and 22 h postchallenge with Ovalbumin. Data represent means ± SEM in pg/ml; significant differences of postnatal treated animals vs. controls are indicated by asterisks ($p < 0.05$), significant differences of IIA vs. MD are indicated by rhombs ($p < 0.05$), and differences between the several time-points are indicated by section symbols ($p < 0.05$).

2.4. Experimental Allergic Encephalomyelitis (EAE)

2.4.1. Disease Model of EAE

EAE is an experimental autoimmune inflammatory disease that serves as an animal model for multiple sclerosis (Swanborg, 1995). Again, the HPA axis is believed to play a major role in determining susceptibility to autoimmune processes (Sternberg *et al.*, 1989), but more recently other regulatory systems such as the autonomous nervous system have also come into focus (Bedoui *et al.*, 2004). In several studies, an inverse correlation was found between disease susceptibility and the HPA axis responsiveness (Kavelaars *et al.*, 1997). We investigated the impact of MD on EAE and tried to treat the deprivation-induced effects with imipramine or a prophylactic strategy with additional stimulation following the deprivation procedure (Stephan *et al.*, 2002b).

2.4.2. Imipramine as a Sufficient Treatment of MD-induced Aggravation of EAE

The clinical course of the EAE is illustrated in Figure 12.5. We observed that MD shortened the interval to onset of EAE and increased the sum of clinical scores as well as the average clinical scores. In other words, MD dramatically aggravated the course of the EAE. Interestingly, chronic imipramine treatment ameliorated the average clinical score compared with maternally deprived rats. Unfortunately, the underlying mechanisms remain unknown. We were not able to detect significant differences in cor-

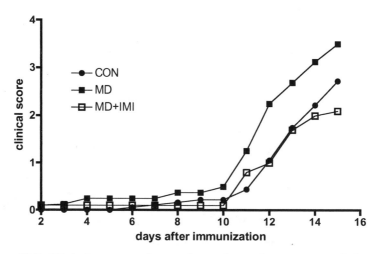

FIGURE 12.5. Clinical course of experimental autoimmune encephalomyelitis (EAE) differs in rats subjected to either repeated maternal deprivation (MD), repeated maternal deprinvation + imipramine (MD+IMI) or control (CON). The clinical disease was scored on the following scale: 0.5, partial loss of tail tone; 1.0, complete tail atony; 2.0, hind limb weakness; 3.0, hind limb paralysis; 4.0, moribund.

ticosterone and cytokine levels in the peripheral blood during the course of the disease.

3. Early Life Stress Increases Disease Susceptibility in Adulthood

Early life stress modulates neural and neuroendocrine mechanisms in host defence and autoimmunity, thereby predisposing to adult disease susceptibility. Here, we reported on opposing effects of MD and HA on a diversity of diseases like asthma, multiple sclerosis, and periodontitis. Summarizing, MD increases, whereas HA decreases, disease susceptibility to all investigated disease models. Furthermore, HA prolongs the pain threshold, decreases anxiety-like behavior, and improves motor function and explorative behaviors, whereas MD induces more or less opposite-like effects. Denenberg and Karas (1959) summarized different experiments and stated that handled animals weighed most, learned best, and lived longest. The question is, how are these effects mediated?

First, postnatal handling and maternal deprivation have been shown to program adult HPA responsiveness (Ladd *et al.*, 2000). However, based on our data, we are tempted to conclude that the demonstrated effects are far beyond oppositional directed changes of the HPA axis but rather due to complex modifications of several regulatory systems that contribute to an altered stress responsiveness and disease susceptibility in adulthood. Otherwise, it would not be allegeable that HA decreases whereas MD increases disease susceptibility of all investigated diseases, because corticosterone should exert protective effects on asthma as well as on multiple sclerosis, while it should aggravate the course of periodontitis. In addition, ACTH levels increased in handled rats during the course of the asthma-like inflammation, thereby protecting from increased symptoms. Therefore, the hypothesis that MD leads to a higher stress responsiveness and HPA axis activity *has to be revised*. More recently, other regulatory systems such as the gonadal axis and the peripheral nervous system have also come into focus (Straub and Cutolo, 2001; Bedoui *et al.*, 2003) and are worthy of further investigation.

Second, postnatal experiences are powerful enough to induce long-term immunological changes, as indicated by different disease susceptibility, but little is known about possible mediators. Before we started the various experiments, we hypothesized MD induces a Th1-Th2 shift. But we had to abandon our hypothesis, because we observed that EAE was aggravated, though a Th2 shift should have had protective effects. Also, the cytokine profiles have not been consistent during the course of different diseases.

Third, it has been reported that postnatal stress effects CNS plasticity (McEwen, 2001), possibly via different BDNF expression profiles (Kuma *et al.*, 2004; Roceri *et al.*, 2004). Another mediator of altered neurogenesis in adulthood after MD has been suggested by Mirescu and collegues (2004), who reported that maternally deprived rats show "a decrease in cell proliferation and immature neuron production" in the dentate gyrus. They hypothesized "that early adverse experience inhibits structural plasticity via hypersensitivity to glucocorticoids and diminishes the ability of the hippocampus to respond to stress in adulthood." Furthermore, the long-term adaptations in glucocorticoid receptor and mineralocorticoid receptor have been recently decribed by Ladd and colleagues (2004). Moreover, there are many implications of postnatal manipulations—or even rearing conditions—for maturation of regulatory systems, cognition, and growth.

Interestingly, we were able to rescue from aggravated diseases by treating maternally deprived rats with imipramine. Because imipramine is an antidepressant drug, *it raises the question of whether MD is* really a model for schizophrenia—as Ellenbrook and Cools suggested (2004)—or a model for depression-like diseases. During the daily deprivation session, maternally deprived rats huddle together in a corner of the cage and remain there totally inactive. In contrast, handled rats are active, even as young pups, showing directed movements, running around, sniffing, *and licking similar to older pups when their eyes were opened.* After the short period of separation, handled pups and their dams are reunited, which may be interpreted as a reward for their searching behavior. As a hypothesis, one can argue that MD induces a "learned helplessness-like" phenotype while handled rats exhibit "learned competence-like" responses.

Summarizing, the postnatal environment causes a complex interplay of endocrine and immunological parameters, as well as of maturation and plasticity that induces long-term effects that program the individual to show altered disease susceptibility in adulthood.

References

Ader, R., and Grota, L.J. (1970). Rhythmicity in the maternal behaviour of Rattus norvegicus. *Anim. Behav.* 18:144–150.

Anisman, H., Zaharia, M.D., Meaney, M.J., and Merali, Z. (1998). Do early-life events permanently alter behavioral and hormonal responses to stressors? *Int. J. Dev. Neurosci.* 16:149–164.

Bedoui, S., Kawamura, N., Straub, R.H., Pabst, R., Yamamura, T., and von Hörsten S. (2003). Relevance of neuropeptide Y for the neuroimmune crosstalk. *J. Neuroimmunol.* 134:1–11.

Bedoui, S., Miyake, S., Straub, R.H., von Hörsten, S., and Yamamura, T. (2004). More sympathy for autoimmunity with neuropeptide Y? *Trends Immunol.* 25:508–512.

Breivik, T., Thrane, P.S., Murison, R., and Gjermo, P. (1996). Emotional stress effects on immunity, gingivitis and periodontitis. *Eur. J. Oral. Sci.* 104:327–334.

Breivik, T., Thrane, P.S., Gjermo, P., Opstad, P.K., Pabst, R., and von Hörsten, S. (2001). Hypothalamicpituitary-adrenal (HPA) axis activation and periodontal disease. *J. Periodont. Res.* 36:295–300.

Caldji, C., Tannenbaum, B., Sharma, S., Francis, D., Plotsky, P.M., and Meaney, M.J. (1998). Maternal care during infancy regulates the development of neural systems mediating the expression of fearfulness in the rat. *Proc. Natl. Acad. Sci. U.S.A.* 95:5335–5340.

Denenberg, V.H., and Karas, G.G. (1959). Effects of differential infantile handling upon weight gain and mortality in the rat and mouse. *Science* 130:629–630.

Dhabhar, F.S., and McEwen, B.S. (1999). Enhancing versus suppressive effects of stress hormones on skin immune function. *Proc. Natl. Acad. Sci. U.S.A.* 96: 1059–1064.

Eder, W., and von Mutius, E. (2004). Hygiene hypothesis and endotoxin: What is the evidence? *Curr. Opin. Allergy Clin. Immunol.* 4:113–117.

Ellenbroek, B.A., De Bruin, N.M., Van Den Kroonenburg, P.T., Van Luijtelaar, E.L., and Cools, A.R. (2004). The effects of early maternal deprivation on auditory information processing in adult wistar rats. *Biol. Psychiatry* 55:701–707.

Francis, D.D., and Meaney, M.J. (1999). Maternal care and development of stress responses. *Curr. Opin. Neurobiol.* 9:128–134.

Heumann, D., and Roger, T. (2002). Initial responses to endotoxins and Gram-negative bacteria. *Clin. Chim. Acta.* 323:59–72.

Hofer, M.A. (1987). Early social relationships: A psychobiologist's view. *Child. Dev.* 58:633–647.

Hofer, M.A. (1994). Early relationships as regulators of infant physiology and behavior. *Acta. Paediatr. Suppl.* 397:9–18.

Kavelaars, A., Heijen, C.J., Ellenbroek, B., van Loveren, H., and Cools A.R. (1997). Apomorphine-susceptible and apo-morphine-unsusceptible Wistar rats differ in their susceptibility for inflammatory and infectious diseases: A study on rats with groupific differences in structure reactivity of hypothalamo-pituitary-adrenal axis. *J. Neuroimmunol.* 77:211–216.

Kruschinski, C., Skripuletz, T., Bedoui, S., Tschernig, T., Pabst, R., Nassenstein, C., Braun, A., and von Hörsten S. (2005). CD26 (dipeptidyl-peptidase IV)-dependent recruitment of T cells in a rat asthma model. *Clin. Exp. Immunol.* 139:17–24.

Kuma, H., Miki, T., Matsumoto, Y., Gu, H., Li, H.P., Kusaka, T., Satriotomo, I., Okamoto, H., Yokoyama, T., Bedi, K.S., Onishi, S., Suwaki, H., and Takeuchi, Y. (2004). Early maternal deprivation induces alterations in brain-derived neurotrophic factor expression in the developing rat hippocampus. *Neurosci. Lett.* 372:68–73.

Ladd, C.O., Huot, R.L., Thrivikraman, K.V., Nemeroff, C.B., Meaney, M.J., and Plotsky, P.M. (2000). Long-term behavioral and neuroendocrine adaptations to adverse early experience. *Prog. Brain. Res.* 122:81–103.

Ladd, C.O., Huot, R.L., Thrivikraman, K.V., Nemeroff, C.B., and Plotsky, P.M. (2004). Long-term adaptations in glucocorticoid receptor and mineralocorticoid receptor mRNA and negative feedback on the hypothalamo-pituitary-adrenal axis following neonatal maternal separation. *Biol. Psychiatry* 55:367–375.

Larche, M., Robinson, D.S., and Kay, A.B. (2003). The role of T lymphocytes in the pathogenesis of asthma. *J. Allergy Clin. Immunol.* 111:450–463.

Lariviere, W.R., and Melzack, R. (2000). The role of corticotropin-releasing factor in pain and analgesia. *Pain* 84:1–12.

Liu, D., Diorio, J., Day, J.C., Francis, D.D., and Meaney, M.J. (2000). Maternal care, hippocampal synaptogenesis and cognitive development in rats. *Nat. Neurosci.* 3:799–806.

McEwen, B.S. (2001). Plasticity of the hippocampus: Adaptation to chronic stress and allostatic load. *Ann. N.Y. Acad. Sci.* 933:265–277.

McEwen, B.S., and Stellar, E. (1993). Stress and the individual. Mechanisms leading to disease. *Arch. Intern. Med.* 153:2093–2101.

Mirescu, C., Peters, J.D., and Gould, E. (2004). Early life experience alters response of adult neurogenesis to stress. *Nat. Neurosci.* 7:841.

Pryce, C.R., and Feldon, J. (2003). Long-term neurobehavioural impact of the postnatal environment in rats: manipulations, effects and mediating mechanisms. *Neurosci. Biobehav. Rev.* 27:57–71.

Reul, J.M., Stec, I., Soder, M., and Holsboer, F. (1993) Chronic treatment of rats with the antidepressant amitriptyline attenuates the activity of the hypothalamic-pituitary-adrenocortical system. *Endocrinology* 133:312–320.

Roceri, M., Cirulli, F., Pessina, C., Peretto, P., Racagni, G., and Riva, M.A. (2004). Postnatal repeated maternal deprivation produces age-dependent changes of brain-derived neurotrophic factor expression in selected rat brain regions. *Biol. Psychiatry* 55:708–714.

Roozendaal, B. (2000). Glucocorticoids and the regulation of memory consolidation. *Psychoneuroendocrinology* 25:213–238.

Schneider, T., van Velzen, D., Moqbel, R., and Issekutz, A.C. (1997). Kinetics and quantitation of eosinophil and neutrophil recruitment to allergic lung inflammation in a brown Norway rat model. *Am. J. Respir. Cell. Mol. Biol.* 17:702–712.

Schuster, M., Tschernig, T., Krug, N., and Pabst, R. (2000). Lymphocytes migrate from the blood into the bronchoalveolar lavage and lung parenchyma in the asthma model of the brown Norway rat. *Am. J. Respir. Crit. Care. Med.* 161:558–566.

Skripuletz, T., Kruschinski, C., Pabst, R., von Hörsten, S., and Stephan, M. (submitted). Postnatal Handling and Maternal Deprivation determine activity, anxiety as well as exploratory responses in adults depending on both gender and genetics.

Stephan, M., Helfritz, F., Pabst, R., and von Hörsten, S. (2002a). Postnatally induced differences in adult pain sensitivity depend on genetics, gender and specific experiences: Reversal of maternal deprivation effects by additional postnatal tactile stimulation or chronic imipramine treatment. *Behav. Brain Res.* 133:149–158.

Stephan, M., Straub, R.H., Breivik, T., Pabst, R., and von Hörsten, S. (2002b). Postnatal maternal deprivation aggravates experimental autoimmune encephalomyelitis in adult Lewis rats: Reversal by chronic imipramine treatment. *Int. J. Dev. Neurosci.* 20:125–132.

Sternberg, E.M., Young, W.S., Bernardini, R., Calogero, A.E., Chrousos, G.P., Gold, P.W., and Wilder R.L. (1989). A central nervous system defect in biosynthesis of corticotropin-releasing hormone is associated with susceptibility to streptococcal cell wall-induced arthritis in Lewis rats. *Proc. Natl. Acad. Sci. U.S.A.* 86:4771–4775.

Straub, R.H., and Cutolo, M. (2001). Involvement of the hypothalamic-pituitary-adrenal/gonadal axis and the peripheral nervous system in rheumatoid arthritis: Viewpoint based on a systemic pathogenetic role. *Arthritis Rheum.* 44:493–507.

Stöhr, T., Szuran, T., Welzl, H., Pliska, V., Feldon, J., and Pryce, C.R. (2000). Lewis/Fischer rat strain differences in endocrine and behavioural responses to environmental challenge. *Pharmacol. Biochem. Behav.* 67:809–819.

Swanborg, R.H. (1995). Experimental autoimmune encephalomyelitis in rodents as a model for human demyelinating disease. *Clin. Immunol. Immunopathol.* 77: 4–13.

Umetsu, D.T., McIntire, J.J., Akbari, O., Macaubas, C., and DeKruyff, R.H. (2002). Asthma: An epidemic of dysregulated immunity. *Nat. Immunol.* 3:715–720.

13
The Relationship Between Stressful Life Events and Inflammation Among Patients with Multiple Sclerosis

DAVID C. MOHR

1. Introduction

Multiple sclerosis (MS) is a chronic, often disabling disease of the central nervous system (CNS) affecting up to 350,000 people in the United States (Anderson *et al.*, 1992; Jacobson *et al.*, 1997; Noonan *et al.*, 2002). As with many autoimmune diseases, it affects women at roughly twice the rate of men, and the prevalence appears to be increasing (Cooper and Stroehla, 2003; Jacobson *et al.*, 1997). Common symptoms include, but are not limited to, loss of function or feeling in limbs, loss of bowel or bladder control, sexual dysfunction, debilitating fatigue, blindness due to optic neuritis, loss of balance, pain, cognitive dysfunction, and emotional changes (Mohr and Cox, 2001). There is a growing literature suggesting that stress may affect risk of exacerbation in patients with MS (Mohr *et al.*, 2004). This chapter will briefly review pathology and pathogenesis of MS, examine the literature on stress in MS, describe the laboratory studies on response to stress, and propose three hypothesized mechanisms by which stress might affect risk of MS exacerbation. We will also briefly review the literature on psychosocial mediators of the relationship between stressful life events and MS inflammation and discuss future directions for research in humans.

2. Brief Review of MS Pathology and Pathogenesis

There are two distinct clinical disease markers in MS: exacerbation and progression. Exacerbation is defined as a sudden onset or increase in a symptom within 24h, which resolves fully or partially over the course of weeks or months. Progression refers to a steady worsening in the absence of exacerbations. Approximately 80% of patients begin with a relapsing-remitting (RR) course that is characterized by periodic exacerbations but no progression between exacerbations (Noseworthy *et al.*, 2000). Within a decade after diagnosis, more than 40% of patients with RRMS convert to secondary progressive MS (SPMS), which is characterized by the onset of progression between exacerbations and a decrease in the frequency of exacerbations. Approximately 10–15% of patients have a primary progressive course characterized by a steady progression of symptoms in the absence of exacerbations and appear to have a different pathogenesis (Thompson *et al.*, 2000). This paper will focus primarily on the effects of stress on relapsing forms of MS as there is almost no literature on stress in primary progressive MS.

MS is a disease in which the immune system attacks the myelin sheath surrounding the axons of neurons in the CNS. The precise etiology of the disease remains largely unknown. Given the clinical, genetic, neuroimaging, and pathological heterogeneity, the pathogenesis of MS is likely multifactorial, involving genetic susceptibility, environmental factors such as exposure to an antigen, developmental factors, autoimmunity, and neurodegenerative processes (Noseworthy *et al.*, 2000). The pathogenesis of an MS exacerbation likely begins long before the emergence of clinical signs. Advanced neuroimaging studies using magnetization transfer ratio (MTR) imaging indicate that changes in the ratio of bound to unbound water are evident weeks or even months before inflammation and exacerbation are evident through conventional imaging or clinical exam (Filippi *et al.*, 1998; Goodkin *et al.*, 1998).

It is generally believed that inflammation and demyelination in MS are the result of autoreactive immune responses to myelin proteins. These are believed to be caused by molecular mimicry and a failure of self-tolerance. Researchers have found support for a number of infectious agents that may serve as a trigger, including the herpes viruses and Epstein-Barr virus (Sospedra and Martin, 2004). The molecular mimicry theory suggests that T cells activated by the virus can cross-react with autoantigens such as myelin basic protein (MBP), myelin oligodendrocyte glycoprotein (MOG), and proteolipid protein (PLP) (Hohlfeld *et al.*, 1995; Noseworthy *et al.*, 2000). Activated Th1 cells secrete inflammatory cytokines that promote proliferation and adherence to the endothelium of the blood vessels through upregulation of adhesion molecules. Recruitment of immune cells is facilitated by a variety of chemokines. Immune cell adhesion to the endothelium and transmigration of cells across the blood-brain barrier (BBB) into the brain are facilitated by increases in matrix metalloproteinases, mast cells, and chemokines (Bar-Or *et al.*, 2003; Sospedra and Martin, 2004). Antigen-presenting cells within the CNS (astrocytes, microglia, and macrophages) further stimulate the T cells by presenting myelin proteins, which are mistaken by the Th1 cells as the foreign antigen presented initially. This can result in an enhanced immune response whereby proinflammatory cytokines trigger a cascade of events resulting in proliferation of Th1 cells, and ultimately immune-mediated injury to myelin and oligodendrocytes (O'Connor *et al.*, 2001). The proinflammatory cytokines that have been most commonly implicated in this process include interferon gamma (IFN-γ), tumor necrosis factor-alpha (TNF-α), and interleukins (IL) 1β, 6, and 12. Damage to the myelin sheath may occur directly through cytokine-mediated injury, digestion of surface myelin by macrophages, antibody-dependent cytotoxicty, complement-mediated injury, and/or direct injury of oligodendrocytes by CD4+ and CD8+ T cells (Bruck and Stadelmann, 2003). Some remyelination can occur via the local response by oligodendrocyte progenitor cells, however exposed axons may also be further injured and transected by continuing inflammation (Bjartmar et al., 2003). We should note that many studies have also shown autoimmune reactions to MBP, MOG, and PLP in healthy control subjects, suggesting that autoreactivity alone is not sufficient to invoke the disease and that failure to establish tolerance may play a critical role in MS (Sospedra and Martin, 2004).

Although the process of inflammation and demyelination is the hallmark of the early, inflammatory period of the disease, it is increasingly recognized that other neurodegenerative processes become more prominent as the disease progresses (Trapp et al., 1998; Confavreux et al., 2000). While axonal transection in the earlier stages of the disease appears to be primarily due to inflammatory processes, degeneration of axons in later stages of the disease may be due to a lack of trophic support from myelin or myelin-forming cells (Scherer, 1999). These two stages may correspond with the RR disease course seen earlier, in which there are exacerbations with quiescence between stages, and the SP course in which progression in the absence of exacerbation becomes increasingly prominent. To date, the impact of stressful life events on progression or neurodegenerative processes remains largely unexamined. We will therefore focus this review on exacerbations and inflammatory processes.

3. Evidence of the Relationship Between Stressful Life Events and MS Exacerbation

Charcot, who first characterized MS in the 19th century, wrote that grief, vexation, and adverse changes in social circumstance were related to the onset of MS (Charcot, 1877). Since then, numerous clinical studies have been conducted examining the relationship between stressful life events and MS exacerbation.

A growing body of empirical work has examined the relationship between stressful life events and MS exacerbation. A recent meta-analysis of 14 studies examining the effects of stressful life events on MS exacerbation found significantly increased risk of exacerbation associated with stressful life events (d = 0.53) (Mohr et al., 2004). Although this is only a modest effect size, it is clinically relevant, given that the positive effect of the most commonly used disease-modifying medications on exacerbation, interferon betas, is estimated at d = 0.30 to 0.36 (Filippini et al., 2003).

All of the studies in the meta-analysis used either neurologist ratings or confirmation of exacerbation. However, because many symptoms of MS are based on self-report (e.g., fatigue, pain, etc.) or can be affected by distress, it could be argued that ratings of exacerbation might be more likely among distressed patients. Therefore, our group replicated previous work on stressful life events and MS using a more objective neuroimaging marker of MS BBB disruption associated with inflammation, Gd+ MRI. Gadolinium is a contrast agent injected into the bloodstream during the MRI scan, which crosses the BBB at sites of focal MS inflammation, thereby providing images of active inflammation. Gd+ MRI is 5–10 times more sensitive than neurologist determinations of clinical exacerbation in evaluating active MS inflammation (Grossman, 1996). In a longitudinal study of MS patients receiving monthly Gd+ MRI, we have shown that stressful life events, particularly those involving family- and work-related stress, as well as disruptions in routine, are associated with the subsequent development of Gd+ brain lesions (Mohr et al., 2000).

It should be noted that while the studies included in this meta-analysis (Mohr et al., 2004) were statistically homogenous, qualitative review of the study designs suggested differences in severity of the stressor. Thirteen of the studies examined the common stressful life events encountered in daily life in the United States and Europe. Many of these stressors tend to be chronic (i.e., lasting weeks or months) and of mild to moderate intensity, such as job-related stressors or family and interpersonal stressors. These 13 studies showed similar increases in risk of exacerbation associated with stressful life events. However, one study following patients in Tel Aviv, Israel, used a traumatic stressor—being under daily and nightly missile attack in the first Gulf War (Nisipeanu and Korczyn, 1993). These patients showed a decrease in risk of exacerbation during and after the stressful life event. Although it is possible that this isolated finding is due to chance alone, this study suggests that traumatic stressors (e.g., sudden and life-threatening) sustained over a period of weeks my have a protective effect, with respect to MS inflammation. This difference highlights the fact that stressors vary across a number of dimensions, including severity and chronicity, and that these differences likely produce different biological responses, which may have very different effects on inflammation in MS.

4. Laboratory Studies of Acute Stress Responses in MS

The association between stressful life events and MS exacerbation has led some researchers to speculate that the stress response is different among people with MS compared with healthy individuals. In this section, we will review the literature from laboratory studies examining the hypothalamic-pituitary-adrenal (HPA) axis responsiveness, as well as response to acute social stress.

4.1. HPA

It has been known for many decades that life events that are perceived as stressful can result in activation of the HPA axis. Stressful life events result in hypothalamic production of corticotropin releasing hormone (CRH) and arginine-vasopressin (AVP). CRH stimulates the pituitary gland to produce adrenocorticotropic hormone (ACTH). The effect of CRH as an ACTH secretogue is enhanced by AVP. ACTH stimulates the adrenal cortex to produce cortisol, which is the final effector of the HPA axis, exerting an inhibitory effect on hypothalamic production of CRH. Thus, the HPA axis is self-regulating, in part through the inhibitory effect of cortisol.

Inflammation can also activate the HPA axis (Chrousos, 1995). The proinflammatory cytokines IL-6, IL-1, and TNF-α have been shown to stimulate CRH and AVP secretion in the hypothalamus (Akira *et al.*, 1990; Bernardini *et al.*, 1990; Tsigos and Chrousos, 2002), while other proinflammatory cytokines such as IFN-γ may participate indirectly by stimulating the production of cytokines that act on the HPA axis (Chrousos, 1995). This activation of the HPA axis leads to increased cortisol release. Due to the ubiquity of glucocorticoid receptors in cells and tissue involved in the immune response, virtually all components of the immune response can be modulated by cortisol, including leukocyte trafficking and function, production of cytokines and other mediators of inflammation, and inhibition of the effects of immune mediators on target tissues (Chrousos, 1995; Jessop *et al.*, 2001; Elenkov and Chrousos, 2002). Indeed, glucocorticoids are the principal treatment for exacerbation in MS (Kopke *et al.*, 2004). This system allows an organism to adjust to changes in levels of inflammation by increasing or decreasing the output of anti-inflammatory glucocorticoids, as needed.

A growing literature has challenged early assumptions that autoimmune disease is associated with HPA hyporeactivity in response to stress and inflammation (Harbuz, 2002). Much of the research on HPA reactivity in MS has used stimulation through injection of human CRH in patients who are pretreated with dexamethasone (dex/CRH test). MS patients generally show significantly greater hyperreactivity rather than hyporeactivity compared with healthy controls (Grasser *et al.*, 1996; Fassbender *et al.*, 1998; Then Bergh *et al.*, 1999; Schumann *et al.*, 2002), although some studies have reported a subgroup that appears to show hyporesponsiveness (Grasser *et al.*, 1996; Wei and Lightman, 1997; Schumann *et al.*, 2002). Studies generally find that Primary progressive MS (PPMS) patients are the most hyperresponsive MS diagnostic group, followed by SPMS patients. RRMS patients are least hyperresponsive among MS patients, although studies generally find they are more hyperresponsive than healthy controls. As RRMS patients shift to a less inflammatory, more neurodegenerative phase (SPMS), the HPA becomes more hyperreactive. Patients with a primarily neurodegenerative disease (PPMS) are most hyperreactive.

Autopsy studies have found that there are significantly more CRH and CRH/AVP neurons in the hypothalamus of patients with MS compared with healthy controls (Erkut *et al.*, 1995). More recently, Huitinga also found chronic activation of the hypothalamic CHR/AVP cells, as marked by high levels of CRH mRNA, in MS compared with controls (Huitinga *et al.*, 2004). However, both the numbers of CRH/AVP cells and the amount of CRH mRNA were relatively reduced in patients with more severe disease. Huitinga also reported that local activation of microglial cells of macrophages were associated with suppression of CRH/AVP neurons. This suggests that while cortisol can be a potent inhibitor of inflammation, inflammation may also downregulate CRH/AVP neurons responsible for activating the HPA axis.

4.2. Immune Responses to Acute Psychological Stressors in MS

Several studies have examined acute responses to psychological stress in patients with relapsing forms of MS compared with healthy controls. Ackerman (Ackerman *et al.*, 1996, 1998), using a 5-min stressor based on the Trier Social Stress Test (TSST) (Kirschbaum *et al.*, 1993), showed a good response to the stressor as marked by self-reported stress, heart rate, blood pressure, and plasma cortisol. Subjects also showed proliferation across a variety of immune cells and increased production of IL-1β, TNF-α, and IFN-γ after whole-blood stimulation with a nonspecific antigen. Although MS patients produced more IFN-γ overall compared with controls, there were no differences between MS patients and healthy controls in the magnitude of the inflammatory response to stress.

Heesen has conducted two additional studies of response to acute stressors. The most recent study used a 30-min cognitive task as the stressor (Heesen *et al.*, 2005). Although Heesen also found an increase in IFN-γ overall, in contrast with Ackerman, the magnitude of the increase was lower in MS patients compared with healthy controls. There was not significant effect for TNF-α or IL-10. An earlier study found no significant effects for stress on serum levels of TNF-α or IL-6 on either MS patients or healthy controls (Heesen *et al.*, 2002). This may have been due to the stress procedure, which used a cognitive task in front of a camera. Nonsupportive human observers produce much greater stress response compared with cameras or other techniques (Dickerson and Kemeny, 2004).

The literature on immune response to acute stress in MS is too scant to permit any clear conclusions. Studies have used different stressors of variable length and different immune markers of inflammation (serum cytokine levels vs. stimulated production), making comparison of these studies difficult. The most extensively documented social stress paradigm, the TSST, suggested a normal stress response among MS patients, however a more extended, possibly less intensive stressor suggested the possibility of hyporesponsivity on IFN-γ a critical cytokine for MS.

5. Three Hypotheses Regarding the Mechanisms by Which Stress Leads to Exacerbation

Although it is important to understand the effect of MS on acute immune and neuroendocrine responses to stress, it is also clear that the effects of stress on MS exacerbation may well extend over periods much longer than those commonly examined with a TSST. We have recently proposed that hypotheses regarding the relationship between stressful life events and MS exacerbation must take into account the temporal relationship between the stress response and the natural history of the development of MS inflammation and exacerbation (Mohr and Pelletier, 2006). As we have described above, the natural history of an MS exacerbation appears to have a long, if not thoroughly understood, trajectory. For many years, it was assumed that BBB breakdown was a very early event in the development of an MS brain lesion and clinical exacerbation. However, newer neuroimaging techniques such as magnetic transfer ratio (MTR) imaging have shown that changes in the ratio of bound to unbound water begin to occur in white matter tissue several months before the emergence of traditional neuroimaging markers of inflammation such as gadolinium enhancing magnetic resonance imaging (Gd+ MRI) lesions (Filippi *et al.*, 1998; Goodkin *et al.*, 1998). Although the specific nature or processes involved in these changes are not well understood, it is clear that vulnerability begins in the tissue long before active inflammation begins. Once there is BBB breakdown and active inflammation at that site, attempts at regulation of the inflammation occur, including the production of Th2 cytokines such as IL-10. If regulation is not sufficiently successful within a brief period of time, clinical signs of exacerbation occur. Even in the absence of treatment with glucocorticoids, active inflammation at the site of the lesion will subside over a period of weeks or months but may leave some residual, permanent symptoms if sufficient irreparable demyelination or axonal damage resulted from the inflammation. Thus, our current understanding is that the pathogenesis, development, and resolution of MS exacerbation can span many months.

Stressors occur over a period of time, particularly those such as family- or work-related, and stressors resulting from disruptions in routine. The biological mediators of stressors vary, depending on the point in the evolution of the stressor one examines. The onset of a stressor, particularly if the onset is sudden, salient, and intense, is often accompanied by sympathetic activation, increases in epinephrine and norepinephrine, and activation of the HPA axis. As the stressor becomes chronic, the HPA axis can become dysregulated, often resulting in higher levels, and sometimes lower levels, of circulating cortisol (Sapolsky *et al.*, 2000). Resolution or adaptation to the stressor under normal circumstances results in re-regulation of the HPA axis and return of circulating levels of cortisol to baseline. Thus, the neuroendocrine processes that mediate stress change with the evolution of the stressor.

Given that the biological underpinnings of both the stress response and the development of MS exacerbation can change over time, we have proposed that the interaction between the two likely can vary depending on where in the temporal trajectory stressful life events intersect with the development of an MS lesion (Mohr and Pelletier, 2006). Evidence suggests that one point of vulnerability in the development of exacerbation occurs approximately 2 to 8 weeks before the development of clinical exacerbation or the appearance of new Gd+ MRI brain lesions (Mohr *et al.*, 2000; Ackerman *et al.*, 2003; Buljevac *et al.*, 2003). [It should be noted that there may be other points of vulnerability. For example, many patients report a worsening of symptoms hours or days after the occurrence of stressful life events, while epidemiological studies suggest that the stress can create vulnerabilities that increase risk of diagnosis years later (Li *et al.*, 2004). However, these reports remain isolated and in need of replication.]

The literature suggests three potential hypotheses for mechanisms by which stress might increase risk of MS exacerbation, shortly before the development of clinical exacerbation or the appearance of new Gd+ MRI brain lesions, which we shall describe below.

5.1. Stress Resolution Hypothesis

This hypothesis suggests that it is the resolution of the stress rather than the onset of stress that facilitates the development of active inflammation during this prodromal period. Whereas chronic stress is commonly marked by increased levels of cortisol (McEwen, 1998), trials of stress-management programs have reported that cortisol decreases as a result of successful stress management intervention (Antoni *et al.*, 2000). MS patients with relapsing disease also often show evidence of low levels of ongoing inflammation not noticeable by the patient or by usual neuroimaging but detectible by triple-dose Gd+ MRI (Silver *et al.*, 1997; Tortorella *et al.*, 1999). Thus, as cortisol rises after the onset of a stressful situation, the person with MS would receive some increased control over inflammation. However, as the stressor resolves, the concomitant reduction in cortisol would represent a decrease in control over inflammatory processes and could leave the individual at an increased risk for exacerbation. Patients' belief that the cause of exacerbation is a stressful life event would be an attributional error. That is, patients may be less likely to attribute an exacerbation to a positive event, such as the resolution of a stressor, and more likely to attribute it to the onset of the stressor, even if this occurred some weeks earlier.

Although there is currently no specific evidence for or against this hypothesis in humans, the EAE literature does offer some support. Most EAE studies find that stress suppresses disease activity, yet these studies routinely sacrifice the animals after the stress-induction protocol, typically after 1–14 days. In an innovative study, Whitacre kept the animals alive after the stress induction and found that 10 days after the termination of the stress protocol, the clinical signs of EAE returned and in many cases were worse than among the unstressed control animals (Whitacre *et al.*, 1998). Although it must be acknowledged that there are problems generalizing animal models to human disease, these findings nevertheless support the hypothesis that the resolution of a stressor can have a permissive effect on inflammatory processes and suggest that this should be further examined among MS patients.

5.2. Development of Chronic Stress: Glucocorticoid Resistance

The glucocorticoid resistance hypothesis suggests that exposure to chronic stress reduces the number and/or function of glucocorticoid receptors on immune cells, thereby making them less responsive to regulatory control by cortisol. There is some support for the notion that immune cells of patients with MS are less sensitive to the regulatory effects of glucocorticoids than the cells of healthy individuals (Stefferl *et al.*, 2001; DeRijk *et al.*, 2004; van Winsen *et al.*, 2005). Furthermore, this glucocorticoid resistance is seen primarily among RRMS patients and is thus most strongly associated with the earlier, more inflammatory phase of the illness, as opposed to SPMS patients who are shifting to a more progressive, neurodegenerative phase of the disease. Whereas glucocorticoid resistance seen in other autoimmune disorders such as rheumatoid arthritis and systemic lupus erythematosus has been attributed to the chronic treatment with glucocorticoids (Chikanza *et al.*, 1992; Tanaka *et al.*, 1992; Sher *et al.*, 1994), this is not the case in MS. Chronic treatment with glucocorticoids is rare in MS, and acute glucocorticoid treatment has not been associated with glucocorticoid resistance (DeRijk *et al.*, 2004).

We propose that glucocorticoid resistance in MS has two related etiologies, independent of the use of glucocorticoid treatments. Patients who experience chronic stressors are also likely to experience increased levels of cortisol (McEwen, 1998). Although these levels are far lower than those used for pharmacological purposes, they nevertheless have been shown to produce glucocorticoid resistance in humans (Miller *et al.*, 2002). This finding is supported by a larger laboratory literature examining the development of glucocorticoid resistance in mice exposed to social stressors (Avitsur *et al.*, 2001, 2002; Stark *et al.*, 2001, 2002). Thus, one potential source of glucocorticoid resistance in patients with MS are the social stressors that have been shown to predict the onset of exacerbation and new Gd+ MRI markers of brain inflammation several weeks after their onset (Mohr *et al.*, 2000, 2004).

The second potential source of glucocorticoid resistance may result from the chronic, low-grade inflammation seen in relapsing forms of MS (Silver *et al.*, 1997; Tortorella *et al.*, 1999). MS patients may produce chronic, if small elevations in cortisol in an attempt to regulate inflammation and maintain self-tolerance. Indeed, these attempts to manage low-grade inflammation may be responsible for the hypercortisolemia seen in these patients (Wei and Lightman, 1997; Then Bergh *et al.*, 1999; Erkut *et al.*, 2002).

The combined effects of inflammation and stressful life events on the immune system may result in a downregulation of glucocorticoid receptor number and function, thereby reducing the impact of HPA regulation of inflammation. Under conditions of downregulated glucocorticoid receptors, if there is a small increase in autoreactive inflammation, immune cells would be less responsive to the regulatory effects of cortisol. The autoreactive immune cascade would be able to continue uncontrolled until a full-blown exacerbation had occurred. In addition, a small drop in cortisol secondary to the resolution of stressful life events might also reduce glucocorticoid regulation of inflammatory processes, leaving the patient more vulnerable to inflammation and at greater risk of exacerbation. Thus, chronic stress, while not causing exacerbation, may leave patients less able to maintain self-tolerance when autoreactive MS immune processes are initiated.

5.3. Stress Onset: The Mast Cell Hypothesis

In our experience, many patients complain that the effects of stress on MS symptoms can occur within hours. We are not aware of any good empirical work in humans that confirm or disconfirm these reports. Laboratory stressors such as the Trier Social Stress Test (TSST) have been shown to produce significant elevations in proinflammatory cytokines in some studies (Ackerman *et al.*, 1996, 1998), although not in others (Heesen *et al.*, 2002, 2005). (We note that Ackerman reported in a personal communication that the cognitive tasks are not as effective as public speaking in eliciting a neuroendocrine stress response among MS patients, possibly because MS patients do not expect themselves to perform as well on cognitive tasks; the two studies that did not elicit a stress response only used cognitive tasks.) However, even when the TSST produced increases in proinflammatory cytokines, these elevations were similar to those seen in healthy controls. Similarly, acute stressors such as injury have not been associated with exacerbation (Goodin *et al.*, 1999). However, it is possible that there are subtle permissive effects and increased risk under specific circumstances. For example, low levels of glucocorticoids, consistent with endogenous cortisol response to moderate stress, may have numerous effects that can promote inflammation, including increased T-cell proliferation (Wiegers *et al.*, 1993, 1995). Indeed, these permissive effects may be particularly enhanced within the central nervous system (Dinkel *et al.*, 2003). However, the mechanism associated with stress onset that has been most thoroughly investigated with respect to potential MS pathways involves the acute activation of mast cells (Theoharides, 2002; Zappulla *et al.*, 2002).

Mast cells have been referred to as an immune gate to the brain (Theoharides, 2002). Increasing evidence suggests that this gate may be opened by environmental stressors (for reviews, see Theoharides, 2002; Zappulla et al., 2002). Mast cells are multifunctional effector cells of the innate immune system and are distributed broadly throughout human tissue, including vascular endothelium in the brain (Zappulla et al., 2002). For more than a century, it has been known that mast cells are found in MS demyelinated plaques, particularly around the venules and capillaries (Kruger et al., 1990; Kruger, 2001). Mast cells may participate in MS exacerbation by facilitating vascular permeability. Mast cells are known to be critical in the initial retardation of leukocytes rolling along the endothelium and the subsequent firm adhesion and extravasation through the cell wall (Kubes and Granger, 1996; Kubes and Ward, 2000). Extravasation is facilitated by mast cell–produced tryptase, as well as adhesion molecules (Kanbe et al., 1999; Theoharides, 2002; Zappulla et al., 2002). All of these vasodilators have been implicated in BBB permeability in MS (Tuomisto et al., 1983; Spuler et al., 1996; Waubant et al., 1999; Piccio et al., 2002).

Mast cell activity is also triggered by stress. Restraint stress has been shown to increase BBB permeability in rats through mast cell activation (Esposito et al., 2001), and subordination stress has been shown to increase numbers of mast cells across a number of brain regions (Cirulli et al., 1998). A principal mediator of stress-related mast cell activation is CRH (Theoharides et al., 1998). Hypothalamic CRH is a primary hormonal response to stressful life events. However, CRH is also present at sites of inflammation. Stress-related BBB breakdown via mast cell activation has repeatedly been shown to be facilitated by immune CRH (Theoharides et al., 1998; Singh et al., 1999; Esposito et al., 2001). Recent studies further indicate that increases in hypothalamic CRH consistent with stress responses can increase immune CRH and induce mast cell degranulation, thereby increasing BBB permeability (Esposito et al., 2002). This would suggest that stress onset might have a permissive effect on MS exacerbation by facilitating BBB permeability.

6. Psychosocial Factors

If indeed there is an association between stressful life events and the occurrence of MS inflammation and exacerbation of a magnitude that is clinically significant (Mohr et al., 2004), the next heuristic question would be, "What can we do to reduce the effects of stressful life events on MS exacerbation. One way of approaching this would be to reduce the effects of stressful events through better stress management. To date, there are no good studies that have provided data that can confirm or disconfirm the hypothesis that stress-management programs can reduce exacerbation. The two largest trials of psychosocial interventions have compared treatments that were all reasonably effective (Mohr *et al.*, 2001a; Mohr *et al.*, 2005). However, several studies have begun examining potential mediators and moderators relevant to stress management and psychosocial intervention.

6.1. Psychosocial Mediators

In a longitudinal study examining the effects of treatment for depression on immune function in MS patients, we showed that reductions in depression were associated with declines in T-cell production of IFN-γ (Mohr *et al.*, 2001b). Importantly, the reductions in IFN-γ production were seen not only for a nonspecific antigen but also for myelin oligodendrocyte glycoprotein, suggesting that treatment for depression can have an effect on highly specific factors in the pathogenesis of MS inflammation and exacerbation. We also examined the effects of change in each of the affective, cognitive, and vegetative symptoms (Louks *et al.*, 1989). While there was a trend toward an association between improvement in affective symptoms and decreased IFN-γ production ($r = 0.41$, $p = 0.11$), there was a strong relationship between improved cognitive symptoms (e.g., self-accusation, guilt, sense of failure) and reductions in IFN-γ production ($r = 0.61$, $p = 0.02$) (Mohr *et al.*, 1999). There was no significant effect for change in vegetative symptoms ($p = 0.25$). Although we cannot exclude the role of affective distress, these findings suggest that cognitions may play an important role mediating IFN-γ production in MS.

6.2. Psychosocial Moderators

The patient's ability to manage stressful life events appears to affect the relationship between stressful life events and exacerbation and inflammation. Cross-sectional work has shown that patients in exacerbation tend to report that they have more emotion-focused coping and lower social support compared with patients who are not in exacerbation (Warren *et al.*, 1991). We examined the effect of coping prospectively on the relationship between stressful life events and the development of Gd+ MRI. We did not find main effects for coping style on the incidence of new Gd+ MRI brain lesions. However, greater use of distraction to cope with stressors was predictive of a significantly weaker effect of stressors on the development of new brain lesions (Mohr *et al.*, 2002). Similarly, there was a trend for instrumental coping in the same direction, while there was a trend for greater emotional preoccupation (a ruminative style) to be predictive of increased effects of stressors on the development of new brain lesions. This suggests that coping may be an important moderator of the relationship between stressful life events and MS inflammation.

Social support has also been implicated as a factor in MS exacerbation (Warren *et al.*, 1991). As noted above, several studies suggest that stressors that impact the risk of MS exacerbation are social in nature, including family and work stressors (Sibley, 1997; Mohr et al., 2000). Certainly, these kinds of stressors can be considered markers of erosion in the patient's social network. However, we have found more specific evidence that social support may have a buffering effect. The effect of depression on T-cell production of IFN-γ noted above was significantly moderated by social support. Specifically, the relationship between depression and IFN-γ production was particularly strong among patients with low levels of support but was virtually nonexistent among patients with high social support (Mohr and Genain, 2004).

Such moderators are potentially important as they could alter the risk of exacerbation after stressful life events in any of the three hypothesized mechanisms or temporal relationships. They are also critical, because adaptive coping may prevent the occurrence of some types of stressors, reduce the distress associated with stressors that cannot be avoided, and influence cognitions that, as noted above, may be related to MS inflammation (Gottlieb, 1997; Mohr et al., 1997). Psychosocial moderators are also potentially useful as they may be modifiable through psychosocial intervention.

7. Summary and Future Directions

There is growing evidence of a relationship between the occurrence of stressful life events and exacerbation in patients who have MS (Mohr et al., 2004). The HPA axis appears to be hyperreactive to CRH challenge, but there is no consistent evidence that immune or neuroendocrine reactions to laboratory stressors are different in MS patients compared with healthy controls. We have proposed three potential mechanisms by which stress could increase risk of exacerbation among patients with MS, including the stress resolution hypothesis, the glucocorticoid resistance hypothesis, and the mast cell hypothesis.

To date, even the best studies of the effects of stressful life events on MS exacerbation and inflammation have been prospective, longitudinal designs, which cannot rule out the potential influences of third variables. One such potential confounding variable is change in normal-appearing white matter that can occur months before the appearance of conventional neuroimaging markers of MS disease activity (Filippi et al., 1998; Goodkin et al., 1998) could arguably increase perceived stress.

The strongest design to investigate the question of causality in the relationship between stress and MS exacerbation, as well as the immune and neuroendocrine mediators, is through a randomized controlled trial of a stress-management program for MS using treatment as usual or a minimal treatment control. The intervention arm of a trial would effectively serve as an experimental manipulation. There is ample evidence that stress-management programs can improve psychological moderators of the rela-

268   D.C. Mohr

tionship between stress and MS exacerbation or inflammation, including coping and social support. A randomized controlled trial of a stress-management program would have the additional advantage of potentially providing a novel treatment tool in the management of MS.

References

Ackerman, K.D., Martino, M., Heyman, R., Moyna, N.M., and Rabin, B.S. (1996). Immunologic response to acute psychological stress in MS patients and controls. *J. Neuroimmunol.* 68(1–2):85–94.

Ackerman, K.D., Martino, M., Heyman, R., Moyna, N.M., and Rabin, B.S. (1998). Stressor-induced alteration of cytokine production in multiple sclerosis patients and controls. *Psychosom. Med.* 60(4):484–491.

Ackerman, K.D., Stover, A., Heyman, R., Anderson, B.P., Houck, P.R., Frank, E., Rabin, B.S., and Baum, A. (2003). 2002 Robert Ader New Investigator Award. Relationship of cardiovascular reactivity, stressful life events, and multiple sclerosis disease activity. *Brain Behav. Immun.* 17(3):141–151.

Akira, S., Hirano, T., Taga, T., and Kishimoto, T. (1990). Biology of multifunctional cytokines: IL 6 and related molecules (IL 1 and TNF). *FASEB J.* 4(11):2860–2867.

Anderson, D.W., Ellenberg, J.H., Leventhal, C.M., Reingold, S.C., Rodriguez, M., and Silberberg, D.H. (1992). Revised estimate of the prevalence of multiple sclerosis in the United States. *Ann. Neurol.* 31(3):333–336.

Antoni, M.H., Cruess, S., Cruess, D., Kumar, M., Lutgendorf, S., Ironson, G., Dettmer, E., Williams, J., Klimas, N., Fletcher, M.A., and Schneiderman, N. (2000). Cognitive-behavioral stress management reduces distress and 24-hour urinary cortisol output among HIV-infected gay men. *Ann. Behav. Med.* 22:29–37.

Avitsur, R., Stark, J.L., and Sheridan, J.F. (2001). Social stress induces glucocorticoid resistance in subordinate animals. *Hormones Behav.* 39:247–257.

Avitsur, R., Stark, J.L., Dhabhar, F.S., Padgett, D.A., and Sheridan, J.F. (2002). Social disruption-induced glucocorticoid resistance: kinetics and site specificity. *J. Neuroimmunol.* 124(1–2):54–61.

Bar-Or, A., Nuttall, R.K., Duddy, M., Alter, A., Kim, H.J., Ifergan, I., Pennington, C.J., Bourgoin, P., Edwards, D.R., and Yong, V.W. (2003). Analyses of all matrix metalloproteinase members in leukocytes emphasize monocytes as major inflammatory mediators in multiple sclerosis. *Brain* 126(Pt 12):2738–2749.

Bernardini, R., Kamilaris, T.C., Calogero, A.E., Johnson, E.O., Gomez, M.T., Gold, P.W., and Chrousos, G.P. (1990). Interactions between tumor necrosis factor-alpha, hypothalamic corticotropin-releasing hormone, and adrenocorticotropin secretion in the rat. *Endocrinology* 126(6):2876–2881.

Bjartmar, C., Wujek, J.R., and Trapp, B.D. (2003). Axonal loss in the pathology of MS: consequences for understanding the progressive phase of the disease. *J. Neurol. Sci.* 206(2):165–171.

Bruck, W., and Stadelmann, C. (2003). Inflammation and degeneration in multiple sclerosis. *Neurol. Sci.* 24(Suppl 5):S265–267.

Buljevac, D., Hop, W.C., Reedeker, W., Janssens, A.C., van der Meche, F.G., van Doorn, P.A., and Hintzen, R.Q. (2003). Self reported stressful life events and exacerbations in multiple sclerosis: prospective study. *BMJ* 327(7416):646.

Charcot, J.M. (1877). *Lectures on Diseases of the Nervous System.* G. Sigerson, trans. London: New Sydenham Society.

Chikanza, I.C., Petrou, P., Kingsley, G., Chrousos, G., and Panayi, G.S. (1992). Defective hypothalamic response to immune and inflammatory stimuli in patients with rheumatoid arthritis. *Arthritis Rheum.* 35(11):1281–1288.

Chrousos, G.P. (1995). The hypothalamic-pituitary-adrenal axis and immune-mediated inflammation. *N. Engl. J. Med.* 332(20):1351–1362.

Cirulli, F., Pistillo, L., de Acetis, L., Alleva, E., and Aloe, L. (1998). Increased number of mast cells in the central nervous system of adult male mice following chronic subordination stress. *Brain Behav. Immun.* 12(2):123–133.

Confavreux, C., Vukusic, S., Moreau, T., and Adeleine, P. (2000). Relapses and progression of disability in multiple sclerosis. *N. Engl. J. Med.* 343(20):1430–1438.

Cooper, G.S., and Stroehla, B.C. (2003). The epidemiology of autoimmune diseases. *Autoimmun. Rev.* 2(3):119–125.

DeRijk, R.H., Eskandari, F., and Sternberg, E.M. (2004). Corticosteroid resistance in a subpopulation of multiple sclerosis patients as measured by ex vivo dexamethasone inhibition of LPS induced IL-6 production. *J. Neuroimmunol.* 151(1–2):180–188.

Dickerson, S.S., and Kemeny, M.E. (2004). Acute stressors and cortisol responses: a theoretical integration and synthesis of laboratory research. *Psychol. Bull.* 130(3):355–391.

Dinkel, K., MacPherson, A., and Sapolsky, R.M. (2003). Novel glucocorticoid effects on acute inflammation in the CNS. *J. Neurochem.* 84(4):705–716.

Elenkov, I.J., and Chrousos, G.P. (2002). Stress hormones, proinflammatory and anti-inflammatory cytokines, and autoimmunity. *Ann. N.Y. Acad. Sci.* 966:290–303.

Erkut, Z.A., Hofman, M.A., Ravid, R., and Swaab, D.F. (1995). Increased activity of hypothalamic corticotropin-releasing hormone neurons in multiple sclerosis. *J. Neuroimmunol.* 62(1):27–33.

Erkut, Z.A., Endert, E., Huitinga, I., and Swaab, D.F. (2002). Cortisol is increased in postmortem cerebrospinal fluid of multiple sclerosis patients: Relationship with cytokines and sepsis. *Mult. Scler.* 8(3):229–236.

Esposito, P., Gheorghe, D., Kandere, K., Pang, X., Connolly, R., Jacobson, S., and Theoharides, T.C. (2001). Acute stress increases permeability of the blood-brain-barrier through activation of brain mast cells. *Brain Res.* 888:117–127.

Esposito, P., Chandler, N., Kandere, K., Basu, S., Jacobson, S., Connolly, R., Tutor, D., and Theoharides, T.C. (2002). Corticotropin-releasing hormone and brain mast cells regulate blood-brain-barrier permeability induced by acute stress. *J. Pharmacol. Exp. Ther.* 303(3):1061–1066.

Fassbender, K., Schmidt, R., Mossner, R., Kischka, U., Kuhnen, J., Schwartz, A., and Hennerici, M. (1998). Mood disorders and dysfunction of the hypothalamic-pituitary-adrenal axis in multiple sclerosis: association with cerebral inflammation. *Arch. Neurol.* 55(1):66–72.

Filippi, M., Rocca, M.A., Martino, G., Morsfield, M.A., and Comi, G. (1998). Magnetization transfer changes in normal appearing white matter precede the appearance of enhancing lesions in patients with multiple sclerosis. *Ann. Neurol.* 43:809–814.

Filippini, G., Munari, L., Incorvaia, B., Ebers, G.C., Polman, C., D'Amico, R., and Rice, G.P. (2003). Interferons in relapsing remitting multiple sclerosis: a systematic review. *Lancet* 361(9357):545–552.

Goodin, D.S., Ebers, G.C., Johnson, K.P., Rodriguez, M., Sibley, W.A., and Wolinsky, J.S. (1999). The relationship of MS to physical trauma and psychological stress: Report of the therapeutics and technology assessment subcommittee of the American Academy of Neurology. *Neurology* 52:1737–1745.

270   D.C. Mohr

Goodkin, D.E., Rooney, W.D., Sloan, R., Bacchetti, P., Gee, L., Vermathen, M.,Waubant, E., Abundo, M., Majumdar, S., Nelson, S., and Weiner, M.W. (1998). A serial study of new MS lesions and the white matter from which they arise. *Neurology* 51:1689–1697.

Gottlieb, B.H., ed. (1997). *Coping with Chronic Stress*. New York: Plenum.

Grasser, A., Moller, A., Backmund, H., Yassouridis, A., and Holsboer, F. (1996). Heterogeneity of hypothalamic-pituitary-adrenal system response to a combined dexamethasone-CRH test in multiple sclerosis. *Exp. Clin. Endocrinol. Diabetes* 104(1):31–37.

Grossman, R.I. (1996). Magnetic resonance imaging: Current status and strategies of improving multiple sclerosis clinical trial design. In D.E. Goodkin and R.A. Rudick (eds.), *Multiple Sclerosis: Advances in Clinical Trial Design, Treatment and Future Perspectives*. London: Springer, pp. 161–186.

Harbuz, M. (2002). Neuroendocrine function and chronic inflammatory stress. *Exp. Physiol.* 87(5):519–525.

Heesen, C., Schulz, H., Schmidt, M., Gold, S., Tessmer, W., and Schulz, K.H. (2002). Endocrine and cytokine responses to acute psychological stress in multiple sclerosis. *Brain Behav. Immun.* 16(3):282–287.

Heesen, C., Koehler, G., Gross, R., Tessmer, W., Schulz, K.H., and Gold, S.M. (2005). Altered cytokine responses to cognitive stress in multiple sclerosis patients with fatigue. *Mult. Scler.* 11(1):51–57.

Hohlfeld, R., Meinl, E., Weber, F., Zipp, F., Schmidt, S., Sotgiu, S., Goebels, N., Voltz, R., Spuler, S., and Iglesias, A. (1995). The role of autoimmune T lymphocytes in the pathogenesis of multiple sclerosis. *Neurology* 45(6 Suppl 6):S33–38.

Huitinga, I., Erkut, Z.A., van Beurden, D., and Swaab, D.F. (2004). Impaired hypothalamus-pituitary-adrenal axis activity and more severe multiple sclerosis with hypothalamic lesions. *Ann. Neurol.* 55(1):37–45.

Jacobson, D.L., Gange, S.J., Rose, N.R., and Graham, N.M. (1997). Epidemiology and estimated population burden of selected autoimmune diseases in the United States. *Clin. Immunol. Immunopathol.* 84(3):223–243.

Jessop, D.S., Harbuz, M.S., and Lightman, S.L. (2001). CRH in chronic inflammatory stress. *Peptides* 22(5):803–807.

Kanbe, N., Tanaka, A., Kanbe, M., Itakura, A., Kurosawa, M., and Matsuda, H. (1999). Human mast cells produce matrix metalloproteinase 9. *Eur. J. Immunol.* 29(8): 2645–2649.

Kirschbaum, C., Pirke, K.M., and Hellhammer, D.H. (1993). The "Trier Social Stress Test"—a tool for investigating psychobiological stress responses in a laboratory setting. *Neuropsychobiology* 28(1–2):76–81.

Kopke, S., Heesen, C., Kasper, J., and Muhlhauser, I. (2004). Steroid treatment for relapses in multiple sclerosis—the evidence urges shared decision-making. *Acta Neurol. Scand.* 110(1):1–5.

Kruger, P.G. (2001). Mast cells and multiple sclerosis: A quantitative analysis. *Neuropathol. Appl. Neurobiol.* 27(4):275–280.

Kruger, P.G., Bo, L., Myhr, K.M., Karlsen, A.E., Taule, A., Nyland, H.I., and Mork, S. (1990). Mast cells and multiple sclerosis: A light and electron microscopic study of mast cells in multiple sclerosis emphasizing staining procedures. *Acta Neurol. Scand.* 81(1):31–36.

Kubes, P., and Granger, D.N. (1996). Leukocyte-endothelial cell interactions evoked by mast cells. *Cardiovasc. Res.* 32(4):699–708.

Kubes, P., and Ward, P.A. (2000). Leukocyte recruitment and the acute inflammatory response. *Brain Pathol.* 10(1):127–135.

Li, J., Johansen, C., Bronnum-Hansen, H., Stenager, E., Koch-Henriksen, N., and Olsen, J. (2004). The risk of multiple sclerosis in bereaved parents: A nationwide cohort study in Denmark. *Neurology* 62(5):726–729.

Louks, J., Hayne, C., and Smith, J. (1989). Replicated factor structure of the Beck Depression Inventory. *J. Nerv. Ment. Dis.* 177(8):473–479.

McEwen, B.S. (1998). Protective and damaging effects of stress mediators. *N. Engl. J. Med.* 338:171–179.

Miller, G.E., Cohen, S., and Ritchey, A.K. (2002). Chronic psychological stress and the regulation of pro-inflammatory cytokines: a glucocorticoid-resistance model. *Health Psychol.* 21(6):531–541.

Mohr, D.C., and Cox, D. (2001). Multiple sclerosis: Empirical literature for the clinical health psychologist. *J. Clin. Psychol.* 57(4):479–499.

Mohr, D.C., and Genain, C. (2004). Social support as a buffer in the relationship between treatment for depression and T-cell production of interferon gamma in patients with multiple sclerosis. *J. Psychosom. Res.* 57(2):155–158.

Mohr, D.C., and Pelletier, D. (2006). A temporal framework for understanding the effects of stressful life events on inflammation in patients with multiple sclerosis. *Brain Behav. Immun.* 20(1): 27–36.

Mohr, D.C., Goodkin, D.E., Gatto, N., and Van der Wende, J. (1997). Depression, coping and level of neurological impairment in multiple sclerosis. *Mult. Scler.* 3(4):254–258.

Mohr, D.C., Boudewyn, A., and Genain, C. (1999). Relationship between treatment for depression and interferon-gamma in patients with multiple sclerosis. *Psychosom. Med.* 61:112.

Mohr, D.C., Goodkin, D.E., Bacchetti, P., Boudewyn, A.C., Huang, L., Marrietta, P., Cheuk, W., and Dee, B. (2000). Psychological stress and the subsequent appearance of new brain MRI lesions in MS. *Neurology* 55(1):55–61.

Mohr, D.C., Boudewyn, A.C., Goodkin, D.E., Bostrom, A., and Epstein, L. (2001a). Comparative outcomes for individual cognitive-behavior therapy, supportive-expressive group psychotherapy, and sertraline for the treatment of depression in multiple sclerosis. *J. Consult. Clin. Psychol.* 69(6):942–949.

Mohr, D.C., Goodkin, D.E., Islar, J., Hauser, S.L., and Genain, C.P. (2001b). Treatment of depression is associated with suppression of nonspecific and antigen-specific T(H)1 responses in multiple sclerosis. *Arch. Neurol.* 58(7): 1081–1086.

Mohr, D.C., Goodkin, D.E., Nelson, S., Cox, D., and Weiner, M. (2002). Moderating effects of coping on the relationship between stress and the development of new brain lesions in multiple sclerosis. *Psychosom. Med.* 64(5): 803–809.

Mohr, D.C., Hart, S.L., Julian, L., Cox, D., and Pelletier, D. (2004). Association between stressful life events and exacerbation in multiple sclerosis: a meta-analysis. *BMJ* 328(7442):731.

Mohr, D.C., Hart, S.L., Honos-Webb, L., Julian, L., Catledge, C., Vella, L., and Tasch, E.T. Telephone-administered psychotherapy for depression. *Arch. Gen. Psychiatry.* 62(9), 1007–1014.

Nisipeanu, P., and Korczyn, A.D. (1993). Psychological stress as risk factor for exacerbations in multiple sclerosis. *Neurology* 43(7):1311–1312.

Noonan, C.W., Kathman, S.J., and White, M.C. (2002). Prevalence estimates for MS in the United States and evidence of an increasing trend for women. *Neurology* 58(1):136–138.

Noseworthy, J.H., Lucchinetti, C., Rodriguez, M., and Weinshenker, B.G. (2000). Multiple sclerosis. *N. Engl. J. Med.* 343(13):938–952.

O'Connor, K.C., Bar-Or, A., and Hafler, D.A. (2001). The neuroimmunology of multiple sclerosis: possible roles of T and B lymphocytes in immunopathogenesis. *J. Clin. Immunol.* 21(2):81–92.

Piccio, L., Rossi, B., Scarpini, E., Laudanna, C., Giagulli, C., Issekutz, A.C., Vestweber, D., Butcher, E.C., and Constantin, G. (2002). Molecular mechanisms involved in lymphocyte recruitment in inflamed brain microvessels: critical roles for P-selectin glycoprotein ligand-1 and heterotrimeric G(i)-linked receptors. *J. Immunol.* 168(4):1940–1949.

Sapolsky, R.M., Romero, L.M., and Munck, A.U. (2000). How do glucocorticoids influence stress responses? Integrating permissive, suppressive, stimulatory, and preparative actions. *Endocr. Rev.* 21(1):55–89.

Scherer, S. (1999). Axonal pathology in demyelinating diseases. *Ann. Neurol.* 45(1):6–7.

Schumann, E.M., Kumpfel, T., Then Bergh, F., Trenkwalder, C., Holsboer, F., and Auer, D.P. (2002). Activity of the hypothalamic-pituitary-adrenal axis in multiple sclerosis: correlations with gadolinium-enhancing lesions and ventricular volume. *Ann. Neurol.* 51(6):763–767.

Sher, E.R., Leung, D.Y., Surs, W., Kam, J.C., Zieg, G., Kamada, A.K., and Szefler, S.J. (1994). Steroid-resistant asthma. Cellular mechanisms contributing to inadequate response to glucocorticoid therapy. *J. Clin. Invest.* 93(1):33–39.

Sibley, W.A. (1997). Risk factors in multiple sclerosis. In C.S. Raine, H.F. McFarland, and W.W. Tourtellotte (eds.), *Multiple Sclerosis: Clinical and Pathogenetic Basis.* London: Chapman and Hall, pp. 141–148.

Silver, N.C., Good, C.D., Barker, G.J., MacManus, D.G., Thompson, A.J., Moseley, I.F., MacDonald, W.I., and Miller, D.H. (1997). Sensitivity of contrast enhanced MRI in multiple sclerosis. Effects of gadolinium dose, magnetization transfer contrast and delayed imaging. *Neurology* 120:1149–1161.

Singh, L.K., Pang, X., Alexacos, N., Letourneau, R., and Theoharides, T.C. (1999). Acute immobilization stress triggers skin mast cell degranulation via corticotropin releasing hormone, neurotensin, and substance P: A link to neurogenic skin disorders. *Brain Behav. Immun.* 13(3):225–239.

Sospedra, M., and Martin, R. (2004). Immunology of multiple sclerosis. *Annu. Rev. Immunol.* 23:683–747.

Spuler, S., Yousry, T., Scheller, A., Voltz, R., Holler, E., Hartmann, M., Wick, M., and Hohlfeld, R. (1996). Multiple sclerosis: Prospective analysis of TNF-alpha and 55 kDa TNF receptor in CSF and serum in correlation with clinical and MRI activity. *J. Neuroimmunol.* 66(1–2):57–64.

Stark, J.L., Avitsur, R., Padgett, D.A., Campbell, K.A., Beck, F.M., and Sheridan, J.F. (2001). Social stress induces glucocorticoid resistance in macrophages. *Am. J. Physiol. Regul. Integr. Comp. Physiol.* 280(6):R1799–1805.

Stark, J.L., Avitsur, R., Hunzeker, J., Padgett, D.A., and Sheridan, J.F. (2002). Interleukin-6 and the development of social disruption-induced glucocorticoid resistance. *J. Neuroimmunol.* 124(1–2):9–15.

Stefferl, A., Storch, M.K., Linington, C., Stadelmann, C., Lassmann, H., Pohl, T., Holsboer, F., Tilders, F.J., and Reul, J.M. (2001). Disease progression in chronic relapsing experimental allergic encephalomyelitis is associated with reduced inflammation-driven production of corticosterone. *Endocrinology* 142(8): 3616–3624.

Tanaka, H., Akama, H., Ichikawa, Y., Makino, I., and Homma, M. (1992). Glucocorticoid receptor in patients with lupus nephritis: relationship between receptor levels in mononuclear leukocytes and effect of glucocorticoid therapy. *J. Rheumatol.* 19(6):878–883.

Then Bergh, F., Kumpfel, T., Trenkwalder, C., Rupprecht, R., and Holsboer, F. (1999). Dysregulation of the hypothalamo-pituitary-adrenal axis is related to the clinical course of MS. *Neurology* 53(4):772–777.

Theoharides, T.C., Singh, L., Boucher, W., Pang, X., Letourneau, R., Webster, E., and Chrousos, G.P. (1998). Corticotropin-releasing hormone induces skin mast cell degranulation and increased vacular permeability, a possible explanation for its pro-inflammatory effects. *Endocrinology* 139:403–413.

Theoharides, T.C. (2002). Mast cells and stress—a psychoneuroimmunological perspective. *J. Clin. Psychopharmacol.* 22(2):103–108.

Thompson, A.J., Montalban, X., Barkhof, F., Brochet, B., Filippi, M., Miller, D.H., Polman, C.H., Stevenson, V.L., and McDonald, W.I. (2000). Diagnostic criteria for primary progressive multiple sclerosis: A position paper. *Ann. Neurol.* 47(6):831–835.

Tortorella, C., Codella, M., Rocca, M.A., Gasperini, C., Capra, R., Bastianello, S., and Filippi, M. (1999). Disease activity in multiple sclerosis studied by weekly triple-dose magnetic resonance imaging. *J. Neurol.* 246(8):689–692.

Trapp, B.D., Peterson, J., Ransohoff, R.M., Rudick, R., Mork, S., and Bo, L. (1998). Axonal transection in the lesions of multiple sclerosis. *N. Engl. J. Med.* 338(5): 278–285.

Tsigos, C., and Chrousos, G. (2002). Hypothalamic-pituitary-adrenal axis, neuroendocrine factors and stress. *J. Psychosom. Res.* 53(4):865.

Tuomisto, L., Kilpelainen, H., and Riekkinen, P. (1983). Histamine and histamine-N-methyltransferase in the CSF of patients with multiple sclerosis. *Agents Actions* 13(2–3):255–257.

van Winsen, L.M., Muris, D.F., Polman, C.H., Dijkstra, C.D., van den Berg, T.K., and Uitdehaag, B.M. (2005). Sensitivity to glucocorticoids is decreased in relapsing remitting multiple sclerosis. *J. Clin. Endocrinol. Metab.* 90(2):734–740.

Warren, S., Warren, K.G., and Cockerill, R. (1991). Emotional stress and coping in multiple sclerosis (MS) exacerbations. *J. Psychosom. Res.* 35(1):37–47.

Waubant, E., Goodkin, D.E., Gee, L., Bacchetti, P., Sloan, R., Stewart, T., Andersson, P-B., Stabler, G., and Miller, K. (1999). Serum MMP-9 and TIMP-1 levels are related to MRI activity in relapsing multiple sclerosis. *Neurology* 53:1397–1401.

Wei, T., and Lightman, S.L. (1997). The neuroendocrine axis in patients with multiple sclerosis. *Brain* 120(Pt 6):1067–1076.

Whitacre, C.C., Dowdell, K., and Griffin, A.C. (1998). Neuroendocrine influences on experimental autoimmune encephalomyelitis. *Ann. N.Y. Acad. Sci.* 840:705–716.

Wiegers, G.J., Croiset, G., Reul, J.M., Holsboer, F., and de Kloet, E.R. (1993). Differential effects of corticosteroids on rat peripheral blood T-lymphocyte mitogenesis in vivo and in vitro. *Am. J. Physiol.* 265(6 Pt 1):E825–830.

Wiegers, G.J., Labeur, M.S., Stec, I.E., Klinkert, W.E., Holsboer, F., and Reul, J.M. (1995). Glucocorticoids accelerate anti-T cell receptor-induced T cell growth. *J. Immunol.* 155(4):1893–1902.

Zappulla, J.P., Arock, M., Mars, L.T., and Liblau, R.S. (2002). Mast cells: New targets for multiple sclerosis therapy? *J. Neuroimmunol.* 131(1–2):5–20.

Index

Printed in the USA